The Limits of Scientific Reason

Continental Philosophy in Austral-Asia

Series Editors: Simone Bignall, Senior Lecturer, Indigenous Strategy and Engagement, Flinders University; P. Diego Bubbio, Senior Research Fellow in Philosophy, University of Western Sydney; Joanne Faulkner, Lecturer in Philosophy and Women's and Gender Studies, University of New South Wales; and Paul Patton, Scientia Professor, University of New South Wales

The *Continental Philosophy in Austral-Asia* series transports a tradition of thought understood as belonging to one place – "the continent" – to places that were transformed in its image through colonisation: Australia, New Zealand, East Asia, and South Asia. The series aims to explore and showcase the diverse ways in which European philosophy has been interpreted and put to use according to the contexts and questions particular to life in even further, stranger, and more "exotic" continents.

Titles in the Series

Young and Free: [Post]colonial Ontologies of Childhood, Memory and History in Australia, Joanne Faulkner

The Phenomenology of Gravidity: Reframing the Maternal in Merleau-Ponty, Levinas and Derrida, Jane Lymer

Deleuze and the Humanities: East and West, edited by Rosi Braidotti, Kin Yuen Wong, and Amy K. S. Chan

Unsettling Food Politics: Agriculture, Dispossession and Sovereignty in Australia, Christopher Mayes

Reimagining Sympathy, Recognising Difference, Millicent Churcher

Towards Continental Philosophy: Reason and Imagination in the Thought of Max Deutscher, Max Deutscher

The Limits of Scientific Reason: Habermas, Foucault, and Science as a Social Institution, John McIntyre

The Limits of Scientific Reason

Habermas, Foucault, and Science as a Social Institution

John McIntyre

ROWMAN & LITTLEFIELD
Lanham • Boulder • New York • London

Published by Rowman & Littlefield
An imprint of The Rowman & Littlefield Publishing Group, Inc.
4501 Forbes Boulevard, Suite 200, Lanham, Maryland 20706
www.rowman.com

86-90 Paul Street, London EC2A 4NE

Copyright © 2021 by The Rowman & Littlefield Publishing Group, Inc.

All rights reserved. No part of this book may be reproduced in any form or by any electronic or mechanical means, including information storage and retrieval systems, without written permission from the publisher, except by a reviewer who may quote passages in a review.

British Library Cataloguing in Publication Information Available

Library of Congress Cataloging-in-Publication Data

Names: McIntyre, John, 1952– author.
Title: The limits of scientific reason : Habermas, Foucault, and science as a social institution / John McIntyre.
Description: Lanham : Rowman & Littlefield, [2021] | Series: Continental philosophy in Austral-Asia | Includes bibliographical references and index. | Summary: "This is the first book to focus on science as a social institution based on a comprehensive analysis of the thought of Foucault and Habermas. A key aspect of this book is its standpoint which critiques science, whilst simultaneously interrogating philosophical critique which must in a certain sense accommodate science, and its effect on modernity"— Provided by publisher.
Identifiers: LCCN 2021020914 (print) | LCCN 2021020915 (ebook) | ISBN 9781538157787 (hardback) | ISBN 9781538157800 (paperback) | ISBN 9781538157794 (epub)
Subjects: LCSH: Habermas, Jürgen. | Foucault, Michel, 1926-1984. | Science—Social aspects. | Science—Philosophy. | Science and civilization. | Civilization, Modern.
Classification: LCC Q175.5 .M39545 2021 (print) | LCC Q175.5 (ebook) | DDC 306.4/5—dc23
LC record available at https://lccn.loc.gov/2021020914
LC ebook record available at https://lccn.loc.gov/2021020915

Contents

Abbreviations	vii
Preface	ix

1 Modernity's Nagging Question — 1
  Science and Society — 1
  The Aim and Contents of the Book — 7
  Philosophy and Its Contexts — 13
  Habermas and Foucault: Lives and Motivations — 19
  Modernity Science and Philosophy — 33

2 Habermas's Critique of Positivism — 39
  Habermas's Response to Positivism — 40
  Knowledge and Human Interests — 50
  Habermas's Theoretical Partitions — 62

3 Science, Modernity, and Communicative Action — 67
  Habermas's Linguistic Turn — 67
  Lifeworld, System, and the Rationalisation of Society — 73
  The Diagnosis of Modernity — 80
  Insights and *Aporias* — 85
  Reinterpreting Habermas — 94

4 Science and Deliberative Democracy — 101
  Between Facts and Norms — 102
  Philosophy and Science — 109
  The Future of Human Nature — 114
  Free Will and Determinism — 122
  Concluding Thoughts — 129

5	Foucault's Archaeology of Scientific Knowledge	133
	Foucault's Radicalisation of Critique	133
	Madness	139
	Archaeology and the History of Science	144
	Order and the Sciences	150
	Concluding Thoughts	163
6	Science and Power	167
	From Archaeology to Genealogy	168
	The Emergence and Dissemination of Modern Power/Knowledge	172
	The Constitution of the Subject	179
	The Natural Sciences	185
	The Normalisation of Society	188
	Biopower and Governmentality	191
	Normative Confusions	200
7	Science and the Genealogy of the Subject	207
	Foucault's Broader Framework	207
	Ethics, Aesthetics, and Spirituality	214
	The Genealogy of the Subject	220
	Philosophy and Science after Kant	228
8	Science, Philosophy, and Modernity	233
	The Reconcilability of Habermas and Foucault	234
	Reflexivity and Its Modern Radicalisation	237
	Discovery and Self-Transformation	239
	Normative Foundations and Confusions	249
	Wrapping Up the Debate	254
	Concluding Reflections	261
Bibliography		267
Index		279
About the Author		289

Abbreviations

The following abbreviations are used for frequently cited works by Habermas and Foucault. Full bibliographic details can be found in "Bibliography" section. The dates shown are the dates of editions used for this book. The original dates of publication are listed in Bibliography.

FOUCAULT

Books by Foucault

AK	1989 *The Archaeology of Knowledge*
BC	1994 *The Birth of the Clinic*
DP	1995 *Discipline and Punish*
HM	2006 *The History of Madness*
HS1	1979 *History of Sexuality* vol. 1
HS2	1990 *History of Sexuality* vol. 2
OT	2002 *The Order of Things*

Foucault's interviews, lectures, and essays

EW1	1997a Essential Works volume 1
EW2	2002 Essential Works volume 2
EW3	2000 Essential Works volume 3
FR	1984 The Foucault Reader
FL	1989 Foucault Live
PK	1980 Power/Knowledge
PT	1997b The Politics of Truth
PPC	1988b Michel Foucault: Politics, Philosophy Culture

Lecture Courses at College de France

AB	2003b Abnormal 1974–1975
BB	2008b The Birth of Biopolitics 1978–1979
CT	2012 The Courage of Truth 1983–1984
GOL	2014 On the Government of the Living 1979–1980
GSO	2011 The Government of Self and Others 1982–1983
HS	2005 The Hermeneutics of the Subject 1981–1982
PP	2008a Psychiatric Power 1973–1974
SD	2003a Society Must be Defended 1975–1976
STP	2007 Security, Territory and Population 1977–1978
WK	2013 Lectures on the Will to Know 1970–1971

HABERMAS

Books by Habermas

BFN	1992 *Between Facts and Norms*
BNR	2009 *Between Naturalism and Religion*
CES	1979 *Communication and the Evolution of Society*
FHN	2003b *The Future of Human Nature*
KHI	1972 *Knowledge and Human Interests*
LC	1988b *Legitimation Crisis*
LSS	1988a *On the Logic of the Social Sciences*
MCCA	1990a *Moral Consciousness and Communicative Action*
OPC	2002b *On the Pragmatics of Communication*
OPSI	2001 *On the Pragmatics of Social Interaction*
PD	1976 *The Positivist Dispute in German Sociology*
PDM	1990b *The Philosophical Discourse of Modernity*
PMT	1992 *Postmetaphysical Thinking*
RR	2002a *Religion and Rationality*
STPS	1989a *The Structural Transformation of the Public Sphere*
TCA1	1983 *The Theory of Communicative Action* vol. 1
TCA2	1989b *The Theory of Communicative Action* vol. 2
TJ	2003a *Truth and Justification*
TP	1971a *Theory and Practice*
TRS	1971b *Towards a Rational Society*

Preface

Some books have a long gestation. The conception of this book was half a lifetime ago in what now seems another world, one of the most remote regions in the Papua New Guinea highlands. Working as a horticultural extension officer, I had been drawn to a region of precipitous mountains, drenching rain, and breathtaking beauty not only by a youthful sense of adventure but also by a vision of science as the means to a better world. After gleaning some insight into the everyday concerns of my indigenous hosts I increasingly began to suspect that something was amiss with this vision. This suspicion was heightened by the many conversations with visiting Western trained experts—agronomists, economists, anthropologists, engineers, plant geneticists, and so forth—some of whom shared my reservations. It wasn't that the scientific standpoint was false or that it didn't hold out genuine emancipatory possibilities. It was rather that to be rigorous, science needed to abstract from the complexity and richness of its messy context, from life as actually lived.

Several decades later, I found myself where I should have started—undertaking a doctoral programme in the Philosophy Department at the University of Sydney. My aim was now to pick up the scattered threads of my youthful suspicions and follow them wherever they may lead. This endeavour led to this book which, in the course of its gestation, profited enormously from conversations, discussions, and exchanges with many people. I will mention just a few. Dave MacArthur first introduced me to a critique of scientific naturalism and the possibilities of more liberal forms of naturalism. I was lucky to have as my doctoral supervisor John Grumley, who broadened the scope of my research by encouraging me to reflect more deeply on the historical and social contexts of science. I am also indebted to Paul Redding who at several critical points was able to suggest fresh perspectives. The perceptive comments of Chris Falzon on the directions a book could take helped me draw together a work that seemed to pull in far too many directions. While Timothy O'Leary's comments spurred me to articulate connections with contemporary issues, Samuel Talcott drew my attention to the importance of Foucault's relationship to Georges Canguilhem. Peter Krockenberger, a wordsmith attuned to the nuance of philosophical language, helped me craft a more polished work. Finally my partner Teresa Mok and our son Angus provided encouragement, support, and remarkable patience.

Chapter 1

Modernity's Nagging Question

A nagging question hangs over modernity—the question concerning science. Never before has the dream of consciously steering history been more compelling. And never before has so much hung on consciousness that is not fully conscious. Science—neuroscience, genetic science, big data—will steer our future, but without more adequate reflection, will we steer science? As the dominant form of rationality in modern societies, science requires reflecting on. Granting to a claim the honorific "scientific" usually obviates the need to consider "non-scientific" accounts. If such non-scientific accounts do come under consideration, they still aspire to not contradict the claims of science. To say that some part of "science" is wrong is to say that what we thought was science wasn't really science, or certainly wasn't good science, the very meaning of which is tied to truth. Like truth, "science" has a future-oriented context-transcendent moment which allows it to surpass itself, and only science can surpass science. Underlying this self-assured authority, there is a sense of disquiet, a suspicion that something is missing, particularly from the scientific account of human beings and their societies. It is not that science is incomplete or that it may not be true. Rather, the anxiety arises from the seemingly ever-expanding scope of science's jurisdiction and its relation to other discourses and practices. It seems that science's authority, if accepted without adequate reflection, could limit freedom and facilitate domination.

SCIENCE AND SOCIETY

This book is concerned with science as an institution. By "institution", I mean the many and varied norms, both formal and informal, which guide shared understandings of what is proper, right, or true.[1] By "science" I mean a family of institutional practices and discourses endorsed by society to produce authoritative truths about ourselves, our societies, and the world. These truths issue from formalised methods and procedures which ensure a degree of objectivity, a distance between the inquirer and their objects as though allowing objects room to speak for themselves, to show themselves unchanged by how we conceive or deal with them. Since what counts as science varies historically, due to science's continual self-revisions, I will avoid questions of demarcation between what we today regard as "science" and "non-science". And rather than draw firm distinctions between the entwined discourses of the human, social, life, and natural sciences, I will adopt a functionalist approach in which I consider the roles played by those authoritative bodies of knowledge most frequently called "science".[2]

While I frequently follow the anglophone convention of "science" in the singular, we must not impute to it the explanatory reach of a continuous, homogeneous, and ultimately unified project which is, in principle unlimited. The thesis of a unity of science that leaves no remainder is ideological, serving as a principle of demarcation, whereby methods and knowledges recognised as belonging to the project of science are set against non-scientific practices of knowledge production, which are consequently devalued. Despite lingering hopes of finding some fundamental essence, concept or method uniting disparate sciences, the disunity of science has become increasingly acknowledged.[3] Today the sciences remain a heterogeneous cluster of diverse practices and knowledges, each with its own set of problems, objects, and methods, unified by a vague set of family resemblances.

By pursuing its truths within its own various specialised conventions, science withdraws from the everyday experience of those whose understanding must be filtered through an array of social discourses (see Wynne 2001). Science is nonetheless shaped by those discourses which in turn it surreptitiously seeps back into, to form an unassailable background against which notions of the self, the good life, and the good society are prescribed. Such unassailability frequently manifests as a naïve scientism that uncritically applauds every scientific

---

[1] Berger and Luckmann (1966) and Searle (2005) provide helpful discussions of institutions.
[2] The term "science" is fuzzy, indeterminate, and frequently contested (see Haack 2003). The distinctions between different sciences shift historically and are as much bridged by continuities as marked by differences. For example, there are marked affinities between "historical" sciences such as cosmology and evolutionary biology and conventional historical inquiry. There are no sharp boundaries between psychology and philosophy of mind, nor between metaphysics and cosmology (Haack 2008, 2012).
[3] Brian Greene's discussion of string theory provides a contemporary example of a quest for unification and an ultimate account of the world (Greene 2000) Members of the Stanford School such as Nancy Cartwright, John Dupre, Peter Galison and Ian Hacking have argued for the plurality and disunity of the sciences.

pronouncement as a further step on the march of progress.[4] In this vein, figures such as Richard Dawkins have popularised science as the struggle against backwardness and superstition (see Dawkins 1997). We frequently see claims from recently emerged areas such as biotechnology, data science or neuroscience that promise to transform the natural and social worlds by their revolutionary insights and technological spin-offs.[5] However, the potential threats posed by science are not seen as needing more careful reflection because, it is implied, the solution to problems only requires a more scientifically rational society guided by a more complete or rigorous science.

In reaction against such unreflective scientific boosterism, we have seen the growth of scepticism about science in recent decades, often backed by the resistance of counter-movements defending ways of life against the apparently inexorable force of instrumental modes of thinking. Science is no longer seen as the bedrock of truth and guarantor of progress.[6] However, the counter-movements that challenge science often draw on the very terms of the rationality they resist, suggesting a limit to what is thinkable and sayable beyond the form of rationality whose preeminent expression is science. Thus we see the scope of political discourse increasingly narrowed by the constraints of a form of scientistic economic jargon which renders its underlying value-laden presuppositions invisible. Into this confused mix, "postmodern" discourses have injected dismissive anti-science attitudes (often based on misreadings of Foucault or Kuhn) claiming that science is no more than another mask for power. In a public realm in which science is both misunderstood and abused in this way, vested interests have been able to cynically distort debate by resorting to pseudo-science or splitting hairs using other specialised areas of science. Thus economic arguments are pitted against environmental arguments, both based on their respective "sciences", as though either could settle the matter. This "weaponisation" of science only further erodes the fragile tissue of trust on which society, and indeed science, depends.

Without being overdramatic, we could call our present situation a crisis of faith in the scientific worldview. I use the term "faith" very deliberately, remaining aware of the irony that faith, or belief without sufficient evidence, to a certain extent underpins the science which proclaims its grounding in evidence alone. By "crisis", I mean a situation which demands a response we can't provide because we can't adequately grasp the situation or the type of response required. In particular instances, the aims, methods, or findings of

---

[4] Habermas provides a succinct definition of "scientism": "science's belief in itself: that is, the conviction that we can no longer understand science as one form of possible knowledge, but rather must identify knowledge with science" (KHI, 4).
[5] Smith (2019) provides an account of the intoxicating dreams of data scientists in the development of artificial intelligence.
[6] See Nowotny et al. (2001) for an account of the shifting relationship between society and science.

science can be clearly problematic and have unintended social consequences.[7] However, exactly what is problematic cannot always be easily recognised or articulated.[8] While we should not deny the achievements of science, we should reject any simplistic idea of scientific progress based on the reassuring dream of an inexorably evolving rational whole. The assumption that society's rational progress is secured by scientific procedures which in principle we could all discern if only we had sufficient time and inclination, is unwarranted and provides one of modernity's important masks.

One question that frequently lies behind concerns about science is of the role that social, economic, and political power plays, not merely in the practical employment of scientific findings but also in the very framing of problems *as* problems and the constitution of categories of objects of scientific investigation. Given its relations to power, can we insist on the freedom of science from values, even as an ideal, or does such insistence itself bolster a partisan position? Like questions about the limits of scientific knowledge, such issues are far from settled and are frequently revisited when new frontiers of science encroach on our lives. Developments in life sciences like genetic engineering and neuroscience have consequences that far outpace any reflective or considered understanding of them. We often rely on science to address problems created by science. However, by treating contemporary society and the natural world as an ongoing experiment, science continues to generate unforeseen consequences. The ensuing controversies focus not so much on the restricted truth claims of the sciences involved, but on what is thought to escape capture in scientific accounts. Notwithstanding disquiet, the play of power frequently ensures the unconsidered endorsement of science and adoption of scientific technology. Concerns are then countered by claims that social power is distinct from science, only coming into play with its subsequent interpretations and implementation. These are then partitioned off from "science itself", which is seen as a pure, neutral, and self-correcting project oriented solely to truth as its ultimate value. Objections that science fails to provide a full or adequate account of the world are then dismissed with the promise that as it proceeds it will bring more accurate, detailed, and comprehensive knowledge.

---

[7] For example, Thomas Szasz's classic work *The Myth of Mental Illness* criticised the propensity of psychiatry to reach into all aspects of life (Szasz 1974). Watters criticises the export of psychiatric diagnoses to societies where such "disorders" make no sense (Watters 2010). More recently, Harrington explores psychiatry's repeatedly frustrated struggle to understand mental disorders in biomedical terms, arguing that the "biological revolution" has had little to do with breakthroughs in , the understanding and treatment of mental illness, and that the field has fallen into a state of crisis (Harrington 2019).

[8] The difficulties in recognising and articulating what is problematic in science are apparent in Boudrey and Pigliucci's (2017) book-length treatment of the question of what exactly "scientism" is and what is wrong with it.

Science is no longer a vocation or a calling as it once was, but a career. Certainly ideas of inquiry into the immutable essence of the cosmos, the contemplation of truth remote from human affairs, did linger until the 19th century. But today science has become enfolded within institutions of power and wealth generation and lacks the transparency it once had. Today scientific research is inextricably tied to its technological and economic exploitation, supported by massive state and private investment and directed by feedback from industry and the economy (Weber 2004; Habermas 1971b, 53–54). Modern natural science grasps nature as it does only because nature has already come to light as a set of calculable, orderable forces, that is to say, technologically. The tight interlocking of science with its "subsequent" technological applications bears out Heidegger's reversal of the relationship usually assumed: "Modern technology is not applied natural science, far more is modern natural science the application of the essence of technology" (Heidegger 1977). Today the sciences—whether we are talking of economics, genetics, psychology, or some quite different science—are incorporated into society primarily in the form of their technical applications which, directed towards the expansion of power and control, feed back into scientific research. Given the tight connections between science and technology, I will not be making too firm a distinction between them, which is not to deny that important distinctions can be made.

Science ceaselessly challenges external constraints on the expansion of its own capacities. It orders and calculates in order to make entities more fully and extensively calculable, thus continually expanding the domain of its research. Over the past two centuries, the power of technical control over nature has been increasingly extending directly into society. For every isolatable aspect of society whose relations can be analysed in terms of presupposed goals, a new discipline emerges in the social sciences. Science and society mutually condition each other in an ongoing generative symbiosis. As members of society, scientists share the assumptions and values of other members of society. These assumptions and values structure the selection of what counts as problematic from a vast background of seemingly unproblematic phenomena. The particular questions asked of a problematic area, and the choice of particular sciences to answer them, also reflect these assumptions and values. Without any assumptions or values, every possibility would remain open and science wouldn't get off the ground. By holding certain aspects of reality as beyond question, particular questions can be addressed to particular sciences. However, this is already to prescribe the nature of possible answers. Thus, for example, we might see science addressing questions in terms of the economy, employment levels, the crime rate, or standards of living and not in terms of redistributive justice, ecological holism, or living well with less.

To inform practical decisions, a multiplicity of specialised sciences must be reduced to a single voice, often filtered through a social science like economics. However, the reality examined by the social sciences is very different from the reality that the natural sciences have in view.[9] Social kinds are not as mind-independent as natural kinds, which exist prior to the concepts which pick them out. They emerge along with the concepts applied to them. In the process of recognising, invoking, or acknowledging them, we endorse, maintain, or reinforce them. Objects, kinds, actions, concepts, and institutions all interact to constitute social reality. Here we see science, itself shaped by society, also shaping society. When President George W. Bush rejected the Kyoto Agreement because studies showed it would damage the U.S. economy, he affirmed the reality of the economy. When British Prime Minister Margaret Thatcher denied the existence of society, she diminished the reality of society. Both the economy and society are constituted by sets of norms which are only real when recognised. By framing problems within certain assumptions and values, we can recruit science and weaponise its authority to narrow or shift the scope of the real.

We must be circumspect in drawing a line between natural and social reality. The distinction between what is and what is not part of the given, non-interpretable "natural" world is fuzzy. History shows us that these categories are not stable. Things become more real as they emerge as scientific objects and become connected to other phenomena embedded in a broader field of material scientific culture and practice.[10] "Nature" is a political concept which marks a difference in how we deal with things and people and what sort of standing they have, rather than one side of a distinction between two clear-cut kinds of reality. Thus the compositional and calculational practices that underpin the modern scientific worldview compel an understanding not only of the earth as a collection of resources to be extracted, conserved or exploited, substituted for or traded off against one another, but also of ourselves as "human resources". Problems like climate change, resource depletion, mass species extinction, and the global refugee crisis highlight the crisis I allude to. While the sciences and their applied derivatives offer "solutions", these frequently fail to address how risks and impacts are distributed across and within societies. By bracketing "externalities" as mere technical problems, technical "solutions" are easy to find. Ideas such as the colonisation of space or the geo-engineering of the planet become thinkable, while ideas of living with less or sharing more are dismissed by our modern cynicism. Judged as naive or romantic they retreat into that vast realm of the unsayable and unthinkable.

---

[9] Hacking (1999, 103–109) provides a good introduction to this topic.
[10] Daston shows how scientific objects are both real and historical by discussing the ways in which objects come into being and pass away as objects of scientific study (Daston 2000).

Scientific discourses are disseminated via journals, government campaigns, advertising, popular science magazines, newspapers, and educational courses. They insinuate their way into social media and face-to-face discussions, eventually being taken up as the furniture of common sense which quietly adjusts and pushes aside preconceptions it can no longer accommodate. Although science's authority is based on its purported separation from partisan interests, it is never as pure as imagined. The power vested in its authority is used strategically to disqualify common sense, religion, and local, indigenous, and traditional knowledges. Notions of human dignity, of the value of the non-human, of individual and collective responsibilities are swept aside by interests which mobilise scientific authority to anchor their claims and narrow the discussion by trumping counter-claims. On the basis of such authority, there is a temptation to think that it is only science, rather than religion, art, philosophy, or common sense, that offers reliable access to truth and secure knowledge. Theories of secularisation and modernisation have simply assumed that religion will fade away under the clear light of an expanding scientific worldview.[11] Stephen Hawking has already pronounced that philosophy is dead, superseded by science (Hawking and Mlodinow 2010, 5). Philosophy is not yet dead but increasingly domesticated within the academy, where it struggles to justify its utility alongside economics, engineering, and other useful disciplines. In recent years we have seen the emergence of "applied" and even "experimental" philosophy, presumably dovetailing neatly into place amongst the sciences.

THE AIM AND CONTENTS OF THE BOOK

The Aim of This Book

This book aims to reveal the nature of the social effects that ensue when scientific discourses migrate from their insulated institutional settings into society more broadly, to interact with power, discourses, institutions, and social practices. I draw on the work of Michel Foucault (1926–1984) and Jürgen Habermas (1929–), two giants of twentieth-century continental philosophy, to analyse the opacities of scientific modernity. My aim is to show the relevance of Habermas and Foucault to our present situation. While neither directly addresses, for example, climate change, artificial intelligence, or species extinctions, their thought is highly relevant to these threats, as it also is to questions of genetic manipulation, neuroscience, and the place

---

[11] A seminal secularisation theory is Max Weber's *The Protestant Ethic and the Spirit of Capitalism* (Weber 2002).

of religion in a modern scientific world. Existential threats such as climate change are not mere accidents which could not be seen coming. They have been known about by scientists for decades.[12] Yet neither leaders nor citizens, even in wealthy, educated, technologically advanced modern democracies, have acted. In recent years, this crisis in rationality has been exacerbated as democracy and other institutions have been undermined by internet newsfeeds, internet searches, and self-organising social media generated by algorithms directed to optimising economic interests which serve to segregate people into more and more closed communities (Smith 2019). The thought of both Habermas and Foucault is both relevant and urgent in relation to these and other pressing current threats.

The failure of our modern scientific reason as it migrates from its specialised institutions into the everyday lifeworld of the citizenry is central to the concerns of our two thinkers. Both open up new possibilities for thought which take seriously the categories which structure our everyday thinking, such as morality, free will, normativity, and value. In certain instances, for example, Foucault's analysis of neoliberalism, they show remarkable prescience. Although Foucault died in 1984 and the latest work of Habermas which I will consider dates from 2005, these philosophers are capable of provoking fresh ways of thinking that are vitally needed today. I also want to show the common ground between Foucault and Habermas as thinkers in the Enlightenment tradition, along with the differences that emerge from Foucault's radicalisation of this tradition, a radicalisation implicit within the Enlightenment project itself. Finally, I want to articulate a view of philosophy as a distinct realm of inquiry which, although respecting the findings of science, cannot be constrained by scientific method or ontologies.

While Habermas and Foucault both criticise various forms of scientism, neither wants to dispute the truth claims of science, or rather what science narrowly defines as truth in terms of its particular rationality. As Foucault says, "It is not a matter of a battle 'on behalf' of the truth, but a battle about the status of truth and the economic and political role it plays" (PK, 132). There are clearly areas of convergence between Foucault's and Habermas's work. There are also stark differences. These differences led to the so-called debate in the 1980s and 1890s.[13] I will discuss the most significant of these differences and argue that we needn't address them in the terms in which they are frequently framed, as an either/or situation requiring a choice between

---

[12] In 1988, James Hanson, director of the Goddard Instute of Space Studies, testified to the U.S. Congress that NASA was 99 per cent certain that greenhouse gases were causing global warming.
[13] The documents of the Habermas-Foucault debate are mostly collected in Kelly (1992) and Ashenden and Owen (1999).

alternatives. Rather we should consider the "point" or overall intention of each project, what each is *doing*, rather than merely *saying*. We can then see their differing approaches more productively as two distinct moments within the broader framework of modern reflexive rationality, where they remain in tension but, in a certain sense, complementary. To see this requires particular ways of reading Habermas and Foucault. Habermas must be read with his transcendentalism dialled down, such that universalism is linked to contingency by an ongoing contextualisation, while allowing his transcendental ideals to retain their regulative function. Foucault must be read as adopting a stance within modernity's reflexive reason that requires him to step back from normative commitments to open a different perspective, though this is not to say that he is "less normative" than Habermas.

This book can most broadly be described as a meta-critique—a critique of Habermas's and Foucault's critiques of science—which ultimately bears on science itself as a social institution and on philosophy as self-reflection. Unlike a philosophy of science that enquires into the truth or empirical adequacy of science, a critique of science as a social institution considers how what are taken as the truths of science interact with other social institutions and discourses to shape societies which in turn shape science. Rather than approaching this task by simply evaluating current science in terms of an internal logic oriented to truth or empirical adequacy, I will tease out Foucault's and Habermas's analyses in their own broad terms, that is, science as part of a history of social institutions, discourses, and concepts. This means, for example, not analysing society's ills from a position which accepts institutions such as asylums, schools, hospitals, and prisons, and the current scientific discourses which justify them as simply given. Foucault and Habermas view such institutions, discourses, and concepts as products of more profound historical transformations which themselves require analysis. To this end, Habermas will develop his dual model of "lifeworld" and "system", along with a historical analysis of their interactions which brings various social pathologies into view.[14] Foucault will initially analyse discourses in terms of *epistemes*, the unconscious structures which underlie scientific knowledge. Later, he will analyse knowledge as inextricably linked to power and to be found in the finest interstices of social relations.

The question of human nature is clearly central to this work. This is the question that Foucault places firmly on the table in *The Order of Things*.

---

[14] By "lifeworld" (*Lebenswelt*), Habermas means the implicit, intuitively present, familiar web of presuppositions that have to be satisfied for an utterance to be meaningful or valid, or invalid. By "system", he means society viewed in terms of various rule-bound structures, such as the market, which go beyond the intentions of individual actors to enable society to maintain itself by producing stable social patterns.

Throughout all phases of his work, the notion of an *essential* human nature as an object of science remains under challenge. It also comes under consideration in Habermas's analysis of neuroscience and genetic engineering, which argues that fundamental distinctions and categories of thought are necessarily presupposed and cannot be eliminated. A related question will be the possibility of a science of society that, having addressed this essential human nature to discern its needs and legitimate aspirations, can lay down in advance what is required for the organisation of the good society. Foucault and Habermas reject such sciences. Rather than a *science* of society, they offer critical *diagnoses*. In these diagnoses, philosophical enquiry must work in tandem with the sciences.

My task will be to make sense of both Habermas's systematicity and Foucault's discontinuity in relation to their overall projects. This will require fleshing out with chronological narratives of Habermas's and Foucault's lives, details of their contexts, and the experiences which animated and shaped their projects. In my final chapter, I place Habermas and Foucault in relation to each other and find that while Foucault's position is more radical than Habermas's, together they represent the restless movement of modern reflexive rationality. We will see Habermas employ science to build a non-metaphysical foundation for a progressive account of reason, employing context-transcending norms to produce an explanatorily powerful, yet fallible, diagnosis of modernity, as well as conceptual tools of analysis and prescription. We will see Foucault suspend science's authority and adopt agnosticism towards its truth, to display its inextricable entanglement with power. The two strategies are not opposed. We need to both posit context-transcending ideals and reveal their illusory status as emerging from power-laden contexts. For both thinkers, this is a political project in which they seek to change consciousness by revealing it to be conditioned by a vast range of historical and social factors.

Outline of Chapters

Following an outline of chapters given next, the remainder of this chapter sketches the philosophical context framing the discussion of science as a social institution, followed by an outline of the influences and motivations that animate Habermas's and Foucault's projects. I then sketch the reflexive nature of modern reason, which will serve as a touchstone throughout the book, to locate these projects within a broader framework. I conclude by discussing how Habermas's and Foucault's projects reflect their understandings of philosophy's role in relation to science, another touchstone of the book.

The following three chapters chart the development of Habermas's theoretical framework in relation to science. Chapter 2 analyses Habermas's early criticisms of positivism as a form of ideology threatening free and open societies. I discuss *Knowledge and Human Interests,* in which this critique found its comprehensive theoretical articulation, and I argue for its contemporary relevance in addressing the new forms of positivism backed by the philosophical dogmas of reductive scientific naturalism. Chapter 3 examines *The Theory of Communicative Action*, in which Habermas employs science—systems theory and social evolutionary theories—to develop a normative philosophical basis for social critique. Maintaining that a dual perspective covering both functional systems and lifeworlds is required for social theory, he steers a path between the objectivism of positivism and the relativism of hermeneutics. Although the resulting theoretical edifice brings modernity's ills into sharp relief, I argue that the complexity of relations between lifeworld and system cannot be rendered in such broad strokes. I conclude by arguing for a reading of Habermas which escapes entrapment by his ambitious theoretical structures. Chapter 4 examines Habermas's work since *The Theory of Communicative Action* to argue that in his major works of this period, his systematic explanatory ambitions still inhibit an adequate account of the complex interactions between scientific, pre-scientific, and non-scientific discourses and their entwinement with politics and power. After considering Habermas's view of the relationship between philosophy and science, I turn to three essays in which he defends everyday intuitions from an unconsidered and overly hasty takeover by the imperatives of scientific technology. I argue that these essays do justice to the complexity of issues by drawing on the structures provided by his major theoretical work, without being entrapped by their problematic commitments.

Chapters 5–7 trace Foucault's project in terms of the conventional periodisation. Chapter 5 introduces Foucault's project as a radicalised critique of Habermas's more conventional critique, linking Foucault's project to the Enlightenment tradition by its *ethos* of ongoing critique and self-transformation. I will show how this radicalisation appears in Foucault's "archaeological" works, which claim that thought and speech in given historical epochs are constrained by anonymous unconscious contingent structures. After examining *The History of Madness*, an early work foreshadowing all Foucault's later themes and preoccupations, I situate his archaeology in relation to more conventional histories. I then discuss *The Order of Things*, a critique of the human sciences in which Foucault claims that thought is caught in an "anthropological sleep". Against his critics, I argue that Foucault's archaeology is not a failed attempt to provide a robust scientific or philosophical theory, but a critical orientation towards the present which opens the human sciences to question by revealing their contingent construction. Chapter 6 draws on *Discipline and Punish*, *The History of Sexuality* vol.1, and his lectures at the *Collège de France* to discuss Foucault's genealogical works. By arguing that power and knowledge condition each other and are mutually generative, Foucault's genealogy opposes the confidence of positivist histories of scientific progress, where truth is assumed to be the correspondence to an existing, though unknown, state of affairs towards which knowledge inexorably moves. Drawing on Foucault's analyses of the human sciences, I discuss how knowledge interacts with practices, concepts, subjects, and objects to produce new objects which don't pre-date their theoretical articulation. We see that Foucault is particularly resistant to moves to naturalise human nature, to dictate some essential biological or transcendental determinant of what we are. I discuss Foucault's notion of "biopower" and its transformation into the concept of "governmentality". Having set out my interpretation of Foucault's archaeological and genealogical works, I will defend Foucault against criticism that he tacitly appeals to an implicit justificatory framework, contrary to his value-free stance. Chapter 7 elucidates the final period of Foucault's work by drawing on *History of Sexuality,* vol. 2 and his lectures at the *Collège* from 1981. I argue that Foucault had always wanted to open a perspective beyond the limits he analysed. By shifting to a notion of the subject who is more than a passive object of power/ knowledge, Foucault's ethics provide the framework to articulate this perspective. By acting upon itself, the ethical subject resists scientific power/ knowledge. The shift in emphasis away from power to subjectivity accommodates the perspectives of all Foucault's previous work. Foucault embraces self-transformation as the *telos* of philosophical life, not directed towards the authoritative truths of science, but towards a subjectivity that problematises whatever is taken as universal and necessary. I argue that Foucault's philosophical critique makes new possibilities visible and invites, rather than prescribes or legislates, political action.

Chapter 8 returns to the themes sketched in chapter 1 to draw together the work of Foucault and Habermas in ways that bear directly on the question of science. By introducing a broader conception of normativity, I articulate a framework of modern reflexive reason in which we can see Foucault and Habermas adopt specific strategies to suit their aims. By characterising these strategies in terms of the imbricated perspectives of observer and participant, necessity and contingency, discovery and invention, I argue that these two strategies represent two distinct genres of intellectual endeavour, neither assessable by the standards of the other. While Habermas wants conclusions and so draws the line at questioning that which must be necessarily presupposed, Foucault questions what seems to be necessarily presupposed. However, both respond to the contemporary ideological distortions induced by scientific naturalism. Both see philosophy as able to articulate social problems not visible from within the specialised perspectives and limited scope of science.

PHILOSOPHY AND ITS CONTEXTS

Within philosophy, the debate over the limits and possibilities of science has taken many twists and turns. Weber set the scene for contemporary anxieties, arguing that in modern rationalised culture, science is accompanied by a particular form of rationality, "means-ends" rationality, and a narrowing of imaginative possibilities (Weber 1978). In contrast to Weber's broad sociological perspective, the logical empiricists sought to capture the rationality of science in terms of a logic that largely abstracted from the social and historical contexts of the sciences. Popper, in this respect consistent with the logical positivists, maintained a heroic optimism regarding the natural sciences, in which he saw an intrinsic potential for human liberation from natural and social constraints (Popper 1963). Against such positive appraisals, Horkheimer and Adorno drew on Weber to argue that the growth of scientific rationality represented a growing means of domination, a new mythology in which instrumental rationality comes to be seen as reason per se (Adorno and Horkheimer 1997). Several decades after this gloomy prognosis, the picture was further complicated by the emergence of new fields of inquiry. Within the English-speaking world, a renewed sociology of scientific knowledge (SSK) emerged, opposing positivistic philosophy of science with theses like holism about testing, incommensurability, and the theory-dependence of observation.[15] Thomas Kuhn attacked the logical empiricists obsessive focus on the logical context of justification to the detriment of the social and historical contexts of scientific discovery. In a further development, the "strong programme" at the University of Edinburgh insisted on sociological explanation of not only false beliefs but all scientific beliefs (Bloor et al. 1996). Although sociologists were quick to point out that political interests didn't straightforwardly determine scientific beliefs, some of their formulations led to accusations of relativism regarding belief and justification. Shapin and Schaffer (1985) developed a sophisticated version of the strong programme, arguing that solutions to the problem of knowledge are inseparable from solutions to the problem of the social order. Criticism again highlighted a problematic relativism.[16] Criticising SSK for its reductionism and

---

[15] See, for example, Kuhn (1996). While Kuhn's historical analysis at the time of its publication (1962) might have seemed a radically new approach, similar arguments had been elaborated in continental Europe thirty years earlier. See, for example, Fleck (1979), as well as the works of Gaston Bachelard and Georges Canguilhem discussed in this chapter.

[16] Krausz (1989) and Hollis and Lukes (1982) provide a background to the problems of relativism in various contexts, including in SSK.

human-centredness, "actor network theory" (ANT)[17] puts forward a theoretical and methodological approach in which the social and natural worlds exist in constantly shifting networks of relationships. Rather than affirming the radical autonomy of humans, it sees objects, ideas, and processes as being just as important in creating social situations.

As a generalisation, we could say that the various strands of the history, sociology, and anthropology of scientific knowledge have depicted the "internal logic" of scientific progress as deeply imbricated with "external influences", and have often sought an explanatory basis for scientific reason in the social and political contexts of science. William Rehg refers to the resulting tension between the logical perspectives favoured by philosophers of science and social-institutional perspectives favoured by sociologists as "Kuhn's gap", a gap that appeared with Kuhn's analysis of science (Rehg, 2009, 33).[18] Whilst this tension is not necessarily problematic, much more needs to be said, and it is here that we will see Foucault meticulously lay out an "analytic" of power and its relationship to knowledge. However, this work requires interpretation and is far from offering easy solutions. It is therefore not surprising that debates around the claims of SSK and the work of Kuhn have, in the public imagination, led to doubts about the neutrality and objectivity of science and, in more extreme forms, the truths of our best science. One consequence is that today we see climate change contrarians present themselves as the overthrowers of a false climate change paradigm.[19]

In the new millennium, we see further shifts, in part generated by breakthroughs in the life sciences, particularly genetics and neuroscience. A popular resurgence of enthusiasm for science, frequently allied with a radical hostility to philosophy and religion, has been promoted by such figures as Sam Harris, Richard Dawkins, and Stephen Hawking. This "new scientism" finds support, or at least no criticism, from the philosophical consensus on scientific naturalism, which has become the dominant paradigm in Anglo-American philosophy.[20] This consensus initially crystallised around the work of figures such as W. V. Quine and Wilfred Sellars, and can be characterised by three themes (De Caro and MacArthur 2004, 7):

1. *Ontological theme*—a commitment to an exclusively scientific conception of nature;

---

[17] ANT is often seen as displacing the strong programme. Owing an intellectual debt to Foucault and French epistemology, ANT was developed by Bruno Latour and several others.
[18] Rehg seeks to bridge "Kuhn's Gap" by an argumentation theory of science that does justice to both logical and contextual factors (Rehg 2009)
[19] For an example of a climate contrarian's use of Kuhn, see Stockwell (2013).
[20] An example of new scientism is Rosenberg's (2011) employment of "scientism" not as a pejorative, but as an honorific term denoting the claim that "the persistent questions" people ask about the nature of reality, the purpose of things, the foundations of value and morality, the way the mind works, the basis of personal identity, and the course of human history, could all be answered by the resources of science.

2. *Epistemological theme*—scientific inquiries are our only source of real knowledge, all other forms of knowledge being either reducible to scientific knowledge or illegitimate;
3. *Metaphilosophical theme*—philosophical inquiry is continuous with science.

These themes give rise to the "placement problem" of how phenomena such as moral, epistemic, and aesthetic norms; meaning; universals; consciousness; and intentionality can "fit into" the world described by physics, chemistry, and biology (Price 2004). According to the scientific naturalist, if the phenomena which these concepts pick out are to be accepted as part of nature, they need to be reduced to the scientific image of nature. Otherwise they are considered non-genuine, second-rate phenomena.

To grasp the implications of these themes, consider the first. Ontology is not a merely academic concern, but profoundly political. To proclaim, with Margaret Thatcher, that there is no such thing as society is not to neutrally describe a putative state of affairs, but to legitimate a particular regime of social organisation. However, it is not only political questions that reductive scientific naturalism bears upon, but also questions that go to the core of how we understand ourselves. Although challenged as far back as Harry Frankfurt's 1971 essay, it seems that reductive naturalistic conceptions of the person remain thoroughly entrenched in Anglo-American philosophy today (Frankfurt 1988, 11–26). Such conceptions betray an undue deference to natural science, which simply does not have persons in view. Such is this deference, that philosophy must follow brain science, which "after more than a century of looking for it . . . [has] long since concluded that there is no conceivable place for a self to be located in the physical brain, and that it simply doesn't exist" (Hughes, cited in Kompridis 2009).[21]

The problem is that scientific naturalism strips away the very vocabulary we need to make sense of ourselves as persons, a problem compounded by an understanding of philosophy so in awe of science that it discounts anything seen to problematise natural science's all-encompassing jurisdiction. Thus, in a major publication on the philosophy of science, Foucault, who remains one of the most, if not *the* most, cited authors in the social sciences, is lumped together with Derrida and Lyotard as a "postmodern philosopher". In a short one-paragraph parody, a few sentences of Foucault's lengthy exposition of Nietzsche's will to truth are quoted and Nietzsche's view attributed to Foucault as his own, devoid of any context that could illuminate Foucault's

---

[21] Kompridis (2009) provides the *Stanford Encyclopedia of Philosophy's* entry on personal identity as an example of entrenched reductive naturalism. See Olsen (2021).

actual position. Habermas, notwithstanding his considerable influence across the social sciences, is not mentioned in the 600 pages of this tome.[22]

A frequent strategy to defend scientific naturalism is to invoke the fact/value dichotomy. Although few philosophers today would defend an absolute dichotomy between the descriptive and the evaluative, the idea that a clear separation can be made between scientific facts and political or social values seems deeply ingrained in our wider culture.[23] The question of whether something is really factual or merely valued has become stubbornly resistant to challenge or debate. While drawing attention to the fact/value dichotomy in the face of concealed values may be reasonable, the dichotomy's extension to knowledge per se gestures towards a realm of knowledge absolutely free of values. Such metaphysical realism supports the idea that natural science is value-free and that its statements aim to faithfully picture objects, properties, and relations that exist just as they are, independently of our thoughts about them or our perceptions of them. In this spirit, Bernard Williams argues for an "absolute conception of reality" of a world that is "to the maximum degree independent of our perspective and its peculiarities" and upon which natural science converges (Williams 1985, 139). Such a view seems committed to the thought that the correct application of scientific concepts is dictated by reality itself. Against such views, philosophers such as Richard Rorty challenge the very idea of our thoughts and language being beholden to facts that make them true (Rorty 1998, 122).

Philosophy cannot be considered a pure form of inquiry any more than science. Whilst ideally, philosophy actively engages with and responds to scientific advances, it is also unconsciously shaped by contexts such as today's maelstrom of scientific and technological innovation driven by the opaque imperatives of power, commerce, and technology. This opacity often precludes serious consideration of potential threats posed by scientific technologies to the recognitive relations that bind society and support institutions like democracy or science itself. Yet, with these momentous changes in the background, we see proponents of evolutionary psychology blithely proclaim genetic-evolutionary "explanations" of art, morality, or a fixed truth of human nature.[24] Presumably an ideal society is in the offing. Some neuroscientists

---

[22] Baghramiam, M. in Psillos and Curd (eds.) 2008, 271.
[23] In philosophy, we saw an absolute dichotomy in A. J. Ayer's meta-ethical emotivism, according to which ethical propositions are normative and, as such, are pseudo-concepts because they don't represent facts (Ayer 1952, 108). Karl Popper sought to harness the dichotomy to disqualify social sciences which don't adopt "a purely objective method" (Popper in Habermas et al. 1976, 102).
[24] As just one example, Wilson's "evolutionary ethics" argues that scientific investigation can pick out "moral sentiments" which, it is assumed, are all an ethical theory requires. Wilson doesn't tell us which sentiments are moral, nor does it follow that an evolutionary explanation can tell us this either (Wilson 1998). This example is from Haack (2009).

deny the possibility of free will, without appreciating the irony of pulling the rug out from under their own feet. Although it can be argued that such claims are neither "real" science nor philosophy, their backing by eminent scientists and philosophers gives them authoritative weight.[25] What starts as rigorous science morphs into careless philosophy, insinuating its way into common knowledge. A reinvigorated critique of scientific naturalism is clearly needed.

For over a century, pragmatist philosophers have argued against such reductionist tendencies. Rather than human understanding mirroring reality as a faithful representation, they argue for a notion of interactive coping with a world which is, at the same time, shaped by this coping activity. The facts of science are then seen to be infused with values that permeate experience. Rather than standing apart from other activities, science is understood as one amongst many human activities to which we grant a certain authority in specific contexts. From a pragmatist point of view, it might be *useful* to distinguish between evaluative and theoretical considerations or, in specific contexts, prioritise one over the other. However, for Putnam, "nothing metaphysical follows from the existence of a fact/value distinction in this modest sense." The most unnerving aspect of the dichotomy for Putnam is how it has "corrupted our thinking about both ethical reasoning and description of the world" (Putnam 2002, 19).

In recent years, scientific naturalism has been challenged by a form of naturalism which draws on pragmatism. Described variously as "pluralistic" (Dupré), "catholic" (Strawson), or "liberal" (McDowell), what De Caro and MacArthur call "liberal naturalism" seeks a place for non-scientific objects in a broader conception of nature (De Caro and MacArthur 2004). Rather than tailoring ontology to the posits of the sciences, this expanded naturalism recognises the prima facie irreducible reality of what Wilfred Sellars called "the manifest image" (Sellars 1963). This form of naturalism can encompass, for example, value in artworks, the moral dimensions of persons, or distinctively first-person experiences such as rational deliberation. Liberal naturalism aims to do justice to the non-scientific subjectivity of the subject. While causal accounts may serve to predict and control mental states, the problem is that normativity just drops out of casual explanations. We can't think of reasons as just happening to lead to sound conclusions. One *ought* to reach conclusions in the light of reasons. What one *ought* to do is simply not explicable in terms of what

---

[25] The view that free will is an illusion is strongly implied, if not argued for, in the work of neuroscientists such as V. S. Ramachandran and philosophers such as Paul Churchland (Ramachandran 2004; Churchland 1984). It has been argued that evolutionary psychologist Steven Pinker harnesses arguments and strawmen to present a deterministic view of human nature which matches his political preferences (Pinker 2002, see the criticisms by Orr 2003; Eriksen 2007; Bateson 2002; Dupre 2003).

will, must, or could happen.[26] Truth, knowledge, and justification are all normative. Liberal naturalism seeks to provide a non-reductive, non-supernatural account of the rational and conceptual normativity to which we are responsive.

While liberal naturalism is attractive, it doesn't go far enough. Specifically, such positions address themselves primarily to questions of what can count as "real", seeking to expand the category beyond the limits of the natural sciences. Habermas and Foucault also want to expand the realm of what counts as "real", "existing", or "true". But, importantly, they focus not on what counts as real or what should be included as real, but on how what is taken as "real" functions and its effects on our social practices. Reality thus becomes political. This is possible because Habermas and Foucault address exactly how science operates as a social institution—how it disseminates into society, how it is entwined with power, and how it mediates and is mediated by other social institutions and regions of rationality. The disciplines of history and sociology of science also adopt this approach. Thus, Shapin and Schaffer examine the work, habits, rituals, and social structures of science to reveal that politics is as tied up with what scientists do and the contexts of those doings as it is with what they say (Shapin and Schaffer 1985). Such historical-sociological accounts show how what are taken as the key scientific givens—facts, interpretations, experiments, truths—ground political and social orders. However, such studies again don't go far enough. They are insufficiently normative. Itself a science, sociology steps back from normative commitments in order to offer descriptions. Certainly, sociological descriptions can undermine whiggish notions of science as inexorably converging on Truth, but the descriptive standpoint offers no hint of a motivation to think that things should be otherwise than they are.

It is here that Foucault and Habermas make their contributions. As politically engaged philosophers, they reject the notion of philosophy as an armchair discipline analysing a priori concepts. Habermas draws on science to mobilise his normative stance. I will argue that Foucault's work is also normative, although not in the way that his critics think. Foucault pursues a practice of history writing that seems resolutely descriptive, while occasionally lapsing into emotionally charged and seemingly condemnatory accounts, giving ample grounds for his critics to denounce his "normative confusions" in helping himself to positions he has ruled out (Fraser 1981). His normative orientation only becomes clear in his late work, in which he spells out his vision of philosophy as an *activity* that opposes science by opening otherwise

---

[26] As Brandom (2004) argues, knowledge is a non-scientific concept upon which the sciences cannot shed any light. Knowledge is intelligible as standing in the space of reasons insofar as it can be taken as a status achieved in the space of giving and asking for reasons. It cannot be accounted for in causal objectivist terms.

unforeseeable possibilities and freedoms. If Foucault's work is easily confused with sociology and history of science, the theoretical structures of Habermas's systematic edifice are also easily confused with the metaphysics of scientific naturalism. One of his most astute critics, Thomas McCarthy, criticises Habermas for importing systems theory, surely a form of science, into his philosophy (McCarthy 1991, 180). Yet Habermas's comprehensive theoretical framework can also be understood, as McCarthy argues, pragmatically and contextually and can be employed to analyse and criticise social arrangements in a way that opens new possibilities beyond scientific reason (Hoy and McCarthy 1994, 72). What we see with both philosophers is a vision of philosophy as having a distinctive and semi-autonomous role in its orientation towards emancipation and the good life.

HABERMAS AND FOUCAULT: LIVES AND MOTIVATIONS

Habermas's Life and Motivations

Habermas lived through Germany's tumultuous years of war and defeat, the Allied occupation and war crimes trials, the drafting of a new constitution, social, economic, and industrial reconstruction, and eventual reunification. Much of this time played out against the backdrop of the Cold War and the threat of nuclear annihilation. As a child, a series of operations on his cleft palate "sharpened awareness of the deep dependence of one person on others", leading him to "those approaches that emphasise the intersubjective constitution of the mind" (BNR, 14). At school, his nasal articulation and distorted pronunciation "directed [his] attention to the otherwise unobtrusive intermediary world of symbols that cannot be grasped like physical objects" and the role of linguistic communication in the genesis of moral and social norms governing communal life (BNR, 15-7). As a teenager he was drafted into the Hitler Youth Movement. After the war, aged sixteen, he was shocked to learn of Nazi atrocities. "Suddenly our own history was shown in a light that changed all essential aspects at one stroke. You realised that you had lived in a criminal political system" (Muller Doohm, 25). He was appalled to witness his elders making excuses for perpetrators (Habermas 1983, 57). Habermas's confrontation with the legacy of the Nazi past became a fundamental theme of his adult life which spurred his thoughts towards the conditions required for stable and just societies (BNR, 17). Heidegger was a major influence on the young Habermas, but in 1953, he was again shocked to find that Heidegger's republished 1935 lecture *Introduction to Metaphysics* retained the original reference to "the inner truth and greatness of National Socialism" without any revision or commentary (BNR, 18). Responding with a newspaper article criticising Heidegger, Habermas took up the role that he has consistently maintained as a public intellectual engaged in political and social questions (Habermas 1953).

This experience also motivated a search for conceptual resources beyond German thought, particularly within the Anglo-American pragmatist and democratic traditions. In 1956, Habermas became Adorno's research assistant at the Institute for Social Research at Frankfurt. Here he encountered the Frankfurt School's integration of social sciences with philosophy in the form of critical theory. During this period, Habermas started writing his *Habilitationsschrift* on the public sphere.[27] Criticised by Horkheimer, who saw him as too radical, he left the School in 1959 to seek supervision from Wolfgang Adendroth at Marburg. After gaining his *Habilitation* in 1961, he took up the position of Professor of Philosophy at Heidelberg. In 1964, he returned to the University of Frankfurt to take up Horkheimer's former chair as Professor of Philosophy and Sociology. From 1971, Habermas was a director of the Max Planck Institute at Starnberg, returning to Frankfurt in 1983 to teach philosophy. He retired in 1994 but taught in the United States for several more years. He maintained a prominent role as a public intellectual on matters to do with German foreign policy during the administration of President G. W. Bush and more recently on the politics of the EU crisis.

The most important intellectual influence on Habermas is the broadly conceived Marxist tradition that grew out of German idealism and inspired Horkheimer's notion of critical theory. Much of the development of Western Marxism can be seen as a shift from a "scientific" Marxism that concentrated on the productive base as the engine of change, towards a focus on the "superstructural" elements of culture as autonomous processes. This "revisionism" involved bringing other thinkers and traditions to bear while still conserving a normative orientation. In this spirit, Habermas's first major work, *The Structural Transformation of the Public Sphere* (hereafter STPS), represents a further development of, or arguably a break from, the earlier conception of critical theory.[28] Like earlier critical theory, it provides an analysis—drawing together insights from history, sociology, literature, psychology, economics, and philosophy—in order to diagnose a situation and articulate emancipatory possibilities. More distant from the war and with additional critical resources, Habermas is better placed than his predecessors to develop critical theory as a response to the challenges of late capitalism. To best orient us within Habermas's vast body of work, I will provide a brief outline of STPS, because in it, we clearly see his abiding commitment to a form of enlightened democratic rationality. This commitment has sustained Habermas's

---

[27] Published in 1962.
[28] Johnson (2009) provides a helpful analysis of Habermas's work on the public sphere.

work throughout his career. In a 2005 interview, he said, "Democracy is like a red thread through my work from the beginning until what I am presently interested in" (Habermas 2007a). Habermas has sought to honour this commitment by articulating a response to what he has seen as the most pressing theoretical and practical issues of the day.

The historical account traced by STPS charts the emergence in the seventeenth and eighteenth centuries of a public sphere of discourse and its subsequent transformations. Habermas maintains that this public sphere, although compromised, can still provide normative guidance today. By "public sphere", Habermas means that within which "something approaching a public opinion can be formed" (cited in Edgar 2005, 31). This would include the exchanges of opinions that occur in face-to-face communication in public spaces such as lecture halls, salons, coffee houses, and clubs, or through journals and newspapers, all of which enabled the formation of more rational opinions. The public sphere was both an institution and a politically effective idea, driven by the need for an arena in which matters of common concern could be rationally discussed and public policy critically assessed in terms of a concept of the public interest. Although actually beset by contradictions, the public sphere is nonetheless implicitly based on the counterfactual norms which underpin democratic discourse.

Habermas articulates this repressed normative core. First, a type of discourse was conducted that "disregarded status altogether." The authority of the better argument could assert itself against any form of status—rank, birth, prestige of public office, or economic power. Second, "the problematisation of areas that until then had not been questioned was possible." Church and state authorities, which previously held a monopoly on the interpretation of matters of public concern, could now be challenged. Third, the public is seen as inclusive, at least *in principle*: "However exclusive the public might be in any given instance, it could never close itself off entirely and become consolidated as a clique" (STPS, 37). These norms, applied to informal collective discourses, grounded claims of legitimate authority based solely on the rational claim of the best argument. Although seldom actualised, these norms are powerfully consequential as ideals against which actual practice can be judged. Despite tensions induced by ideological distortions, the growth of the public sphere throughout the eighteenth century was accompanied by a series of reforms. Public opinion forged through argumentation could now challenge the authority and power traditionally vested in the state. Throughout the eighteenth century in Britain, there were increasing appeals to "the sense of the people", "the common voice", and the "public spirit". Such appeals represent the first moves by which Parliament became responsive to the people, whose collective opinion legitimated parliamentary lawmaking (STPS, 31).

Habermas's entire project will rely on the derivation of the counterfactual normativity of the early bourgeois public sphere.

However, this early bourgeois public sphere began to transform as those left out challenged their exclusion from its supposedly universal openness and reciprocity. In the nineteenth century, the masses comprising the working class established themselves as political agents and formed trade unions to demand political and social rights. At the same time, capitalism's boom-bust cycle increasingly stripped any appearance of "natural" justice from the free market and, from the beginning of the twentieth century, the small-scale entrepreneur was displaced by larger companies. Increasingly, the concentration of capital in large oligopolies undermined the notion of perfect competition. The state responded by expanding its activity and regulating the economy to become a major provider of goods and services. With economic rationalisation and technological development, the state and society became interlocked and the public sphere lost its role of the go-between mediating between public and private. New and larger powers increasingly came to contest and control this public space. After the First World War, government regulation and power increased in order to regulate capitalism as it moved from laissez-faire to a more monopolistic form. The public sphere was transformed from a sphere of independent entrepreneurs attempting to articulate their own standpoints, to a domain vulnerable to the vicissitudes of monopoly capitalism, where big business, governments, and trade unions could advance their own agendas in the political domain. These transformations, culminating in the welfare state, undermined the initiative and autonomy of the bourgeoisie, who came to see themselves as subservient to government economic support and regulation. The vitality of the public sphere was thus diminished as autonomy was narrowed to the exercise of private preferences in leisure activities and lifestyle choices. With the rise of formally organised political parties, the politician became a professional, drawing upon advertising and public relations to elicit acclamation. The bourgeois public sphere, originally a conduit between society and the state, had assumed that rational debate could be free of economic interests and that all members of a universal public sphere shared common interests. Now, in advanced capitalism, rational debate based on this commonality is replaced by the haggling and bargaining of special interest groups.

In many respects, Habermas's argument runs parallel to Adorno's and Horkheimer's *Dialectic of Enlightenment* of 1944. In line with their critique of Enlightenment transformed back into myth, Habermas's bourgeois public sphere is transformed into manipulated publicity in which governments and corporations secure plebiscitary acclamation. The difference is that Habermas does not abandon the Enlightenment ideals. Rather, he insists that the values articulated in the norms of the bourgeois public sphere have an enduring and functioning significance, both regulative and critical. If the goals of society lie in free communication, we must theoretically reconstruct its abstract norms, which transcend their embodiment in any actual society. For Habermas, the bourgeois public sphere remains a crucial normative resource for democratic discourse. His critical theory differs from that of his predecessors in other ways as well. Since Adorno's and Horkheimer's version sought to enable individuals to resist homogenisation by current social reality, freedom was characterised only negatively, by the capacity to resist. Habermas, however, links freedom inextricably to democracy and the renewal of its institutions and thus aims to identify the social and institutional conditions of democratic institutions which foster autonomy. In Habermas's earliest work, we see the commitment to democracy reflected in his articulation of the norms of the bourgeois public sphere and his interest in domination-free communication, which would later form the core standards of his moral-political theory. We will also see that this commitment to democracy and undistorted communication forms the background to Habermas's various responses to science.

After STPS, Habermas took critical theory in a different direction, still aimed at understanding the basis for democratic discourse, but now directed against the then-dominant positivism which he saw as a truncated form of reason that distorted communication.[29] This move must be seen in the context of the generally uncritical attitude towards scientific and technological progress prevalent in the early 1960s, a time when the state played a major role in the economy, to the extent that "technocracy", the idea that political decisions were merely technical and could be made by scientific experts, seemed a real possibility. Habermas responded by contributing to "the positivist dispute", an exchange initiated by Adorno and Popper in which Habermas and Adorno defended a dialectical social science against the reductive tendencies of positivism (Habermas et al. 1976, TP). Habermas's critique of positivism found its definitive formulation in *Knowledge and Human Interests* (hereafter KHI), which put forward a Kantian epistemology where three deep-seated anthropological interests serve as the quasi-transcendental conditions of three distinct domains of knowledge. Habermas surveys the "abandoned stages of reflection" to reconstruct how Kant's differentiated notion of reason had given way to positivism's view of the natural sciences as being the paradigm of knowledge per se. Drawing on Kant, Hegel, Marx, Peirce, Dilthey, and Freud, this critique reflects on both the function of knowledge in the reproduction of social life and the historical conditions in which the subject of knowledge is

---

[29] By "positivism", I mean the philosophical attitude which restricts what can count as "knowledge" to what is manifest in sense experience, thus excluding value judgements and normative statements. Chapter 2 provides further discussion. Also see Kolakowski (1972).

shaped. It represents Habermas's first attempt to provide a systematic framework for a critical social theory. As positivism lost its dominance, Habermas began developing a more comprehensive social theory drawing on various developmental accounts of morality and communication from the social sciences. However, the basis of his early concern with positivism—that it suppresses reflection and free communication—is sustained throughout his entire career and is seen in his later engagement with Luhmann's systems theory and his more recent discussions of scientific naturalism (LC, 130–42; BNR, 151–181).

After KHI, Habermas's Kantian critical theory of cognitive interests gave way to his theory of communicative action. He had long insisted that reason resides at the very heart of human communication, which is the engine for human emancipation (KHI, 314). Habermas turns to empirical sciences to reconstruct the irreversible and progressive stages of childhood learning which he extrapolates to a developmental sequence explaining the stages of social evolution. By focusing on communicative structures, Habermas's "post-metaphysical" theory of rationality and social integration, based on reconstructions of the competences and normative presuppositions underlying communication, avoids any claim to ultimate foundations. Habermas's 1981 *magnum opus, The Theory of Communicative Action* (hereafter TCA), moves critical theory away from the strong transcendental framework of KHI towards a more modest, fallibilist, empirical account of the philosophical claim to universality and rationality. He will characterise this naturalistic approach as a form of "soft naturalism" where "weakly transcendental" conditions are directed to invariant structures and conditions and raise universal, but defeasible, claims which govern the familiar processes of communication underlying social cooperation. Drawing on the speech act theory of J. L. Austin and John Searle, he shows how, in everyday life, speakers make claims that transcend specific conversational contexts. The expectations raised in such speechacts imply a set of rules which define the meaning of a domination-free dialogue undistorted by power. This provides Habermas with a rational standard against which he can discern various pathologies of modernity.

The 1980s also marked a period when Habermas engaged in debates about contemporary issues such as how Germany's Nazi past should be remembered, immigration, and neo-Nazism. His commitment to a universal humanism led to an interest in cosmopolitanism and multiculturalism which played out in questions about the relationship between constitutional law and democracy, the role of the nation state, and the "clash of civilisations" thesis. From his earliest work, Habermas was committed to the reconstruction of his own society as a liberal democracy capable of resisting totalitarianism. This commitment continued in his later works on communicative action, legal philosophy, international law, and post-secular society. While conscious of philosophy's ultimately broader purview, Habermas recognises the emancipatory potential of science and incorporates it into his theorising to provide perspectives not available to philosophy. He is, however, concerned that science increasingly links itself to economic imperatives, overstepping its proper domain and short-circuiting democratic deliberations. For example, he argues that technologies such as human genetic engineering are advancing so rapidly and are so tied to economic imperatives that they are bypassing rational scrutiny and informed public discussion (FHN, 18). Criticising all forms of fundamentalism, Habermas reconsiders the relationship between reason and religion which, in TCA, he had treated from a sociological perspective as an archaic mode of integration. From the 1980s, Habermas explores the relationship between religious and philosophical modes of discourse and the role of religion in politics. He concedes that religious discourse still harbours potentials for meaning from which philosophy can learn (RR, 77, 162). With the heroic vision of modernity of 1960s positivism now long discredited, Habermas defends a more nuanced account of the "incomplete project of modernity" and its rational foundations against anti-modernists, traditionalists, conservatives, and "postmodernists" (with whom he includes Foucault).

In his 1992 *Between Facts and Norms* (hereafter BFN), Habermas sets out his theory of deliberative democracy, reworking the issues first raised in STPS by applying his complex idealised discourse theory to the political-legal institutions of complex modern societies. In one sense, Habermas is staunchly Hegelian, locating his theory's normative dimension in existing institutions. Unlike Foucault, who sees Kant as the founder of modernity, Habermas thinks it was Hegel who first saw modernity as a philosophical problem. While distancing himself from the excesses of Hegelian metaphysics, Habermas seeks to recover the hidden traces of immanent reason so as to provide a better grounding for a critical social theory of modernity. Philosophy has come to recognise that consciousness has no unmediated access to being and the subject is conditioned by a multiplicity of historical, economic, social, psychological, and cultural factors. Philosophy must therefore direct its attention to formal conditions of rationality, such as those found in language.

Habermas's critical philosophical theory is not distinctive because it endorses any particular scientific theory or method. In fact, he uses social sciences in methodologically and theoretically pluralistic ways. What is distinctive about his critical philosophy is that it unites normative and empirical inquiry. The motivating commitment that lies behind all Habermas's work is the idea of human emancipation through the establishment and maintenance of democratic institutions, which requires interactive open-ended discourse as free as possible from all constraints, including the internal constraints which Kant referred to as "immaturity". Despite the many directions in which Habermas's practical engagements and sources of inspiration have taken him, he has never strayed from this commitment. However, in this vast body of work, science is always pivotal in two respects. First, in his criticisms of positivism and reductive naturalism, what Habermas criticises is not the content of science or its truth or falsity, but the philosophical misunderstanding of science as the most authoritative form of knowledge which can be unproblematically used for practical ends. This is the basis of his critiques not only of positivism but also of systems theory, technocracy, genetic engineering, neuroscience, economic rationalism, and many other areas. Second, in these very critiques, Habermas harnesses sciences like linguistics, anthropology, sociology, and psychology as resources enabling an "external" perspective which is capable of being harnessed as a normative standpoint to cut across conditioned philosophical reason.

Foucault's Life and Motivations

Foucault was born in 1926, the son of a surgeon. He was academically gifted but troubled. Depressed and obsessed with thoughts of self-mutilation and suicide, his father sent him to a psychiatrist. Eribon cites a doctor who knew Foucault and suggested that his condition arose from the distress of coming to terms with his homosexuality, an unsurprising response given the homophobia of 1940s France (Eribon 1991, 26).[30] It is also not surprising that, in addition to studies in history and philosophy, Foucault developed an "obsessive interest" in psychiatry, psychology, and psychoanalysis (Eribon 1991, 27). Working in mental hospitals during his studies, he qualified in psychology as well as philosophy. His writings and political engagements continued to reflect a concern for marginal groups and the forms of power to which they were subjected. It was surely these experiences that prompted his insights into the inextricable relationship of power with scientific knowledge. While the meanings of "power" and "knowledge" differ, Foucault claims they cannot be analysed separately. Both are inextricably tied to the subject, which must also be drawn into this constellation. Thus Foucault refers to particular forms of subjectivity, such as homosexuality, madness, or delinquency, as constituted by particular regimes of power/knowledge.

Rabinow and Dreyfus situate Foucault's development between phenomenology and the two reactions to it, structuralism and hermeneutics (Rabinow and Dreyfus 1983, xix–xxi). Although this contextual triangulation gives an

---

[30] Miller notes "that rumours of homosexuality could and did break academic careers. Prejudice was backed by legislation" (Miller 1993, 30).

adequate account of his relationship to the dominant tendencies in French intellectual life, it does not include what Foucault himself saw as his dominant influences. In particular, it fails to include the history and philosophy of science, an independent sub-dominant stream in French intellectual life. On Foucault's account, as structuralism came to be seen as superior to phenomenology, there was a series of attempts to combine structuralist or Freudian thought with Marxism and produce a "structural-Freudo-Marxism" (FL, 350). While such attempts dominated the scene until the late 1960s, Foucault comments that there was a group of students in this period "who did not follow this movement [but instead] . . . participated in the history of science." This group was aligned with the philosopher and historian of the life sciences Georges Canguilhem (1904–1995), "who had a decidedly influential effect on young French university life. Many of his students were neither Marxists, nor Freudians, nor Structuralists . . . I'm speaking about myself here" (FL, 350). Canguilhem was the director of Foucault's doctoral thesis and the publisher of Foucault's *The Birth of the Clinic*. He remained a staunch supporter in the face of Foucault's existentialist critics (Macey 1995). Addressing Canguilhem, Foucault claimed that his own work "carries the imprint of your mark . . . my 'counter-positions' . . . are only possible beginning from what you have done, from this layer of analysis that you introduced, from this 'eidetic epistemology' that you have invented" (Eribon 1991, 103). Canguilhem's method for investigating the history of sciences was the starting point for Foucault's own works. Talcott argues convincingly that, without Canguilhem, Foucault would have been very different (Talcott 2019).[31] To give a sense of Canguilhem's formative influence on Foucault's project, I will provide a brief sketch of Canguilhem and his approach.

After studying philosophy, Canguilhem undertook medical studies which he completed in 1944 with a thesis that would establish his reputation. Originally published in 1966, the English edition appeared in 1978 as *The Normal and the Pathological* and included an introduction by Foucault (Canguilhem 1991).[32] Canguilhem later completed a thesis in philosophy at the University of Paris in 1955, on the formation of the concept of reflex, under the renowned philosopher of the natural sciences Gaston Bachelard, whom he would later replace as director of the *Institut d'Histoire des Sciences et des Techniques* (Lecourt 2008, 56).

Canguilhem recognised the need for a particular historical approach to the life sciences which was different to both science itself and to other types

---

[31] I am indebted to Samuel Talcott for drawing to my attention the importance of Foucault's relationship with Canguilhem. The following account of Canguilhem in part draws on Talcott (2019).

[32] A modified version of Foucault's introductory essay appeared in 1985 (EW2, 465–77).

of history. The historian cannot assume that universal truth simply develops according to its own logic once scientific procedures have broken with everyday experience. Canguilhem rejects this positivistic philosophy of history, which he sees as "based on a generalisation of the notion that theory ineluctably succeeds theory as the true supplants the false" (Canguilhem 1994, 42). He employs "epistemology" to discern the thread of reason in the past, including in what is now considered false. This rational thread links the past with present science, even though both the concepts and the standards of rationality may have changed. The task of the epistemologist is to "abstract from the history of science in the manifest sense (i.e. a more or less systematic series of pronouncements claiming to state the truth) in order to uncover the history of science in the latent sense (i.e. the order of conceptual progress that is visible only after the fact and of which the present notion of scientific truth is the provisional point of culmination)" (Canguilhem 1988, 8–9). Thus the history of a science is the account of its normative ruptures which come about in its search for norms which improve its access to truth. As Foucault argues, the sciences continually break from their past and are only instructive insofar as they have rectified their own procedures and errors (EW2, 471). Therefore, since this successive transformation of truth-orienting norms entails the reshaping of its very history, writing the history of a science is a recurring task (EW2, 472).

In order to discern continuities or ruptures in the history of science, the historian must work at the level of the concept. As Gutting points out, this project is distinct from the histories of terms, phenomena, or theories. (Gutting 1989, 32) The problem with histories of terms is that they naively presuppose that their content remains fixed, while histories of phenomena fail to take account of interpretation. Rather than histories of theories, Canguilhem wants to distinguish between theories and their interpretation, a distinction not always made by analytic philosophers inclined to view the meanings of concepts used to interpret data as deriving from the theories which explain that data.[33] Canguilhem separates the *theories* which subsequently explain data from the *concepts* that enable our initial grasp of phenomena. The advantage of Canguilhem's finer-grained distinction here is that it can make sense of "theoretical polyvalence", where the same concept plays different roles in different theories. Canguilhem's history of concepts is well illustrated by his history of the reflex, in which he cut the link between Descartes's theory and the modern theory of the reflex by separating the truthfulness of concepts from the truthfulness of the theories in which they are embedded (Gutting 1989, 34–37). Canguilhem's history of concepts can revive forgotten names as

---

[33] Feyerabend, for example, maintained that the entire meaning of a concept is given by the role it plays in the statements of a theory (Feyerabend 2010).

well as throw traditional chronologies into disorder by revealing unexpected connections. However, unlike traditional chronologies, it is conscious of its historical nature and so doesn't see itself as the final or definitive account.

For Canguilhem, the history of science must include a particular form of pseudo-science which he calls "scientific ideology".[34] This is not necessarily ideology in Marx's sense, nor the unfalsifiable assertions which Popper saw in the work of Marx and Freud, nor is it superstition in opposition to science. Scientific ideologies have "an explicit ambition to be science, in imitation of some already constituted model of what science is" (Canguilhem 1988, 33). Although they stray beyond the norms of science and make claims beyond the capacity of contemporary science to establish, scientific ideologies do not reject science. Rather, they take a science as their model of knowledge and pretend to scientific status. Since scientific ideologies can give birth to concepts which are preserved in the later scientific theories which replace them, they can be "both an obstruction to and a necessary precondition of progress" (Canguilhem 1988, 32). While Bachelard made a firm distinction between science, with its theoretically constructed objects, and non-science with its given commonsense objects, Canguilhem's intermediate concept of a scientific ideology represents the non-science preserved in the science. Foucault, as we will see, regards ideologies as inseparable from science, although this need not detract from scientificity.

In *The Normal and the Pathological*, Canguilhem criticised the view of diseases as being merely quantitative modifications of normal states by arguing that norms represent *desired states of affairs* rather than averages (Canguilhem 1991).[35] Norms refer to *values*, which can't be reduced to an objective concept determined scientifically. Abnormality involves *subjective* experience—loss, sadness, disempowerment—and the values attached. Anomaly does not. What follows is that the concept of biological norms "cannot be reduced to an objective concept determinable by scientific methods. Strictly speaking, then, there is no biological science of the normal. There is a science of biological situations *called* normal" (Canguilhem 1991, 228). In other words, it is not science that picks out what counts as normal. If biological normality and abnormality are values that are posited socially, then sociology and biology and social and biological norms are inextricably bound together (Canguilhem 1991, 159). Canguilhem is particularly concerned to emphasise the hidden values embedded in what we take to be given biological determinants (Canguilhem 1991,

---

[34] Canguilhem acknowledged that this notion was influenced by the work of Althusser and Foucault (Canguilhem 1988, 9).
[35] Foucault regarded *The Normal and the Pathological* as "without any doubt the most important and the most significant of Canguilhem's work" (Canguilhem 1991, xviii).

161). He shares with Foucault a notion of normativity as a dynamic interactive web pervading all social and biological life.

Canguilhem's notion of norms cannot be separated from the problem of error which his work addressed in an attempt to correct a long tradition in the history of medicine that viewed truth as the successive denunciation of previous errors.[36] For Canguilhem, error is proper to living things themselves before being the object of the scientist. While error in its usual sense pertains to the human pursuit of knowledge, it has its origins in the experience of failure in the costly trial and error of living (Canguilhem 1991, 130). Acting at first without full consciousness, the living seek to attain their goals by whatever means seem fruitful. Awareness of a problem arises only when confronted with failure, when the risk taken has led to suffering or illness (Canguilhem 1991, 222). We seek knowledge in the face of such error or failure. As such, this error is creative. However, the pursuit of knowledge in itself is an erroneous one, since it is not oriented to solving any particular problem but is solely motivated by the will to know, that is, to grasp the true and avoid the false. Respecting the mutual exclusivity of these values, the search for knowledge seeks to expel all error from itself, even though it is only error which makes possible the will to know. Such positivism denies the value of error, the value of erring for living beings, and demands that humans conform to the rigid contours of the true. Foucault recognised one of Canguilhem's most important contributions to be his attempt to rethink the relationship between the subject, life, and knowledge on the basis of the problem of error (EW2, 477).

Today, Foucault tells us, the question of Enlightenment seeks to examine "a reason whose structural autonomy carries the history of dogmatisms and despotisms along with it—a reason, therefore, that has a liberating effect only provided it manages to liberate itself" (EW2, 469). In Foucault's view, it was Canguilhem who best posed this question of Enlightenment to the generations coming of age after the war. In his introduction to Canguilhem's *The Normal and the Pathological*, Foucault asks, "Why . . . following its own logic, [has the philosophy and history of science] turned out to be so profoundly tied to the present?" His response is that it "avails itself of one of the themes which was introduced surreptitiously into late 18th century philosophy: for the first time rational thought was put in question not only as to its nature, its foundation, its powers and its rights, but also as to its history and its geography; as to its immediate past and its present reality" (Foucault in Canguilhem 1991, 9).

In other words, by viewing reason in terms of the contingent contexts which constitute it, French history and philosophy of science enabled an external standpoint from which to mobilise critique. It is this tradition that

---

[36] Although the problem of error is not explicitly addressed in his major publications before 1966, Talcott shows that this problem was central to Canguilhem's thought throughout his career (Talcott 2019).

Foucault sees as taking up the Kantian Enlightenment project of the critique of reason, not in terms of its nature, function, or limits, but by engaging with the present and revealing its contingency. Foucault carried on this critique in his critical histories of the human sciences and their consequences for the contemporary world. He can therefore place himself within this tradition of history and philosophy of science, extending back to Comte and transmitted through Bachelard and Canguilhem, a tradition "serving to support the philosophical question of the Enlightenment" (Foucault in Canguilhem 1991, xi). Foucault's historical relation to the Enlightenment is a theme I will return to in chapter 5. For now, it is important to note that while Foucault frequently responded to the dominant philosophical fashions of his time, he saw himself as fundamentally committed to and engaged with the French tradition of history and philosophy of science. The themes of Canguilhem's work I have sketched earlier appear throughout Foucault's work, whether as borrowings, extensions, or counter-positions.

Although Foucault's orientation was shaped by the French tradition of the history of science, it was also given a particular inflection by other philosophers, most notably Nietzsche and Heidegger. Foucault was interested in Nietzsche as a genealogist who problematised truth as being intimately entwined with relations of power, who understood subjectivity as a construct and saw a complexity of forces as the lowly origin of our concepts. In many respects, Foucault went beyond Nietzsche, with a more sophisticated and detailed analysis. "The only valid tribute to thought such as Nietzsche's is precisely to use it, to deform it, to make it groan and protest" (EW3, 54). About Heidegger's influence, in a late interview, Foucault commented, "Heidegger has always been for me the essential philosopher. . . . My entire philosophical development was determined by my reading of Heidegger. I nevertheless recognise that Nietzsche outweighed him . . . . It is probable that if I had not read Heidegger, I would not have read Nietzsche." Foucault's other references to Heidegger are brief. Stuart Elden argues that although Nietzsche's influence on Foucault is immense, it is continually mediated by Heidegger (Foucault, cited in Elden 2001, 1).

Foucault was constantly adjusting his theoretical positions, at times pronouncing bold provocations and at other times revising or retracting previous positions. His ongoing retrospective reinterpretations of his work gave the sense of an experimenter wanting to test ideas, to put them into circulation before committing further to them. In an interview in 1977, Foucault announced, "What else was I talking about, in *Madness and Civilisation* or *The Birth of the Clinic*, but power? Yet I'm perfectly aware that I scarcely even used the word and never had such a field of analysis at my disposal" (PK, 115). Five years later he was saying, "It is not power, but the subject, which is the general theme of my research" (Rabinow & Dreyfus 1983, 209). This was typical of the ongoing revision of Foucault's self-understanding and work. In some cases, he deliberately bracketed questions for methodological reasons. In other cases, he seemed to later recognise what was implicit earlier. However, the boldness of Foucault's revisions and self-reinterpretations suggests a desire to move on, to look back at his previous selves from new perspectives. In an interview in 1978, he said, "I'm an experimenter in the sense that I write in order to change myself and in order not to think the same thing as before" (EW3, 240). Foucault's reinterpretations of his work may offer clues, but it is ultimately for others to judge whether or not to take him at his word. Clearly, the conventional periodisation of his work—archaeology, genealogy, ethics—gives rise to questions about consistency and coherence.

A number of interpretive schemas offer an overall sense and direction of Foucault's project.[37] None of them is entirely satisfactory, since they neglect what Foucault himself thought about fitting into *any* general interpretive schema. In *The History of Sexuality* vol. 2, he describes philosophical activity as "the critical work that thought brings to bear upon itself . . . the endeavour to know to what extent it might be possible to think differently, instead of legitimating what is already known" (HS2, 9).[38] Seen in these terms, Foucault's philosophical project cannot easily be judged against fixed theoretical foundations. By squeezing his *oeuvre* into a general interpretive schema, we risk distorting the specificity of each of his works. Since his work does not aspire to systematicity, we must not place too much weight on harmonising his work within an overall schema. The danger is that such schemas typically interpret work in terms of the problems or questions of established disciplines, but for Foucault, it is precisely established disciplines that are the problem. To present his work within established paradigms is to deny, in advance, the thrust of Foucault's entire project to rethink the most fundamental presuppositions. Foucault distanced himself from contemporary understandings of philosophy as a socially disengaged, theoretical, and universalist enterprise seeking ultimate truths. He saw himself as part of a stream of philosophical thought more or less marginalised in modernity. As I will discuss in chapter 7, he becomes clearer about his place within this stream as his work progresses, eventually coming to identify philosophy as a form of *ascesis* or spiritual exercise, a way of life in which one is transformed.

At the very least, Foucault's project can legitimately be seen as an ongoing exploration which continues today to challenge who we are, to change us and what we think. The central focus of this ongoing experiment in thought is the

---

[37] For example, Rabinow and Dreyfus (1983), Han (2002), Gutting (1989).
[38] Cf. Foucault (1989, 19).

constitution of subjects through the three axes of knowledge, power, and the self's constitution of itself. In a 1983 interview, he tells us:

> Three domains of genealogy are possible. First, an historical ontology of ourselves in relation to truth through which we constitute ourselves as subjects of knowledge; second, an historical ontology of ourselves in relation to a field of power through which we constitute ourselves as subjects acting on others; third, an historical ontology in relation to ethics through which we constitute ourselves as moral agents. (Rabinow & Dreyfus 1983, 237)

If we see these three domains as mapping onto the three periods of Foucault's work, we can say that these periods do not consist of a series of corrections on the path to a single approach, but that Foucault adopts whatever methods are required by a specific subject matter. Archaeology produces histories of thought which abstract from the individual subject in order to get at the discursive structures which both constrain and enable the fields in which individual subjects operate. Genealogy traces the origins of practices and institutions in their contingent beginnings to display the operation of power as it shapes and reshapes subjects. Ethics charts the possibility of resistance, where subjects are seen to be not totally determined by knowledge and power. The three methods are perspectives, each with their insights and blind spots. It would, however, be too tidy to claim that the works on archaeology, genealogy, and ethics correspond precisely to the axes of truth, power, and subjectivity. However, it is fair to say that a central focus of Foucault's work is the constitution of subjects in history and although each of the axes—truth, power, and the subject's self-formation—is inextricably entangled with the others, each period of Foucault's work thematises one of these axes in addressing the constitution of subjects. If we recall that science is the most authoritative form of knowledge in the modern era, we can see how science, power, and subjectivity form a tangle that Foucault's genealogies attempt to tease apart. He firstly examines knowledge, then its links to power, then the role of self-constitution. Seen in this light, all the periods of Foucault's work can be brought to bear productively on the question of science as a social institution.

MODERNITY, SCIENCE, AND PHILOSOPHY

Both Foucault and Habermas elucidate the relations between science as an institution and form of knowledge and its environing social and historical context. They both address the complex ways in which science and its contexts interact to condition, express, and transform each other. They examine the role and function of science in relation to aspects of culture and social institutions such as law, morality, the family, and religion. They consider the operation of power in specific forms like economic and administrative power, as well as in processes of socialisation and the formation of particular forms of subjectivity. The relationship of science to other discourses like philosophy and everyday common sense is woven into their accounts.

With such analyses, Habermas and Foucault give substance to the sense of crisis I have discussed.[39] They both offer broad notions of reason in which scientific reason, though increasingly dominant in modernity, does not equal reason per se. By treating science as a historical phenomenon within the larger picture of knowledge and social life, their differentiated analyses highlight the dangers and risks that science poses, while not denying its emancipatory possibilities. Although accepting scientific standards of evidence and objectivity, they both challenge the uncritical acceptance of science as the paradigm of modern knowledge. However, there are differences. Habermas wants to recover the emancipatory aspects latent within Enlightenment humanism, while Foucault wants to radically challenge it. Both articulate their projects in their own terms—Habermas in terms of the public sphere, democracy, and the ideals of communicative action, and Foucault in terms of genealogical histories of the human sciences.

To adequately grasp Habermas's and Foucault's critiques of science, we must locate both within the broader framework of modern reflexive reason. Both see modernity as characterised by reflexivity and craft their critical projects accordingly. Although humans have presumably always been able to adopt a third-person stance towards themselves, both see this as radicalised in modernity. Foucault notes that the modern *cogito* can no longer infer from "I think" to "I am" because it is not clear whether "I" is what I consciously survey, or the unthought that conditions such a survey (OT, 352–4). This instability produces a craving for explanations and a never-ending compulsion to keep uncovering what remains hidden in our nature. Foucault's analysis is directed as much to the sciences as to philosophical reflection. While philosophy motivates thinking the unthought, it is science that provides the content. In the contemporary scene, specialised fields such as neurophysiology and evolutionary psychology most explicitly uncover the conditions of possibility

---

[39] From its initial formulation in Max Horkheimer's 1937 essay "Traditional and Critical Theory", critical theory has been strongly linked to the concept of "crisis", towards which it is directed (Horkheimer 2002) This link was reaffirmed by Habermas's 1973 *Legitimation Crisis* in which crisis not only referred to economic dysfunctions or administrative failures but also rational and communicative crises. His 1982 *Theory of Communicative Action* analysed such crises in communication in terms of the attrition of the normative space of reasons. Foucault, who identified himself with the critical theory tradition, can be seen as similarly engaged in addressing the crises of our times.

for our thinking while leaving more unthought to condition this uncovering. Although never fully grasped, "the whole of modern thought is imbued with the necessity of thinking the unthought" (OT, 356). This necessity to pursue and uncover a background of continually receding conditions of possibility is "profoundly bound up with our modernity" (OT, 357). There is no end to this task, which requires "the permanent reactivation of an attitude . . . a permanent critique of our historical era" (PT, 109).

Unlike Foucault, who highlights this instability, Habermas discerns an underlying "logic" of rational progress. He argues that modernity is the outcome of a progressive series of "distantiations", enabling potentially emancipatory differentiations to accrue in the lifeworld. He sees the differentiated rationality of the decentred subject of modernity which can "burst asunder" the "provinciality" of other forms of subjectivity as a positive development (PDM, 322). But he is also concerned that the fundamental communicative structures of the lifeworld which bind social relations are increasingly threatened by interpretations of science which proclaim the primacy of the scientific observer's perspective (Habermas 2007b, 36). He contrasts the stance of a participant in social practices to a neuroscientific third-person observer's stance. In modernity, these two moments, of the participant and the observer, have been radically divided into two *conflicting* perspectives. Although we are able to debunk another's arguments by casting aspersions on their reasoning capacity, we still to some extent remain within the "space of reasons". However, now the stakes are raised. By adopting the scientific observer's stance and the authoritative causal language of natural science across the board, we deny epistemic normativity not only to blatantly irrational reasoners but, by extension, to our own intrinsically normative rationality.

This difficulty had been recognised by the previous generation of critical theorists who, in highlighting the omnipresence of instrumental reason, left little ground for critique to stand on. Is there *any* position from which we can confidently commit to our judgements? Can we find any "external" perspective sufficiently objective to critique conventional reason? These are questions that Habermas and Foucault inherited in responding to modernity's reflexive turn. Habermas responds by employing the human sciences to gain an "external view" of the communicative structures that have evolved in natural history and which he endorses as necessarily and universally constituting reason. He is worried by modernity's tendency to narrow the scope of this reason, reducing the interpretive human sciences to the law-like regularities of natural science directed towards prediction and control. Foucault radicalises Habermas's external view, by stepping outside the social and historical position of current science's "truth regime" to a position where the human sciences themselves are seen as problematic. In chapter 8, I will characterise these two different responses to modernity's reflexive reason as strategies by which Habermas seeks to discover what is stable, universal, and necessary, while Foucault seeks to transform subjectivity by problematising whatever is presented as stable, universal, and necessary.

The differences between Foucault and Habermas, as articulated in their informal "debate" and continued by commentators in subsequent decades, may suggest diametrical opposition on a range of matters. Foucault is typically criticised for various forms of relativism and for eliminating the subject. Habermas is typically criticised for idealisation, abstract universalism, and naturalism. Foucault and Habermas are frequently seen as two opposing leftist philosophers, one humanist and the other non-humanist, both actively engaged with politics but with differing assumptions and objectives. This impression was encouraged by their intermittent criticisms of each other's work, which sometimes simply reflected misunderstandings of what the other was claiming. What is not always recognised is the extent to which Foucault and Habermas share common ground. They both recognised that fundamental questions about the human condition cannot be answered by purely scientific approaches. They shared a perspective of a critical philosophy that could articulate social problems not visible from within the specialised perspectives and limited scope of science. By virtue of its self-reflexive nature, philosophy must draw on resources outside itself to gain perspectives not otherwise available. For Habermas, philosophy accepts, addresses, and harnesses the insights of the sciences to enable a critical perspective on science. Foucault suspends commitment to science's truths but draws on broad historical contextualisations to reveal scientific categories as not given by nature, but conditioned and subject to change. Yet both Habermas and Foucault would insist that philosophy relinquish all pretensions to playing a foundational role. For both, what is distinctive about philosophy is its anarchic nature, unconstrained by the methods of science which, in what Kuhn calls "normal science", appear fixed and laid down in advance (Kuhn 1996). This shared view of philosophy as standing at some distance from science contrasts with contemporary attitudes to philosophy as merely science's more speculative dimension.

Nevertheless, there are differences between the two. Rather than smoothing them over, I will present the differences between Habermas and Foucault as separate moments of modernity's scientifically informed reason. Habermas seeks "the reconciliation of a modernity that has fallen apart, the idea that without surrendering the differentiations which have made modernity possible . . . one can walk tall in a collectivity that does not have the dubious quality of backward-looking substantial forms of community" (Habermas 1986, 125). Modernity is the "incomplete project" which must not be abandoned (Habermas 1996). Habermas wants to endorse cultural modernity by taking its knowledge and interpreting it for contemporary society, thus showing not only the limitations of science but also its emancipatory potential, something Foucault barely canvasses. Habermas's strategy yields a progressive developmental account where the sciences are linked to cognitive advances in distantiation and differentiation, but this progress is by no means guaranteed or universal and carries considerable cost and risk. Foucault is more wary in his assessment of modernity, especially the human sciences and their ubiquitous power effects. He employs his genealogical critique more radically to highlight science as a contingent activity embodied in multiply conditioned subjects in historic social settings, so undermining any temptation to view scientific reason as reason per se.

Both Habermas and Foucault engage with the sciences, not theoretically, but practically in a way intended to provide some orientation for how we should live. Both identify themselves with the Enlightenment tradition which aims to understand society and change it by freeing human beings from entrapment in systems of dependence or domination. As a second-generation critical theorist, Habermas harnesses science in collaboration with philosophy, also taking on additional resources such as American pragmatism and the analytic philosophy of language. Late in his career, Foucault acknowledged critical theory as a form of reflection close to his own. French epistemology and German critical theory could then be seen as two national traditions which, despite their differences, take up the same question of Enlightenment.[40] Like Habermas, he thought that philosophy must diagnose the particular social and historical formations of modernity in which it finds itself. This involves a critique of modernity's paradigmatic form of reason, science, to expose the false pretensions of various versions of scientism and their oppressive and alienating power effects.

---

[40] In a 1978 interview, referring to the members of the Frankfurt School, Foucault says he "should have read them long before, should have understood them much earlier" (EW3, 274). Foucault locates himself in the tradition of the critique of reason extending through Kant, Nietzsche, Weber, and Horkheimer (PT, 97–120).

Chapter 2

Habermas's Critique of Positivism

A hallmark of Habermas's long career has been his active political engagement as a public intellectual. As early as his 1953 criticisms of Heidegger, he invoked the "role of public critique as guardian" (Habermas 1953). He has continued to speak out, to mention just a few examples, against attacks on minorities, self-assertive patriotism, the idea of a balance of terror in the face of nuclear destruction, and a statute of limitation covering Nazi war crimes (Muller-Doohm, 112–122). During the 1960s, Habermas's concern for open and free discourse was reflected in his response to positivism, which he criticised as an ideological misinterpretation of science and a truncated form of reason that stifled reflection. In this chapter, I will consider Habermas's early critiques of positivism, drawing on essays collected in *Theory and Practice* (hereafter TP), *The Positivist Dispute in German Sociology* (hereafter PD), and *Towards a Rational Society* (hereafter TRS). These works coincided with a period of national reconstruction when the German state played a major role in the economy and when a "technocratic" state, in which major decisions would be made by scientific experts, seemed possible.[1] In these essays, Habermas argues against positivism's understanding of science, and the technocentrism it supports, as obstacles to egalitarian democratic will-formation. I will then turn to Habermas's substantial theory of rationality, *Knowledge and Human Interest* (hereafter KHI). Here he develops a differentiated account of reason that grants a place to the form of rationality inherent in natural science,

---

[1] Sociologist Helmut Schelsky and philosopher Hans Freyer argued that, given technology's increasing power and autonomy, a technocratic state was already underway. "Political norms and laws are replaced by objective exigencies of scientific-technical civilisation" (Schelsky quoted by Habermas TRS, 59).

while not identifying it with rationality per se. I will conclude by discussing the tensions inherent within KHI.

HABERMAS'S RESPONSE TO POSITIVISM

Logical positivism (or, after the Second World War, "logical empiricism") was a broad movement generally committed to empiricism, scientific naturalism, and the unity of science.[2] The logical empiricists emphasised the role of rigorous science in advancing society. They argued that science progresses by generating verifiable theories and that any statement which cannot in principle be confirmed by empirical observation is metaphysical nonsense. Against this, Popper insisted that theories could never be confirmed but only falsified. The mark of a good theory was its falsifiability. Both the positivists and Popper argued that all scientific enquiry, including the social sciences, should be based on empirical observations which can potentially verify (or for Popper, "corroborate") statements, resulting in causal explanations and general laws. This stance entailed that the application of scientific methods to the study of social phenomena required rigorous avoidance of normative considerations, since value judgements did not admit of truth or falsity. All sciences, including social sciences, had to be "value-free". Scientific knowledge could be brought to bear on options, preconditions, or consequences, but the choice of ends itself was ultimately a question of values requiring decisions, not facts. Including normative considerations in social inquiry would result in dogmatism and ideology, not the cumulative progress of objective knowledge. Thus, positivism sought to unmask the normative aspects of arguments as "non-cognitive", "subjective", and "irrational", bringing to light what was seen as a confusion of facts and values. By revealing what it took to be the pseudo-scientific, ideological character of such arguments, positivist critique sought to establish its continuity with the Enlightenment battle against ignorance, superstition, and dogmatism.

Habermas was acutely aware of the encroachment of functionalism and systems theory into social science. He thought that such positivism, blind to its own value-laden presuppositions, uncritically affirmed the *status quo*. Anticipating his shift to a communicative paradigm a decade later, he responded with the idea of "a rationality with unconstrained validity in a domination-free discussion" (cited in Muller-Doohm, 113). He argued that causal explanation is generally inappropriate in the social sciences because social processes typically involve processes of communication which rest

---

[2] For a more detailed account of positivism, see Kolakowski (1972).

upon social actors making sense of their world by interpretations. By adopting the methods of the natural sciences as the model for all knowledge, positivists assume that the social scientist is distinct from the object of inquiry. However, this assumption of value-freedom fails to account for the economic, historical, and social contexts which condition the thoughts, perceptions, and language through which the social scientist's observations are undertaken and articulated (PD, 157–8). By abstracting from these background conditions, positivism restricts science to narrow interests of instrumental control and prediction, which it regards as the only access to reality. The proper aim of social science, in Habermas's view, is not instrumental control and prediction but critique and transformation. Popper's rigid distinction between non-observable values and observable facts consigns critical evaluation of humanity's true interests to the subjective status of decisionism, not susceptible to rational resolution and only resolved through arbitrary assertions of will (PD, 144; TP, 265). Habermas argues that the social scientist can only escape the ideological entrapment entailed by this view if she can grasp "the societal life-context as a totality which determines even research itself" (PD, 134).

Rather than scientific hypotheses being tested directly by empirical observations, Habermas argues that a statement derived from a hypothesis is related to a statement describing empirical conditions. However, it isn't clear how a given observation is sufficiently motivated by experience to confirm a law-like hypothesis (PD 152). Positivists appeal to a purely descriptive language stripped of vagueness, an appeal presupposing that "facts" are directly imprinted on our consciousness, from whence they are seamlessly transformed into protocol statements that enable deductive steps leading to scientific knowledge. However, Habermas tells us, "There is no such thing as immediate knowledge". Since "empirical data are interpretations within the framework of previous theories . . . they themselves share the latter's hypothetical character" (PD, 201–2). For example, "The simple assertion that 'here is a glass of water' could not be proved by a finite series of observations, because the meaning of such general terms as 'glass' or 'water' consists of assumptions about the law-like behaviour of bodies" (PD, 150). The procedure by which an observation is mapped onto a description cannot be formulated as a set of logically consistent self-contained rules. Popper, agreeing with Habermas on this point, argued that there must be a discursive process within the scientific community similar to jury deliberation (PD, 152). Picking up this thought, Habermas suggests there is nothing that *compels*, either logically or empirically, acceptance of any given empirical observation as an instance of a statement of the testing of a hypothesis. A decision must be made which relies on a pre-understanding which cannot be theoretically elaborated, but is secured by a practical interest in the domination of objective processes (PD, 151). Yet, for the positivist, decisions relate to values and are thus relegated to the realm of the subjective and irrational. By positivism's own criteria, "Values are in principle beyond discussion" (TP, 271). This restriction on reason renders positivism incapable of justifying its own interests. If a practical orientation is ultimately beyond rational justification, then positivism's commitment to empiricism is itself subjective and unjustifiable.

However, science is neither subjective nor unjustifiable. Scientists achieve rational consensus by reference to a vast number of implicit beliefs which constitute the scientific lifeworld, within which science's fundamental interest of control and prediction "recedes into the background . . . disappear[ing] from the consciousness of those involved in the research process" (PD, 155). This fundamental interest entails as its sole admissible "value"—apart from methodological or logical values immanent to science—efficiency in the selection of purposive-rational means guaranteed by conditional predictions in the form of technical recommendations. This value cannot be seen as a value, "because it seems simply to coincide with rationality as such" (TP, 264). Positivism thus conceals a commitment to a particular form of rationality behind a veil of value-neutrality. Against this, Habermas points to the social and consensual nature of science and argues that there must be a form of rationality, beyond instrumental rationality, which is not mere rule-following but includes a communicative dimension to decide what counts as an instance of a rule. "Argumentation", he tells us, "differs from mere deduction by always subjecting the principles, according to which it proceeds, to discussion. . . . What can pass as criticism always has to be determined on the basis of criteria which are only found, elucidated and possibly revised again in the process of criticism" (PD, 214).

Habermas rejects the straightforward identification of the methods of natural science with emancipation. Certainly, early Enlightenment philosophers saw the scientific study of nature as combating prejudice and the unjustified authority of despots and the Church (TP, 257). However, Enlightenment reason was inherently practical, with an interest in emancipation from natural and quasi-natural constraints. The decision to act was not external to reason (TP, 254). However, this relation of critical reason to enlightened practice was narrowed by positivism to one of prognosis and control. "As our civilisation becomes increasingly scientific . . . science, technology and administration interlock in a circular process [in which] the relationship of theory to practice can now only assert itself as the purposive-rational application of techniques assured by empirical science" (TP, 254). With positivism, the critique of values becomes universal and the possibility of self-reflection is abandoned. Assuming its own value-neutrality, science becomes blind to its inherent values, thus suppressing social practices not structured in those terms (TP, 269). By refusing to recognise the historical emergence of social forms, positivism treats social relations in terms of ahistorical natural laws, applicable to the rational administration of society (TP, 210). With the normative basis of society occluded, society and its members are treated as causally determined natural objects. Habermas argues that ultimately, the values by which we judge human actions risk erosion in a society understood in such narrow terms (TP, 273–4). Scientific research, however objective, is still embedded in a lifeworld. However, the positivist's analytic-empirical methods, assumed to be value-free, ensure that "simplified basic assumptions [are] chosen in such a way that permits the derivation of empirically meaningful law-like hypotheses" (PD, 133). Thus, positivism ideologically endorses the *status quo* by producing a false picture of society as being subject to unchanging natural laws. By failing to reflect on the social conditions of its emergence, positivist science projects its own values as universal in scope and validity.

Positivism found expression in "technocracy", the idea of creating a "technical state" in which major decisions could be made by experts rather than politicians. This idea should be seen in the context of the generally uncritical attitudes towards scientific and technological progress prevalent in the early 1960s. Notwithstanding the justified fear of nuclear annihilation as the Cold War escalated, faith in science and technological automation led many to believe in the possibility of scientific rationality ushering in an age of prosperity in which scientific techniques could be extended to the administration of society. While today such confidence has been considerably muted, Habermas's response to technocracy remains relevant as an antidote to the ongoing temptations of technocratic thought. His response involved highlighting the relationship of scientific expertise and technology to political practice. At this time, Habermas was reading American pragmatism, in which he found a model for the critical interaction of the technical expert and the politician. Questions of life conduct require rational discussion not focused exclusively on either technical means or the application of traditional norms. This requirement bears on today's problems, where "new technical capacities, erupt[ing] without preparation into existing forms of life-activity and conduct" are being determined by "unreflected goals, rigidified value systems, and obsolete ideologies" (TRS, 55, 60). These goals, value systems, and ideologies must be made explicit and then tested against, and adjusted to, real possibilities. In this way, the ongoing development of technologies will not entail remaining bound to the same values. Value systems would be "partly confirmed, partly rejected, articulated and reformulated, or denuded of their ideologically transfigured and compelling character" by the challenges posed by technology (TRS 67).

One can think of examples from reproductive technology or artificial intelligence which illustrate how traditional values are transformed when confronted with previously unimagined possibilities. Given that technology and the scientific research enabling it respond to our values, themselves shaped by technology, Habermas advocates a "dialectic of potential and will" that could enlighten us about our tradition-bound values in relation to technological possibilities. He argues that this dialectic actually occurs, but without reflection and in accordance with interests for which no public justification is given or demanded. He wants scientific knowledge and practical concerns to interpenetrate and inform one another in a process where technical knowledge is employed to enhance human life. The challenge is to consciously set into motion a politically effective discussion that brings the social potential of technical knowledge into dialogue with values (TRS, 61). In this way, the dialectic of potential and will avoids both technocracy, which reduces normative political decisions to merely administrative questions handled by scientific-administrative experts, and "decisionism", which insists that the burden of moral choice, unable to be eliminated by scientific-administrative expertise, must rest solely upon the subjective preferences of politicians. Both these positions contradict the normative self-understanding of democracy that public decisions are reached by autonomous citizens in processes of unconstrained exchange of opinions.

The tendency towards the ideological use of simplified, decontextualised, and over-extended interpretations of science understandably provokes a demand for the positivistic separation of theory and practice. In this context, Habermas sees the notion of the "neutrality" of the sciences being reasonably employed against the short-circuiting of the connection between scientific expertise and a public vulnerable to manipulation by imbalances in social power and access to information. He would agree that it may be reasonable to invoke a fact/value distinction when faced with climate change denialism or the promotion of technological "progress" serving narrow interests. Nonetheless, as soon as this strategy calls into question the connection between power and reason per se, it "succumbs to the limitations of positivism and an ideology that makes science impervious to self-reflection" (TRS, 70). Habermas is not optimistic about the possibilities for an informed rational dialogue about science, although this is what he advocates. The ideals of rational and open debate have been undermined by powerful corporations as well as the managerial state, with its bureaucratised power reinforced by specialised large-scale research organisations, which excludes the public as a political player (TRS, 76).

Even without a technocracy, "technocratic consciousness" serves an ideological function by posing practical problems exclusively in terms of possible technological solutions. Technocratic consciousness differs from earlier bourgeois ideologies which selected, amplified, and distorted ethically loaded images of the good life and could therefore still be traced back to a basic normative pattern of just interactions. By not projecting such images, "technocratic consciousness is less vulnerable to reflection than bourgeois ideology", in which discrepancies between ideological projections and reality can prompt reflection and critique. It renders inert our interactions in the ordinary languages in which we are socialised and in which "domination and ideology ... can be reflectively detected and broken down" (TRS, 112). Habermas follows Herbert Marcuse in thinking that the key to analysing late capitalism is to understand scientific technology as a legitimating power, a role it assumed as "free market" ideologies were replaced by state interventionism designed to compensate for market failures (TRS, 100–101). Interventions attempt to deal with social problems by eliminating dysfunctions and avoiding risks. However, they are interpreted as addressing technical economic questions, rather than as political questions requiring democratic deliberation in terms of, for example, justice. By the fusion of science, technology, industry, and state administration, "the development of the social system *seems* to be determined by the logic of technical progress". Scientific-technical progress "produces objective exigencies ... which must be obeyed" (TRS, 105). Through the prism of technocratic consciousness, science and technology fulfil the ideological function of legitimating the exercise of political power over a depoliticised public. The basis of legitimation for a rational society becomes the growth of productive forces that issue from quasi-autonomous scientific and technological progress, with existing relations of production appearing as their technically necessary organisational form.

Against such technocratic thought, Habermas insists that the challenge of controlling our species' destiny in the face of the unplanned sociocultural consequences of technological progress cannot be met by purely technological imperatives (TRS, 60). We cannot assume that rationality extends from the capacity for technical control to the practical mastery of historical processes (TP, 275). History cannot be rendered more rational by greater technological control, however sophisticated, "but only by a higher stage of reflection, a consciousness of acting human beings moving forward in the direction of emancipation" (TP, 275–6). Rather than simply presupposing technological solutions, this "higher state of reflection" would address fundamental questions about the goals of society, which Habermas insists cannot be decided in advance of unconstrained democratic deliberation.

Habermas's criticism of technocratic thought does not mean that he opposes scientific-technological development per se. In fact, he embraces a modernity in which science has become the primary productive force. However, he is concerned about the orientation of practical life towards technological imperatives in a society that addresses conflicts, not by discussion, but by depoliticisation. Foreshadowing his later 'lifeworld colonisation' thesis, subjects are both depoliticised and self-objectified in purposive-rational action, thus diminishing their capacity to maintain the intersubjectivity of mutual understanding and communication without domination. Since these effects are concealed behind the expansion of the power of technical control, the new ideology requires a reflection that penetrates beyond particular-class interests to disclose the fundamental interests of humanity as such. In KHI, we will see Habermas spell out these interests comprehensively.

Habermas's analysis of technocratic consciousness as an ideology was most likely influenced by Marcuse's notion of "one-dimensional man", which was also possibly the germ of the colonisation thesis we will see in chapter 3 (Marcuse 1964). While Habermas and Marcuse agreed that technological consciousness played an ideological role in advanced societies, they didn't agree on how this arose. Habermas thinks that technocratic consciousness is problematic because it reflects an imbalance between two logically distinct dimensions of reason—the dominance of "labour" over "interaction". By "labour", Habermas means the growth of productive forces by the technical mastery of nature as well as over certain social processes by the application of technical knowledge. It is a success-oriented form of purposive-rational action aiming to predict and control nature. "Interaction" refers to communicative relations among individuals pursuing common understanding, leading to decreasing repression, increasing role distance, and more flexible applications of norms. Technocratic domination results from an *imbalanced* expansion of the rationality embedded in scientific technology, governed by its logic of labour, into "the 'historical totality' of a lifeworld" which at the same time puts this particular form of reason beyond criticism (TRS, 90).

Marcuse had a different view. He argued that current scientific technology is constituted on the basis of society's control by elites. Technologies are not simply means applied to independently chosen ends; they shape and are shaped by those ends. Science and technology are contingent and could be reconstructed to play different roles in different social systems. He therefore proposed a revolutionary transformation of basic practices leading to a change in the very nature of instrumentality. Rather than treating nature as mere raw material to be exploited for power and profit, with the abolition of class society, human beings could adopt a more responsive attitude towards nature's inherent processes and potentials. Habermas doesn't accept that such an alternative new science is possible. While science and technology are not politically neutral or value-free, as the positivists claimed, they are nonetheless an expression of a fundamental human, not class, engagement with the world. Habermas thinks Marcuse has jumped too quickly to the conclusion that the basic presuppositions of the natural sciences are exclusively "determined by class interest and historical situation" (TRS, 85). In contrast, Habermas posits a fundamental logical structure as being a precondition for the survival of the species as a whole (TRS 87). This structure underlies scientific technology's various forms, which rely on purposive-rational action as a feedback-controlled learning process.[3]

It seems to me that Habermas misunderstands Marcuse's utopian reflection on a new, emancipatory relation to nature. Marcuse is suggesting a technology designed and applied with sensitivity to the inherent potentials of the medium and the context that it presupposes. This would involve it being responsive to nature, rather than merely controlling it by abstracting from its qualitative, ethical, and aesthetic values. Actual science and technology would still be governed, though not fully determined, by the feedback-controlled objectivity that Habermas attributes to it in the abstract. As we will see, the debate with Marcuse reveals Habermas's commitment to a form of essentialism. I shall expand on this point at the conclusion of this chapter, where I will raise the issue of fundamental theoretical structures that persist, in various forms, throughout Habermas's entire project.

During the second half of the 1960s, the technocratic vision was opposed by increasing student activism which was spurred on by geopolitical issues like the nuclear arms race, the Cuban missile crisis, and the Vietnam War, leading to heated debates about the relationship of technology to society. By the end of the decade, rising counter-cultural movements, often embracing romantic anti-technology attitudes, voiced strong critiques of technocracy. The situation was changing rapidly. With the publication of works like Rachel Carson's *Silent Spring*, this critique moved into mainstream society, which became more critical of technocentrism and, joined by scientists, demanded legislative controls and public participation (Carson 2002). Positivism came under increasing challenge not only from the critiques of Habermas and Adorno, but also works as diverse as those of Wittgenstein and Foucault. With the revolution ushered in by figures such as Kuhn, Feyerabend, and the historians and sociologists of science, positivism came to be seen as naïve and dogmatic. As the question of technology receded from the political-cultural horizon, Habermas's interest shifted to a communicative paradigm.

In the retreat from the utopian aspirations of the 1960s, Habermas's influence and reputation came to eclipse Marcuse's. However, it is not surprising to see a renewed interest in Marcuse today. In recent decades, radical

---

[3] Habermas borrows the idea from the anthropologist Arnold Gehlen of modelling technological developments on a series of substitutions of mechanical devices for human functions and faculties—a socially neutral sequence of prosthetic substitutions from the digging stick to the computer (TRS, 87).

criticism of technology has been reinvigorated by the new field of science and technology studies, the environmental movement, and by analyses like Foucault's. I will return to Habermas's treatment of science and technology in the conclusion to this chapter and again in the following chapter, to argue that Habermas has been unable to theorise the value-ladenness, variability, and specificity of technology adequately to incorporate it into his critique of modernity. Notwithstanding these difficulties, Habermas's critiques remain relevant today for three reasons. First, the demise of positivism did not result in any clear consensus, but rather a plethora of approaches to the philosophy of science. Second, for the social sciences, there remain fundamental questions about their appropriate methodological paradigms and their relationship to the natural sciences. And third, positivistic thought remains a potent force. While a degree of scepticism towards science is now commonplace, a strain of unreflective technology-inspired optimism persists. Advanced technologies are frequently proffered as obvious solutions to complex problems—industrial agribusiness to feed the hungry masses, geo-engineering to address climate change, economic models to address social injustice, social media to unleash grassroots democratic forces.[4] Silicon Valley entrepreneurs talk earnestly of the utopia that new "disruptive technologies" will usher in, while basic rights and the democratic public sphere continue to be eroded by their incursions. What these approaches have in common is not only their "top-down" authoritarianism, but also their occlusion of ethical concepts such as justice, fairness, and responsibility, consequently avoiding any cost to those proposing such "solutions". This suggests the ongoing ideological function of the conviction that practical problems are fundamentally technical, a conviction which occludes the deepest roots of our present failures, including the direction, goals, meaning, and social implications of scientific-technological growth.

In recent decades, we have seen a variation of technocratic consciousness. Rather than scientific-technical consideration by experts pre-empting practical deliberation, climate denialism seeks to eliminate practical deliberation by undermining science. Powerful interests finance the ongoing questioning of climate science regarding whether change is happening, whether it is human-induced, and whether the consequences are of concern. Rather than questions about the nature and consequences of climate change being investigated and answered by experts and accepted by the lay public, who then discuss

---

[4] References to such proposals frequently appear under the rubric of "ecological modernism", a technology-based approach oriented to innovation in environmental policy. Opposed to deindustrialisation, it promotes "environmentally sensitive" technologies which increase resource productivity. Critics claim this amounts to "greenwashing". Predicated on the notion of sustainable growth, questions arise about the adequacy of ecological modernism on its own to address gross injustices and about its applicability outside rich nations (see Fisher and Freudenburg).

what should be done, these scientific questions are treated as open to public determination. To the extent that expert advice is accepted, we see a response closer to Habermas's account of technocratic consciousness which prompts a technical solution foreclosing any broader consideration of the values or aims of society. The question of what should be done becomes a merely technical question framed in the reductive language of economic costs and benefits. The problem is that climate change raises political, ethical, and moral questions regarding the fairness of the distribution of the risks and burdens of climate change and the action required to limit it. This "dialectic of potential and will" can only be addressed by genuine public discussion which is not truncated at the outset.

Habermas's critique was undertaken during the ascendancy of the social democratic activist state, when there was more confidence in the capacity of science and technology to make a better world. During this period, Habermas sought a middle ground between entrenched positions. Unlike the positivists, he did not see science as "pure" theory, free from contaminating interests. Nor did he embrace technocratic planning like conservatives did. He also rejected Marcuse's proposal for a new science underpinned by a different relationship with nature and society. Habermas's overriding concern about positivism was as a form of ideology which suppressed and distorted thought and speech by disqualifying anything beyond the ambit of empirical natural science. This concern was sustained and developed throughout his career, including his later engagement with Luhmann's systems theory and his more recent discussions of naturalism.[5] Habermas's early critiques are still relevant to today's versions of technocratic consciousness. They are also significant because they articulate distinctions that formed the basis of his later theoretical developments. By adopting the categories of "labour" and "interaction", Habermas foreshadowed the development of his differentiated account of rationality, firstly grounded in deep-seated anthropological interests and then in the normative structures of communication. We will see, in the following chapter, the categories of "labour" and "interaction" rearticulated as "system" and "lifeworld", while "ideology" will be understood intersubjectively in terms of "systematically distorted communication." Beyond such reformulations, the same commitment to the analysis of obstacles to egalitarian, democratic will-formation in modern societies can be seen.

---

[5] Habermas's debate with Luhmann can be found in German in Habermas and Luhmann (1971). Some helpful comments are found in McCarthy (1991, 155). For Habermas's discussions of scientific naturalism, see PDM, pp. 368; BNR, Ch 6.

KNOWLEDGE AND HUMAN INTERESTS

The concerns which motivated Habermas's critique of positivism found their comprehensive theoretical articulation in KHI, an epistemological thesis of three different ways in which reality is known, each linked to a fundamental interest necessary for the survival of the species. Positivism's dominance had undermined the tradition of epistemological inquiry in which natural science was just one category within a broader conception of knowledge (KHI, 3–5). Kant's core question of how reliable knowledge is possible had been corrupted into the narrow question of natural science and its methodology (KHI, 4). Habermas wants to oppose this by providing a systematic framework in which critical social theory has a place as a distinct form of valid knowledge. However, towards the end of the 1960s, positivism's notion of the unity of scientific method and universal laws was giving way to the relativistic implications of hermeneutics and various forms of historicism and culturalism. In the new front opening up between relativists and universalists, Habermas defends a universalist approach to social inquiry. In a stance opposing both "pure" theory and relativism, Habermas sees knowledge as necessarily bound to interests grounded in the biological and social imperatives of human survival. This differential account gives separate places to natural sciences, human sciences, and philosophical critique. Habermas's broadly Kantian stance sees the observing subject and observed object as not distinctly separate. Embracing the idea of the subject as inherently social and historical, he aims to salvage Kant's claim of universal autonomy and responsibility, but in a way that accounts for the empirical processes of self-formation.

KHI aims to understand philosophy's present situation by reconsidering the "abandoned stages of reflection" located in the movement of philosophical thought from Kant to Marx (KHI, vii). In reconstructing this history, Habermas analyses the connection between knowledge and human interests to reflect the history of reason characterised by his earlier dichotomy between "labour" and "interaction". These categories are now augmented by "domination", yielding three distinct types of knowledge, each oriented by a deeper, non-subjective "transcendental" interest with which we apprehend reality. These human interests—the technical, the practical, and the emancipatory—are not transcendental in the Kantian sense. Habermas's project is empirical, employing reconstructive approaches developed in developmental psychology and linguistics. His cognitive interests are therefore "quasi-transcendental", based in the natural history of the species and tied to the imperatives of sociocultural forms of life, although not reducible to them. Labour and interaction give rise to natural and social science by providing viewpoints on the natural and social worlds, while emancipation from domination is specifically related to critical philosophy. Habermas insists that the sciences can't be understood without reference to these knowledge-constitutive interests. This touches on a theme we will see foregrounded by Foucault. Philosophy has traditionally seen power, interest, or utility as irrelevant, or even an impediment, to knowledge. Opposed to the idea of the absolute purity of knowledge, Habermas salvages the objectivity of the sciences by tying scientific knowledge to deep-seated interests shared by the species as a whole. As we will see, Foucault historicises scientific knowledge by linking it to particular interests and regimes of power.

Because Kant's transcendental subject lacks any evolutionary, developmental, or historical constitution, Habermas draws on Hegel's Jena lectures of 1803–1806 to articulate a quasi-transcendental framework within which knowledge unfolds, using the categories of labour and interaction (KHI, 15–16). However, he avoids the later Hegel's idealism by turning to Marx's articulation of the subject-object relationship and conceiving of humans as subjects actively transforming both themselves and objects by social labour (KHI, 24). Marx views the scientific-technological rationality built into labour as propelling the evolution of society, providing the impetus to transform rigidified ways of life. However, the problem which Habermas sees is that Marx tacitly conceives of reflection, how we grasp the world and ourselves within it, not communicatively but instrumentally, in terms of production, and thus comes close to obliterating the distinction between the natural and human sciences (KHI, 44).[6] A critical social science must avoid this conflation by distinguishing the rationalisation of purposive-rational action, or labour, from the rationalisation of communicative interaction which removes restrictions on self-reflection and communication. Productive forces play a role in emancipatory movements only if they don't replace rationalisation of the institutional framework which occurs by communicative interaction. If communicative interaction is neglected, rationalisation of productive forces will lead to an increase in technical control over objectified social and natural processes. Scientific-technological progress must therefore not bypass, nor substitute for, rationalisation of the medium of symbolic interaction which removes restrictions on communication.

Grounded in three fundamental interests, the viewpoints of three "sciences"—empirical-analytic, historical-hermeneutic, and critical—disclose different aspects of reality.[7] The first two viewpoints respectively employ

---

[6] It is doubtful that Habermas provides a fair interpretation of Marx. See Grumley (1992).
[7] Habermas's German *Wissenschaft* has a broader meaning than "science" has in English. "Empirical-analytic sciences" are the natural sciences, but they also include other sciences to the extent that they aim at producing nomological knowledge. "Historical-hermeneutic sciences" are broadly "the humanities", including the social, historical, and human sciences to the extent that they aim at interpretive understanding of meanings. "Critical sciences" are forms of inquiry oriented towards emancipation, although the concept is not yet developed beyond KHI's preliminary formulation.

distinct categories such as "bodies in motion" or "acting and speaking individuals". The use of such categories "implies an *a priori* relation to action to the extent that 'observable bodies' are simultaneously 'instrumentally manipulable bodies' whereas 'understandable persons' are simultaneously 'participants in linguistically mediated interaction'" (Habermas 1975, 174). The third viewpoint represents the already-constituted objects of the first two interests in terms of emancipation. Habermas's threefold differentiation allows autonomous areas of knowledge not reducible to natural science, which becomes just one form of knowledge among others. I will now consider KHI's three types of science, before examining some criticisms and Habermas's responses.

The Empirical-Analytic Sciences

The survival of the species clearly depends on successfully coping with threats and opportunities in the physical world. The technical interest which guides the empirical-analytic sciences addresses this dependency in the form of a learning process. Basic scientific statements reflect the success or failure of experimental operations within a fixed framework of possible experiences which is given by the invariant relation of human beings to their environment. Formulated into law-like hypotheses, these statements enable prediction and control of nature. Habermas thus argues that the empirical-analytic sciences are therefore ultimately determined by the behavioural system of instrumental action which arises contingently in human evolution and binds scientific knowledge of nature to the technical interest in control over natural processes. However, for Habermas, this is not a biological, psychological, or sociological matter, nor does he want to reduce science to its utility. His analysis is directed at the types of validity and the meaning of empirical scientific statements which are intrinsically related to purposive action. The control and prediction constitutive of empirical-analytic science are seen in its fundamental methodological and rational norms, such as falsifiability, accuracy, and replication.

Habermas turns to Peirce to articulate the category of labour within the context of natural science (KHI, 91–139). Peirce's pragmatism offers a form of transcendentalism that fully embraces fallibilism. For Peirce, a belief is that upon which we are willing to act. We orientate our behaviour according to our beliefs, which become settled as our actions succeed (KHI, 121). Knowledge is not grounded in immediate intuitive contact with the world or rational reflection; it advances through practical engagement, mediated by prior beliefs and practical expectations (KHI, 97). New knowledge is generated as a challenge to settled beliefs, reacting to the "irritation of doubt" when nature doesn't respond as anticipated. If the behaviour that a belief gives rise to succeeds, the belief remains unproblematic. However, when it fails, we seek to restore a settled state by supplementing, modifying, or rejecting that belief. The capacity for rational-purposive control of nature involves an ongoing and cumulative learning process (KHI, 120–1).

According to Habermas, instrumental feedback-controlled action, grounded in the category of labour, makes natural scientific knowledge possible as *a particular form* of knowledge (KHI, 120–6). Labour, our instrumental interaction with the world, enables the discernment of distinctions and the consequent synthesis of experience in terms of specific, contingently acquired, concepts (KHI, 129–30). By instrumental action, we establish which observable regularities are actually the causally necessary powers and properties that define objects. Since causal relations are not something directly sensed, but involve a counterfactual condition, this requires our active intervention.[8] Causal necessity is not observed but is based on what Peirce identifies as habit, or "fixity of belief". Since no belief is certain in the Cartesian sense, "truth" is the capacity of a belief to withstand experimental tests in the long run. The experimental method allows truth to emerge when applied by an indefinite community of inquirers over an indefinite period of time. Thus, according to Peirce, a belief is "true" if this ideal community of inquirers would accept it (KHI, 91–5).

Habermas's Peircean account captures the operation of a background of shared tacit beliefs within science. Taken-for-granted beliefs enable scientists to interact within the scientific community in a manner not fully specifiable in terms of rules, but dependent on implicit cultural understandings embedded in the lifeworlds of scientists. If scientists are successful, beliefs about the right way to practice science are not opened to question, but remain part of this implicit background. It is only when science enters what Kuhn calls a "revolutionary" period, that the paradigms on which scientific theories depend are drawn into the open to be challenged (Kuhn 1996, 6–8). The difference between science and everyday belief formation is that science operates as an institution, a rule-governed disciplined structure that develops and refines pre-scientific knowledge by means that guarantee intersubjective agreement (e.g. by peer review), precise references (e.g. standard terms and measurement conventions), abstraction (e.g. isolating experiments from extraneous complexity), and systematic knowledge (universal assumptions which link theories). Isolated from mundane life processes, scientific feedback is reduced to a few significant forms, directed towards prediction and control. Habermas acknowledges that the technical application of knowledge is by no means *intended* in scientific research processes. However, the

---

[8] Since causal relations involve counterfactual conditionals of the form "If A had not occurred, C would not have occurred", we need to intervene to see whether C would occur without A occurring.

structure of propositions and the types of testing conditions already embody a methodological decision which "has been taken in advance concerning the technical utility of empirical scientific information" (PD, 209). The cumulative progress of natural science is rendered systematic through interlocking research programmes which are increasingly planned in advance.

Inquiry cannot be justified by simply appealing to "reality", because if all experience of reality is mediated by prior belief and hence language, then the accuracy of a hypothesis cannot be tested by an appeal to any immediate knowledge of the real (KHI, 99). To avoid a perspectivism that allows reality to be constituted in many ways, Habermas needs to ensure that the relationship between interest and knowledge is a transcendental one rather than a merely empirical one. For the relationship to be transcendental, there must be a discursive process like the early Hegel's notion of interaction. By such a process, the community of scientists test a hypothesis by determining whether empirical observations conform to a statement of a logical consequence of the hypothesis, a process Popper saw as akin to disputation by a jury. Clearly, what our words mean and how we apply them is not a matter for us to decide alone, but can only be determined intersubjectively (KHI, 108). For Habermas, humans come to share an understanding of reality only to the extent that their engagement with it is mediated by concrete intersubjective relationships with each other. The notion of intersubjectivity is harnessed in KHI to explain the progress of science as a series of increasingly adequate statements about the same reality.

The Historical-Hermeneutic Sciences

Habermas's historical-hermeneutic sciences are guided by a practical interest anchored in sociocultural imperatives to maintain and expand the possibilities of mutual understanding. The intersubjective and normative nature of this practical interest renders it inaccessible to the analytical-empirical sciences (KHI, 176). Rather than drawing an ontological distinction between different objects of inquiry, the same objects can be disclosed under different aspects. Human beings can be studied as objects of an empirical science like biology, or as communicative actors whose actions can be interpreted. Both forms of knowledge, though constituted differently, are achieved with the fallibilistic pragmatism that Peirce described in relation to the empirical-analytic sciences. "Both are set off by disturbances of routinised intercourse whether with nature or with other persons. Both aim at the elimination of doubt and the reestablishment of unproblematic modes of behaviour" (KHI, 175). In the case of empirical-analytical sciences, doubt arises from any frustration of feedback-controlled, purposive-rational action. In the case of historical-hermeneutic sciences, doubt arises from the disturbance of a consensus, the non-alignment of the reciprocal expectations of acting subjects.

While the objects and laws of empirical-analytical science exist independently of the human mind, the objects of historic-hermeneutic sciences are products of human intentional action, of humanity coming to an understanding of its own created meanings. The empirical-analytical sciences bracket their own constitutive activity in order to apply hypotheses to alien material, while the historical-hermeneutic sciences grasp cultural events "from within", since researchers are, after all, products of culture (KHI, 145). To elucidate this intersubjective dimension of understanding, Habermas turns to Dilthey's hermeneutics. According to Dilthey, the empirical-analytic sciences, directed towards "explanation", formulate particular experiences in a way that can be subsumed under universal laws. They "grasp reality with regard to technical control that, under specified conditions, is possible everywhere and at all times" (KHI, 195). The specificity and particularity of the individual experience is lost by abstracting from the richness of the context and smoothing out irregularities, leaving only what is quantifiable and experimentally reproducible (KHI, 162). This is expressed in a "pure" language where context is irrelevant to interpretation and meanings are conventionally standardised. In contrast, the historical-hermeneutic sciences, directed towards "understanding", invoke the particularity of experiences in ordinary language by an intersubjective process. They "grasp interpretations of reality with regard to possible intersubjectivity of action-orientating mutual understanding specific to a given hermeneutic starting point" (KHI, 195). Rather than subsumption under general laws, hermeneutics seeks expressions of unique experiences in all their rich particularity, within the generality of a shared ordinary language. Instead of a clear distinction between language and reality, as assumed by the empirical-analytic sciences, hermeneutic reality is constituted in language as the reality of meanings.

According to Habermas, Dilthey's hermeneutics rested too heavily on the monological psychological concept of "empathy", through which the experience of an historical agent could be re-experienced in the mind of the historian (KHI, 179–82).[9] Habermas argues that rather than attempting to experience the psychic state of the other, the opening of dialogue is required. He therefore turned to his contemporary Hans-Georg Gadamer (see LSS, 162–70; Gadamer 2004, 1990). Rather than founding knowledge on immediate insight, Gadamer sees all understanding as involving interpretation. For Gadamer, ongoing traditions are the medium in which

---

[9] Habermas's interpretation of Dilthey is disputed by a number of commentators who consider that he overplays the role of empathy and discounts Dilthey's intersubjective ideas of expression. See Harrington (2001).

"prejudices" or prejudgements, as the conditions of understanding, are transmitted and developed. Rather than the reproduction of an original meaning, hermeneutical understanding involves the interplay of dialogue between tradition and the interpreter. The meaning of a text is in principle inexhaustible, since it must remain open to future interpretations when the movement of history reveals new viewpoints which cast existing elements in a new light. This is important for Habermas because modern historicism views the continuity of tradition as broken. Scientific-industrial society sees itself as no longer an ongoing unreflective extension of tradition, but as rationally planned and technically mastered. According to this technocratic consciousness, a nomological social science could formulate general laws to which history could appeal for causal explanations of particular events. Habermas saw that, by demonstrating the dependence of interpretive understanding on an essentially historical prior sociocultural sharing of pre-understandings, Gadamer's hermeneutics requires us to reflect on the relations of theory to history. Both Habermas and Gadamer recognised that the idea of a society freed from historically conditioned consciousness, or a social science freed from the context of its interpretation, is illusory. The interpreter cannot assume a purely subject-object relation to her sociocultural tradition, because all her conceptual and normative orientations arise from this tradition. Since she belongs to the object domain being investigated, she cannot dominate it theoretically with objectivating techniques like those of natural science.

Habermas agrees with Gadamer about this embeddedness of the social analyst in historically situated contexts, but he also sees the need for critical reflection to transcend context. Habermas is concerned about Gadamer's emphasis on the authority of tradition as a medium of unreflectively shared pre-understandings. He sees Gadamer as elevating tradition to a position of unquestionable authority comprising the entire holistic background of pre-understood meaning that, by its very nature, cannot be made explicit all at once. Against this apparent conservatism, he argues that tradition is "profoundly transformed as a result of scientific reflection" (LSS, 168). When the everyday speech bequeathed by tradition is systematically distorted, only an "external" framework like science enables sufficient distance for the distortion to come into view. Habermas thinks that to identify hermeneutic inquiry with the continuation of tradition is to emphasise participation and dialogue over distantiation and critique. While accepting the context-boundedness of all understanding, Habermas wants to mitigate Gadamer's relativistic conclusions arising from the situated character of understanding.

To achieve this, Habermas harnesses scientific theories of communicative development and social evolution as standpoints, external to the traditional lifeworld, which transcend the context of normal language competence.

Such theories specifically reveal the empirical conditions of the formation and development of tradition. Although all social institutions depend on language as tradition, language is based on social processes involving power and domination which it does not articulate. If culture is viewed through the prism of an external perspective that brings into view its relation to the social, political, and economic conditions of life, cultural tradition loses its appearance of self-sufficiency. Responding to Gadamer, Habermas notes that everyday linguistic understanding contains possibilities for reflection more critical than the tradition-bound, dialogical reflection that Gadamer has in mind. Certainly, one cannot adopt the stance of a *totally* detached observer of one's own language without at the same time leaving the realm of meaningful experience, but such radical distancing is not necessary for a radical critique because "every language is its own metalanguage" (Habermas 1980, 246). Here Habermas foreshadows his development of the context-transcending norms of communicative action which will be discussed in the following chapter.

The debate between Habermas and Gadamer continued for a decade. The two may have had more in common than either acknowledged. It is beyond my present scope to further examine their arguments or attempt a rapprochement. What is important is that Habermas's arguments point ahead to the future development of his project. He thinks that Gadamer devalues methods of social analysis that go beyond normal linguistic competence. He will develop reconstructive theories of linguistic competence, cognitive and moral development, distorted communication, and social evolution which can provide quasi-causal, though not nomological, explanations of social phenomena. Habermas will integrate this external perspective into his social theory. We see this beginning in his discussion of the critical sciences.

The Critical Sciences

Thus far, we have considered Habermas's epistemological thesis on the analytical-empirical and historical-hermeneutic sciences, that the former are constituted by the technical interest underlying instrumental reason and the latter by the practical interest underlying communicative reason. However, the dependence of these sciences on interests embedded in empirical structures of human life suggests that reason is merely serving these interests. Habermas's critiques of both positivism and hermeneutics suggest that neither form of reason is sufficiently rational. Habermas needs a critical perspective from a position outside of empirically bound interests and he thus posits an emancipatory interest which guides a critically oriented science. This interest directs the subject towards autonomous agency by engagement in a form of self-reflection which recognises distorted and repressed content, "releas[ing] the subject from dependence on hypostatised powers." Guided by this interest, critical science can analyse the other two sciences to "determine when theoretical statements grasp invariant regularities of social action as such and when they express ideologically frozen relations of dependence that can in principle be transformed" (KHI, 310).

The emancipatory interest constitutes reality in ways that bring "second nature" into view.[10] Neither natural science, which only addresses "first nature", nor hermeneutics, which assumes all products of human subjectivity to be accessible to natural language, is sufficient for this task. The emancipatory interest requires self-reflection by a type of reason understood in terms of Kant's idea of Enlightenment as freeing humanity from its self-imposed tutelage by an orientation towards autonomy and maturity. In his 1965 Frankfurt inaugural lecture, Habermas claimed a priori status for this interest, noting that its standards

> are exempted from the singular state in which those of all other cognitive processes require critical evaluation. They possess theoretical certainty [since] the human interest in autonomy and responsibility . . . can be apprehended *a priori*. What raises us out of nature is the only thing whose nature we can know: language. Through its structure, autonomy and responsibility are posited for us. Our first sentence expresses unequivocally the intention of a universal and unconstrained consensus. (KHI, 314–15)

Habermas grounds this emancipatory interest in a developmental account of the species in which communication becomes distorted and constrained by socialisation into traditions which serve to conceal the power structures underlying social relations. This conjectural scientific account provides the critical sciences with a normative standpoint from which distorted communication can be recognised. Habermas offers Freudian psychoanalysis and Marxist ideology critique as paradigmatic examples of critical "sciences" directed towards emancipation. However, neither are completely adequate due to their incorporation of "scientistic misunderstandings", specifically their self-understandings as natural sciences. To formulate an account of an emancipatory form of reason, Habermas must develop novel interpretations that go beyond established disciplines. KHI should therefore be seen as laying out his preliminary reflections as preparation for a new conception of inquiry directed towards critical reflection on taken-for-granted knowledge,

---

[10] By "second nature", I mean that which appears as simply "given" but is revealed to be an intentional product of human agency, unrecognised as such. "First nature" is what is immutably distinct from the perceiving subject.

in the light of the awareness that knowledge is formed within constellations of social power (KHI, 310).

To this end, Habermas treats psychoanalysis as a "depth hermeneutics" to be understood in terms of interpretation (KHI, 256–7). However, rather than treating meaning as accidentally obscured, Habermas draws on Freud's treatment of meaning as systematically distorted, thus calling for a quasi-causal explanation, a task requiring a general theory of cultural evolution (KHI, 217, 257). He looks to Freud's diagnosis of "communal neurosis", which took account of the cultural evolution of the human species. The motive for human society was seen as the overcoming of scarcity, varying historically according to the level of development and organisation of productive forces (Freud 1985). The social institutions which inhibit the satisfaction of otherwise achievable desires could then be viewed as communal neuroses, maintained by social power. With such an account, Freud is able to give interaction and language a more developed role than Marx, who Habermas thought tended to reduce it to labour (McCarthy 1978, 84).

By incorporating Freudian themes into Marxist historical materialism, Habermas reconceptualises concepts like power and ideology. Like individual neuroses, institutionalised power relations result in the rigid reproduction of behaviour. Neurotic motives for action are redirected towards substitute gratifications. By drawing on Freud's treatment of dreams and neurotic symptoms as meaningful, Habermas develops a reading of psychoanalysis as a theory of language. He looks to Alfred Lorenzer's reconstruction of Freud in terms of developmental linguistics which shows that what may initially appear as a purely natural phenomenon, such as memory loss or neurotic behaviour, is actually meaningful (KHI, 256–8). A trauma or desire that is unacceptable to consciousness is made incomprehensible by sealing it within the private language of the unconscious. Like Marx's historical materialism, Lorenzer's reconstruction of Freud provides a means to diagnose such disturbances and understand their deeper meanings. However, the meaningful content can only be grasped by virtue of it being possible to *explain the origin* of the symptomatic scene by reference to the formative conditions of early childhood. This general theory offers an explanatory framework within which the patient's fragmentary experiences may be reconnected. Similarly, the theory of systematically distorted communication which legitimises and conceals social power structures provides an explanatory framework for ideology.

For Habermas, reason's underlying emancipatory ideal is seen in the overcoming of ideologies by the undermining of rigidified distorted communication and the power structures it legitimates. The *telos* of this ideal is "an organisation of social relations according to the principle that the validity of every norm of political consequence be made dependent on a consensus arrived at in communication free from domination" (KHI, 284). This counterfactual critical ideal, an anticipation of a future condition intended to guide social transformation, is the core rational value that grounds human social activity. This ideal, which we will see maintained and transformed in *The Theory of Communicative Action*, is the critical force that can challenge the ideological hypostatisation of sciences as they are drawn into constellations of social power. The technical interest enables the material conditions for political self-reflection. The practical interest provides traditions with stability (KHI, 210–11). However, it is the emancipatory interest that grounds the transcendental status of all three knowledge-constitutive interests. Its recognition entails that an appeal to knowledge-constitutive interests can never be merely a biological or sociological reduction. These interests are not mere facts of human nature. They necessarily entail procedures for the rational assessment of claims regarding both facts and values. Their justification now exceeds mere species survival and is directed towards political maturation.

Habermas's critical science does not exempt itself from the self-formative process upon which it reflects. As his major work rebutting positivism, KHI argues against the idea of pure theory in the form of either science or philosophy (KHI, 311). The three knowledge-constitutive interests are conceptually irreducible to each other or to anything else. However, the fact that an irreducible interest underlies the possibility of natural science doesn't mean that the natural sciences lose their legitimate authority. Their authority is based on the objectivity of their statements, which must be protected "against the pressure and seduction of particular interests." While it can secure its authoritative objectivity by these means, natural science nonetheless "deludes itself about the fundamental interests to which it owes not only its impetus but the conditions of possible objectivity themselves" (KHI, 311). It would seem that, although natural science is entitled to its authority regarding the objective world, it is always tempted to claim a 'God's eye view' as its standpoint of inquiry. However, while the philosophical commitment to pure theory "succumbs to unacknowledged external conditions and becomes ideological", within scientific practice, this ideology can actually protect the progress of scientific method by excluding reflection on knowledge-constitutive interests. "From knowing not what they do methodologically, they are that much surer of their disciplines, that is of methodical progress within an unproblematic framework" (KHI, 315). The problem is that natural science has today become identified with knowledge per se. Whereas philosophy once marked out the legitimate place of the sciences within a broader field of knowledge, "the philosophy of science that has emerged since the mid-nineteenth century as the heir of the theory of knowledge is methodology pursued with a scientistic self-understanding of the sciences." We will see in chapter 4 that Habermas doesn't think that philosophy in this post-metaphysical age can resume its former role as adjudicator of other areas of knowledge, but neither can science take the role of "first philosophy," addressing that which is most fundamental and on which all other knowledge depends (KHI, 4).

After the Theory of Cognitive Interests

In KHI, we see Habermas articulate a position from which each of the sciences can be seen as a partial form of reason, thus enabling a critique directed against the threat posed by an increasingly dominant orientation of practical life towards the imperatives of natural science and technology. These imperatives distort the communicative resources required for maintaining stable, free, and open societies. Given its ambitions, it is not surprising that KHI remains the book Habermas has been most ambivalent about, or that it has attracted considerable critical discussion (Muller-Doohm, 136). Claiming "no more than the role of a prolegomenon" for KHI, Habermas's responses have taken the form of a series of adjustments away from the transcendental analysis of constitutive interests and towards the communicative paradigm.[11] Following publication of KHI, Habermas acknowledged several problems. One significant problem is the conflation of two senses of "reflection" (Habermas 1975, 182–185). In one sense, "reflection" refers to critical insight into the unconsciously produced constraints and distortions which subjects succumb to in their socialisation. Such insight is emancipatory in itself because it reveals pseudo-objectivities. In another sense, "reflection" involves bringing something to consciousness that does not distort consciousness, for example, the implicit anonymous universal rule systems like grammar or logic which competent communicative agents unconsciously follow. This type of reflection, which Habermas calls "rational reconstruction", is directed to the conditions required for knowing, speaking, and acting subjects as such. Their mere discovery is not emancipatory (TP, 23).[12]

However, there is a gap between the "pure" disinterested reason of reconstruction and the emancipatory interest of critique. Rational reconstruction, independent of any ensuing emancipation, seems to be beyond the scope of the three fundamental human interests. Lacking a concrete practical

---

[11] Habermas's responses can be found in TP preface, Habermas (1975, 1982).
[12] Reconstructive sciences are revisable and fallible. They analyse the deep structures of cognition and action which are operative in the activities of individuals. However, unlike transcendental philosophy, they are not ahistorical, but rather frameworks or patterns of rule competencies which evolve in the history of the individual and the species. Thus, individuals are unknowingly the agents of evolution through the learning processes by which they acquire such competences.

orientation itself, it is nevertheless one of the conditions of possibility for critique and practical emancipation. However, emancipation itself requires further engagement and commitment. It is in response to this difficulty that we will see Habermas, in his turn towards the reconstructive science of communicative competences, split the concept of an emancipatory interest so that the unconscious constraints to which subjects surrender in self-formation are demoted to a merely empirical inquiry into possible deviations from a formal developmental logic.

HABERMAS'S THEORETICAL PARTITIONS

The primary target of Habermas's early works was positivism's reduction of knowledge per se to natural science, which it regarded as the only form of genuine knowledge. In KHI, he adopted a strategy of partitioning to grant legitimacy to three fundamentally distinct and irreducible forms of knowledge. Natural science governed by the technical interest is then only one form of reason, distinct from and irreducible to the two other forms of reason grounded in the practical and emancipatory interests, respectively. While this strategy blocks the positivists' reduction of knowledge per se to natural science, it is not clear that knowledge fits into the partitions that Habermas has constructed. In chapter 5, we will see Foucault's archaeology cut across divisions to include a vast web of heterogeneous discourses that overlap and infect each other. Foucault would insist that Habermas's partitions are more invented than discovered.

Prompted by this provocation, I will direct my analysis to Habermas's partitions. The problem is that, although his "cognitive interests" are supposed to strictly align with distinct areas of knowledge, it seems that those areas overlap and merge in ways that render his division arbitrary. However, on Habermas's account, each interest underlying a form of science is logically distinct and irreducible to any other interest. Pursuing success in the natural sciences cannot map directly onto mutual understanding by interaction. The problem with this account is seen in the case of natural science's manifestation in technology. Habermas, I think correctly, doesn't make a firm distinction between science and technology because he sees today's scientific research processes as tightly coupled with technology in such a way that the economic exploitation of technology generates feedback for science.[13] Modern sciences are incorporated into society primarily in the form

---

[13] A number of distinctions between science and technology can be made. For example, science, like technology, aims for precision and generality, but it is always open-ended because there is always more that might be discovered. In contrast, technology has a specific goal in view which it seeks to

of their technical applications (TRS, 52–5). Habermas thus treats scientific technology as politically neutral due to its orientation to a technical interest, not bound to particular social forms but necessary for the survival of the species as a whole. He sees technology as a general form of action, transcending particular political interests by reflecting the species-wide technical interest of the natural sciences in the prediction and control of nature. If technology remains linked solely to this interest, it poses no threat. The problem which Habermas identifies is that technocratic consciousness has infiltrated communicative interaction and distorted the practical interest underlying the historical-hermeneutic sciences. The task is to restore the balance, not attempt to invent a different form of science-technology, as Marcuse argued.

I remain unconvinced. Habermas doesn't adequately address the fact that technological substitutions for bodily exertions are used differently in different social contexts. It is only by being contextually specific that one can explain actual technologies, because they don't arise exclusively from any constant of human nature. Habermas insists that the natural sciences and their associated technologies are guided by technical rules rather than social norms, with failure defined by inefficiency rather than social sanction (TRS, 92–3). I agree that inefficiency is different from social sanction. However, what counts as inefficiency is nevertheless determined by specific social contexts. While technical efficiency is no more than the ratio of inputs to outputs, it must also be understood concretely, for example, in the context of workers' resistance to unjust practices. As Feenberg points out, when applied to specific contexts, the notion of efficiency implies decisions as to what kinds of things serve as inputs and outputs, who can offer and acquire them and on what terms, and so forth (Feenberg 1999, 160). Habermas's neutrality requires the bracketing of such background decisions which necessarily define the criteria for the concept of efficiency.

Even a species-wide interest in survival is not simply given in a way that determines a fixed logic of action. This interest relies on the mediation of fundamental distinctions, such as harmless/harmful or useful/useless, by particular forms of cultural and social life. Scientific research pursues certain questions and not others, employs certain categories and not others, and applies certain distinctions and not others. This suggests that by its very nature science cannot avoid reflecting particular interpretations, rather than simply following a universal neutral ahistorical technical interest directed towards species survival. What I am suggesting is that there are always

---

achieve by the most efficient means, reflecting a range of pragmatic concerns. (See Tiles (2005) for a discussion of this view held by Bachelard.) However, because of the complicit nature of the social interests which inevitably mould science, the critical theory tradition tends to identify technology with science, often calling it "technoscience".

alternatives, none of which are neutral, for satisfying the "universal" interests of "efficient" species survival. For example, this interest might address climate change by massive substitution of technical systems for natural processes. Alternatively, this same interest might reduce emissions by forgoing what a century ago were luxuries but are now deemed essential for the "standard of living". Efficiencies are informed, consciously or unconsciously, by social norms, values, and meanings.

As an analytic ideal, Habermas's technical interest may be universal, but technology can only be *actualised* in particular social and cultural contexts, where it necessarily embodies particular values and interests. Natural science only becomes manifest when instrumentalised in speech acts in particular pragmatic contexts or when "translated" into technological projects where norms, values, and meanings coalesce. Technology is only realised in actual designs, unavoidably imbued with norms, values, and meanings, responding to what particular societies recognise as problems according to their criteria for what counts as a "problem". This blending of the technical and the social is not extrinsic as Habermas assumes, but is a defining feature of science-technology.

Consider buildings as a form of science-technology. They both constrain and enable interactions—the manner in which we stumble across, meet, acknowledge, see and are seen, speak to and are spoken to by certain others. Buildings don't achieve this by a spatial arrangement of forms dictated by a function whose efficient fulfilment neutrally flows from the logic of a deep-seated interest. While buildings fulfil such generic abstract functions, at the same time, they necessarily fulfil very particular functions. They fulfil such functions not only by their physical realisation, but by signifying this function. Thus a door not only manifests the function of controlling and enabling movement between spaces, but *signifies* further functions by employing a "language" which marks socially sanctioned distinctions between servant and master, tradesman and guest, front and back, private and public, and so forth. Latour refers to this process as "delegating" norms to technology in such a way that they are routinely embodied in devices that serve to enforce, dictate, or prompt observance of social norms, both physically and symbolically (Latour 1992).

However, Habermas cannot accept what he sees as a mixture of communicative and instrumental rationalities. While technology appears to relieve physical effort but not communicative effort, this appearance relies on a failure to recognise how abstract universal functions are shaped into concrete value-laden designs. Habermas's commitment to natural science and technology as pure expressions of a certain type of rationality approaches essentialism. He fails to recognise that the under-determination of technology by this abstract functional rationality leaves room for its integration with the particular social environments that support its function. Fundamental imperatives tie technology not only to species-wide interests, but to features of social formations such as capitalism. What characterises capitalist technological commodification is its focus on maximising exchange value, which suppresses other values and meanings which are found in creative relations to work, relationships, or traditions. Norms embodied within technologies can centralise power and control, systematically diminishing agency and participation and hence also communicative rationality. Today, neural networks underpinning advanced artificial intelligence systems are moving into core areas of society. Apart from the obvious risks posed by this technology, we see the delegation of human competences to technology shifting the balance between workers and employees and between citizens and the state, thus diminishing agency and participation and strengthening top-down control.[14] By downplaying the hegemonic social interests which co-determine the direction of scientific research and the actualisation of technology, Habermas limits his critique. The problem is that, although technical systems may have essential attributes which entail an objectifying, success-oriented relation to nature, Habermas wants to build an entire social critique out of these few abstract properties. Although we may be tempted to invoke distinctions between pure and applied research, or between science and technology, Habermas rightly rules such moves out. It is then only by bracketing out the fact that the logic of scientific technologies requires a social context to become actual that he can suggest that technology neutrally follows its own rationality, governed by its technical interest.

Throughout Habermas's work, the employment of abstract structural categories like "interaction" and "labour" enable broad-brush and seemingly robust diagnoses and prescriptions. Habermas is correct to criticise forms of scientific naturalism which rule out what doesn't fit their schemas and which reduce all knowledge to their limited purview. He is also correct not to demonise modern technology or instrumental rationality, as Horkheimer and Adorno did (TRS 86, Adorno and Horkheimer 1997). By positing three irreducible areas of knowledge linked to fundamental interests, Habermas can give different regions of rationality their place while arguing that the pathologies of modernity do not reside in reason per se, but in an imbalance of its differentiated components. Against positivism, he establishes the legitimacy of forms of inquiry outside natural science. Against hermeneutics, he denies that these are merely arbitrary products of social arrangements or convention.

---

[14] Artificial intelligence is currently being trialled in criminal justice to influence whether a convicted person spends time in jail. See https://epic.org/algorithmic-transparency/crim-justice/. It is also used to assist doctors to diagnose illnesses (Jiang et al. 2017).

His early critiques of positivism remain relevant today, when ideas of justice, fairness, and the sustainability of life are challenged by a revived technocratic consciousness and the "solutions" it engenders. Nonetheless, difficulties arise because Habermas's theorising doesn't sufficiently acknowledge the value-ladenness, variability, and specificity of scientific technology in his critique of modernity. KHI's theoretical structures tend towards essentialism, resulting in tensions throughout Habermas's entire project.

Thus far we have seen that, in his early critiques of positivism, Habermas defends a broad conception of reason. This defence was developed in KHI as an empirical thesis of quasi-transcendental interests which ground, and give legitimacy to, three distinct areas of knowledge. Difficulties that became apparent following the publication of KHI motivated a shift from this thesis. There were also other motivations. Habermas was aware that the technical interest he had ascribed to empirical science was looking increasingly implausible. Since Kuhn, there had been a gradual weakening of confidence in the hypothetico-deductivist model of scientific method, along with an increasing awareness of the roles of interpretation and practical choice in scientific practice (Kuhn 1996). In any case, positivism was waning. In 1975, Habermas notes that "the criticism and self-criticism of scientistic assumptions are well underway" (Habermas 1975, 159). He recognises the "confrontation of science theory with the history of science" within Anglo-Saxon traditions and mentions Kuhn, Feyerabend, Lakatos, Toulmin, and Searle. Habermas will now shift from the critique of positivism to a more comprehensive basis for his project. The epistemological approach will give way to the communicative paradigm. We will also see new difficulties emerging in Habermas's theoretical constructions.

Chapter 3

Science, Modernity, and Communicative Action

In the decades following the publication of KHI, Habermas developed a distinctive way to incorporate science into a philosophical basis suitable for a comprehensive social theory. This far-reaching theory, both descriptive and normative, would enable the diagnosis and critique of modern society. It took form as *The Theory of Communicative Action* (hereafter TCA), which appeared as two volumes in 1981. Turning away from KHI's cognitive interests, TCA moves critical theory in a naturalistic, "postmetaphysical" direction by reducing its reliance on transcendental philosophy. The "philosophy of the subject" entailed by cognitive interests had obscured comprehension of the intrinsically intersubjective and dialogical nature of reason. While avoiding transcendentalism, Habermas insists that philosophy can still retain key universalistic claims, now reformulated in conjunction with the human sciences to enable an analysis of the general conditions of rationality manifest in various human capacities. A theoretical reconstruction of the everyday communicative competences of social actors forms the foundation of Habermas' theory of society, rationality, and modernity.

HABERMAS'S LINGUISTIC TURN

Habermas's communicative theory drew on a number of distinct theoretical strands—formal pragmatics, systems theory, developmental psychology, and social evolution. His formal pragmatics employed the speech act theory of Austin and Searle to formulate a general theory of communication by reconstructing the universal pragmatic presuppositions of speech and action (Austin 1962; Searle 1969). This became the basis of Habermas's communicative turn in social theory (OPSI, 3–104). First signalled in his 1970 Christian Gauss lectures at Princeton and further developed in the pivotal 1976 essay, *What Is Universal Pragmatics?*, formal pragmatics aims to render theoretically explicit the pre-theoretical know-how underpinning basic communicative competences. This rational reconstruction of formal competences which make the communicative acts sustaining everyday social life stable, meaningful, and ordered explains the formation and ongoing reproduction of society. By uncovering implicit universal rules common to language per se, formal pragmatics anticipates an ideally rational form of communication directed towards uncoerced consensus (Habermas 1980, 267). By giving formal pragmatics a definitive form in TCA, Habermas's paradigm of communicative interaction preserves the aspirations of the critical theory tradition, suitably modified for contemporary conditions and needs.

Although Habermas continued to dispel positivistic notions of cybernetically self-regulated societies, he also recognised the possibilities of systems theory.[1] In his 1973 *Legitimation Crisis* (hereafter LC), he designated the lifeworld a "sociocultural subsystem", distinguished by its holistic and normative nature, its coordination by natural language, and its resistance to functional specification. In TCA, he draws a firmer distinction between system and lifeworld to delineate certain areas where social systems theory has priority, and other areas where it does not. The lifeworld is not a subsystem, but "a concept complementary to that of communicative action" (TCA2, 119). By adapting systems theory, Habermas is now able to reveal objective interconnections beyond those which are subjectively intended or expressly articulated, thus gaining a perspective unavailable to interpretive sociology, which is incapable of seeing where "language is *also* a medium of domination and social power" (LSS, 172).

Borrowing from Piaget and Kohlberg, Habermas radically recasts historical materialism in terms of the developmental stages of societies (CES, 95). While the cognitive-developmental psychology of Piaget and Kohlberg provides a general theory of socialisation in terms of the acquisition of communicative competences, Weber, Durkheim, and Mead provide theories of social evolution which enable systematic reconstruction of historical situations.[2] By employing these speculative theories, Habermas can view the development of societies as learning processes comprising a series of stages, each of which

---

[1] Systems theory analyses conglomerations of interacting parts that are at least partly self-regulating, bounded by space and time, influenced by their environments, defined by their structures and purposes, and expressed through their functioning. As an example, an organism is a system of organs. Used in fields as divergent as biology and cybernetics, systems theory was developed for sociology in the mid-twentieth century by Talcott Parsons and revitalised in Germany in the 1970s by Niklas Luhmann.

[2] Piaget (1965); Kohlberg (1971); Weber (1978, 2002, 2004); Durkheim (1974); Mead (1934).

depends on the previous stage. Thus able to establish provisional standards by which societies can be analysed, he can retain the normative standpoint essential to critical theory.

By incorporating all these broadly scientific strands, TCA provides a comprehensive theory of modern society which is critically aimed against forms of thought that reduce the possibilities of freedom. Habermas framed his approach as an alternative to Luhmann's social systems theory which argued that modern societies are no longer primarily integrated by norms, but have developed systemic alternatives to deal with complexity.[3] He was also worried by the relativism of the reactions against positivism—Gadamer's hermeneutics, Heidegger's historicism, and the Wittgensteinian sociology of Winch. To retain a critical perspective, Habermas needed to contain the relativistic implications of such reactions, which view understanding as inextricably historically and socially bound. By drawing on generalised empirical knowledge of the human capacity to critique arbitrary structures of prejudice by demanding discursive justifications, he can counter these relativistic tendencies and reduce the context-dependency of understanding (LC, 84). The external counter-intuitive perspective of the sciences can be transformative. "The medium of tradition", Habermas tells us, "has been profoundly transformed as a result of scientific reflection" (LSS, 168). The paradigm of communicative interaction seems to be the perfect alternative to both systems theory and hermeneutics. However, while acutely aware of science's capacity to both liberate people from prejudices and entrench new prejudices, Habermas is not exempt from the hazards presented by this ambiguity, a problem I will come to.[4] In this chapter, I will first discuss Habermas's major theoretical structures—formal pragmatics, the system/lifeworld distinction, and social evolution. Having elucidated this framework, I will consider Habermas's diagnosis of modernity and conclude with some criticisms which I suggest can be adequately addressed by reinterpretation.

Formal Pragmatics

The basic idea of formal pragmatics is that speaking subjects know how to achieve their goals through communication without relying on any explicit rules, criteria, or schemata. In his formal pragmatics, Habermas aims to unify all dimensions of language under the implicit norms which underpin the competences needed to communicate. Drawing on Austin and Searle, he is concerned with the illocutionary component—what language *does*—in the

---

[3] See Habermas and Luhmann (1971) for Habermas's critique of Luhmann.
[4] McCarthy suggests that Habermas is "seduced" by the illusion of a rigorous science (McCarthy 1991, 180).

establishment and reproduction of specific relationships constitutive of everyday social life. In this regard, speech acts are only successful if they have the capacity to generate freely entered interpersonal relationships between subjects. The illocutionary force—that is, the speaker's particular intention to, for example, assert, question, promise, and advise—derives from the speaker's recognisably sincere willingness to enter into a particular social relation, to accept obligations, and to draw the relevant conclusions for action. The role of communication in establishing and reproducing specific relationships constitutive of everyday social life is assured by the "binding and bonding" force of the speech act (OPC, 83–4), which is aimed at facilitating voluntary cooperation by raising three claims to validity, that is, the truth, rightness, and sincerity of a speaker's utterance. Only one of these claims is thematised at a time, depending on whether the speech act is constative, regulative, or an avowal. The other claims recede into a pre-thematic holistic background behind the dominant claim.

Each of Habermas's three validity claims corresponds to one of three "formal world concepts"—the objective, intersubjective, and subjective "worlds". Any of these validity claims can be drawn out of any utterance and subjected to the same intersubjective norms of redemption. Promises, for example, are explicitly illocutionary, but also presuppose propositional content which can be challenged quite apart from the relationship established (CES, 34). Meaning is always framed in terms of these three presupposed areas of validity which apply to language per se. The formal world concepts are transcendental in the weak sense of providing the conditions of possibility of communication aimed at mutual understanding.

What Habermas refers to as "communicative action" is action specifically directed towards unconstrained agreement in language. Its rules govern the way we use language and determine its appropriateness within particular social contexts (OPSI, 73). It is the fundamental medium of social cooperation and involves commitments immanent to speech acts. Constative speech acts, such as reporting, asserting, or narrating, mark the distinction between appearance and reality by thematising the validity claim of truth. They entail "an *obligation to provide grounds* [and] . . . contain the offer to recur if necessary to the *experiential* source from which the speaker draws the *certainty* that his statement is true". Regulative speech acts, such as requesting, commanding, and warning, mark the distinction between what is and what ought to be by thematising the validity claim of rightness. They entail "an *obligation to justify* . . . to indicate if necessary the *normative context* which gives the speaker the *conviction* that his utterance is right". Expressive speech acts, such as revealing, admitting, or deceiving, mark the distinction between the authentic self and the expressions in which it appears by thematising the validity claim of truthfulness. They entail "an *obligation to confirm*, to show in the consequences of his action that he has expressed just that intention which actually guides his behaviour". Thus, the basis of the illocutionary force that leads to the hearer understanding and freely accepting the speaker's intention is rational because "speech act obligations are connected to cognitively testable validity claims" (OPC, 85).

Habermas's account is not simply a description of the mechanisms by which competent agents establish and maintain relationships. Since the illocutionary force has a rational basis, it is also normative and displays not only how relationships *are* conducted but also how they *should be* conducted. Its orientation towards the goal of critical theory—human life free from domination—is thus anticipated in every act of communication. Habermas's entire project rests on this possibility of providing an account of communication that is both theoretical and normative. Formal pragmatics is a continuation of his ambition to chart a comprehensive concept of reason which is a viable alternative to instrumental reason. This form of reason yields knowledge that is empirical rather than a priori and fallible rather than certain, but it is not reducible to an empirical science which causally explains language usage. It is directed to invariant structures and conditions and raises universal, but defeasible, claims. Though it explains the implicit know-how needed to generate speech acts, the rules that form the conditions are not innate, but acquired by socialisation. Like hermeneutics, it relies on tacit understandings and normative expectations regarding speaking, acting, and knowing. These tacit norms are *unavoidable*. We cannot imagine substituting another set of norms.

When part of the background consensus which we always already naively assume becomes problematic, we can still generate rationally motivated consensus by engaging in "discourse", that is, suspending assertive force and treating an argument's propositional content as hypothetical. Discourse is guided by normative considerations much like those Habermas identified in STPS as residual in the bourgeois public sphere (see chapter 1):

1. Every subject capable of speech and action is allowed to participate;
2. Everyone is allowed to introduce or question any assertion whatsoever and express attitudes, desires, and needs;
3. No speaker is prevented by external or internal constraints from exercising her rights as laid down earlier (MCCA, 89).

These norms serve as critical resources by which we recognise illegitimate forms of social organisation. Legitimacy is not merely a matter of the current acceptability of norms, but rather the way in which the norms have been achieved, which can be judged by the "ideal speech situation" (OPSI, 85–103). This standard is a "meta-norm" which specifies the formal properties that discursive argumentation needs for a consensus to be distinguished from a mere compromise. It is not intended to delineate some ideal communication or community, but to specify what is counterfactually presupposed in every communicative interaction. Like Kantian "regulative ideas", it functions as a critical reference point, critical because validity is not defined in terms of particular contexts but transcends all contexts.[5]

In explicating the ideal speech situation, Habermas tempers Adorno's concern that imbalances of power always covertly invade language and thought. He pragmatically accepts the taken-for-granted world as a starting point, from which competent agents make a number of assumptions, almost invariably counterfactual. Agents assume that communication will not be distorted by imbalances of power, that all participants are equally able and willing to redeem the validity claims implicit in their utterances, and that participants in communication relate to each other as responsible subjects (OPSI, 148). Social agents presuppose these counterfactual "fictions" as the critical standard inherent in communicative action (OPSI, 102). Equipped with this standard, participants in communicative action proceed by calling for and providing reasons, as obligated by the commitments implicit within speech acts. However, when the ideal speech situation is disturbed by strategic action oriented towards success rather than mutual understanding, this standard has been abandoned.[6]

There is a further class of deviations from the ideal speech situation which agents cannot recognise. "Systematically distorted communication" constrains actors by covert power below the level of conscious articulation. Habermas thus replaces the concept of ideology employed by the philosophy of the subject with the intersubjective concept of "systematically distorted communication", which manifests as the inability to raise validity claims to intelligibility or truthfulness. This inability is not due to lack of hermeneutic effort, but rather reflects an inability to deal with external situations, say unjust power relations, which manifest as the disruption of communicative

---

[5] Habermas's idealised presuppositions of communication are not free of problems. The formulation of norms is extremely slippery, with Habermas sliding between different versions. (see Benhabib, 287–8). Since using the term "ideal speech situation" in TCA, Habermas has distanced himself from this expression: "Unfortunately, I once called the state in which those idealising conditions would be fulfilled the 'ideal speech situation'; but this term is easily misunderstood because it is too concretistic. . . . At no time, have I intended the 'unlimited communication community' to be understood as more than a necessary presupposition" (Habermas 1990a).
[6] Culler argues that Habermas's distinction between strategic activity and communicative action misinterprets speech act theory and that linguistic communication can't be so easily separated from purposive activity (Culler in Rasmussen, 38–41).

competence (OPSI, 159–64). To grasp the role that Habermas's concepts of communicative action and systematically distorted communication play in his diagnosis of modernity, I will first sketch his concepts of lifeworld and system and his account of their evolutionary development.

LIFEWORLD, SYSTEM, AND THE RATIONALISATION OF SOCIETY

Lifeworld and System

Habermas initially developed formal pragmatics as a means to incorporate communicative intersubjectivity into a comprehensive theory. However, he lacked a substantive sociological theory. TCA addresses this gap by binding formal pragmatics into a critical social theory which incorporates the concepts of lifeworld and system, thus enabling an account of modernity not only in terms of its social, linguistic, and intersubjective nature but also in systemic terms that escape that nature. Maintaining that both concepts, lifeworld and system, are essential to social theory, Habermas steers a path between the objectivism of Luhmann's systems theory and the relativism of Gadamer's hermeneutics, while maintaining the perspectives of both participants and observers of social practices.

Before TCA, Habermas had not developed the concept of the lifeworld beyond its origin in Husserl's phenomenology as a naive familiarity with an unproblematically given background comprising the stock of unthematised meanings, values, and assumptions which individuals within a society share. Now he enriches this concept to fully accommodate its intersubjective linguistic nature and serve as a counterpoint to the concept of system. In TCA, the lifeworld becomes the condition of possibility for communicative action, oriented towards mutual uncoerced agreement, a resource that harmonises intended action orientations by normatively securing communicatively achieved agreements. "The lifeworld is the intuitively present, familiar and transparent vast and incalculable web of presuppositions that have to be satisfied if an actual utterance is to be at all meaningful, that is valid or invalid" (TCA2, 131).

Communicative action proceeds by ongoing redefinitions of "situations" or "segment[s] of the lifeworld context of relevance that is thrown into relief ... and articulated through goals and plans of action" (TCA2, 122–3). Claims initially in the background are made explicit, put on the table, where they can be clarified, accepted, or rejected. Unlike Husserl's concept, Habermas's lifeworld is intrinsically intersubjective, enabling understanding to be reached and intersubjective breakdowns to be repaired by judging thematised claims on the basis of unproblematic background convictions. Situations are ordered by validity claims allocated to one of the three formal world concepts—the subjective, intersubjective, and objective "worlds". The socially shared lifeworld has a quality of certainty which is prior to any possible disagreement. While a segment of the lifeworld can become problematic and acquire the status of contingent reality, it is not possible that everything be called into question at once.

In analysing particular social phenomena from a lifeworld perspective, we adopt the perspective of participants, acting subjects recognising and responding to norms (TCA2, 132). From this perspective, the lifeworld itself remains withdrawn from thematisation and unavailable to theory. In order to explain the reproduction of the lifeworld as a whole, Habermas turns to the lifeworld's structural components, the three functions which the medium of language performs in the lifeworld's reproduction. First, by enabling mutual understanding, language functions to transmit and renew cultural knowledge, allowing reproduction of the contents of culture. Second, language functions to coordinate action, integrating society and stabilising group solidarity. Third, language functions to form personal identities by the socialisation of responsible social actors. Language thus serves to connect new situations with present conditions in the lifeworld dimensions of culture, society, and personality (TCA2, 137–8).

These three structural components are reproduced through the symbolic interactions of everyday communicative practice in cultural reproduction, social integration, and socialisation. This division maps onto the three formal world concepts and their associated validity claims. "Culture" is a resource to challenge and redeem claims of truth in the objective world, thus reproducing commonly accepted beliefs, meanings, and interpretations. "Society" is a resource to challenge and redeem normative rightness within the intersubjective world, thus enabling integration of shared norms and behavioural expectations. "Personality" is a resource to challenge and redeem the claims to truthfulness and authenticity of the subjective world and produce shared competences by socialisation processes. All three structural components reproduce society by operating in a circular fashion, mediating between the particularity of the individual or situation and the general framework of lifeworld resources (TCA2, 141–4).

Into this account Habermas introduces a social evolutionary dimension which I will elaborate shortly. For now, the lifeworld should not be seen as a conservative "given" but as developing towards increasing rationality. Emerging out of unquestioned social solidarity, societies increasingly differentiate claims into the relevant validity spheres of truth, rightness, and truthfulness pertaining to the objective, intersubjective, and subjective "worlds" (TCA2, 146). For example, as science becomes institutionalised, we see an increasing demarcation and separation of the validity sphere of truth as pertaining solely to the objective world known to science. There is a cost for this increasingly differentiated rationality. As actions become less subject to guidance by compulsory norms, both the volume of communication required and the risk of failure to reach consensus increase. As social integration becomes more difficult, systems like the market economy develop and relieve the burden on communicative action (TCA2, 150). While the lifeworld reproduces society communicatively, such systems reproduce its material aspects by the purposive activity of work through which individuals realise their aims. The more the lifeworld is expanded and differentiated, the more its analysis needs to be complemented by a systems analysis, to get at the counter-intuitive aspects of sociation. Goal-oriented activities of social actors are coordinated not only by their intended orientations to mutual agreement but also by functional interconnections not intended or recognised within the horizon of everyday practice. Habermas claims that by maintaining both system and lifeworld perspectives, we see that the more complex systems become, the more provincial lifeworlds become.

Habermas regards the systems perspective as a necessary complement to the lifeworld perspective, which is blind to materially induced distortions of symbolic reproduction (TCA2, 148–50). Systems are viewed from an objectivating observer's perspective, while the lifeworld only comes into view from the perspective of a participant in social practices. Including a systems perspective alongside, but distinct from, a lifeworld perspective, enables two analytic perspectives necessary for social theory (Habermas in Honneth & Joas, 253). Social integration harnesses lifeworld mechanisms that harmonise action orientations by normatively secured consensus. System integration harnesses the various sorts of "invisible hand" mechanisms through which the pursuit of individual interests produces desired patterns which are unintended and often unrecognised by actors (TCA2, 150). For example, a market economy stabilises unintended consequences of actions by functionally intermeshing action consequences, ideally matching supply to demand. Systemic outputs (prices) are also inputs which push systems towards a homeostatic equilibrium, a "goal state". Attempting to avoid socially unacceptable outcomes induced by market processes, governments intervene in this feedback loop with a "steering medium" such as power or money. This encodes "information" into systems which rely on "a purposive rational attitude toward calculable amounts of value, so enabling a generalised, strategic influence on the decisions of other participants while *bypassing* processes of consensus-oriented communication" (TCA2, 183). As I will argue later, scientific discourses, for example, economics, clearly have a preeminent role in constructing, maintaining, adjusting, and harnessing systems, as well as anchoring them to the lifeworld by legitimation.

Habermas insists that symbolic reproduction of the lifeworld remains a matter for communicative action.[7] While non-linguistic mechanisms can free people from the need for excessive negotiation and the risk of consensus failure, this freedom cannot be extended to areas of life that primarily fulfil functions of cultural reproduction, social integration, and socialisation by reaching mutual understanding in language. The structure of mutual understanding is required to recognise the criticisable validity claims embedded in lifeworld contexts defined by cultural traditions, institutional orders, and competences. Communicative action feeds off the resources of the lifeworld by actualising the potential for rationality as particular shared lifeworld contexts lose their background quasi-naturalness by becoming subject to validity claims.

By itself, systems theory can't make the distinctions necessary to grasp the pathologies that emerge in modernity *as* pathologies. Increased complexity achieved at the cost of the lifeworld does not appear *as a cost*. From the systems perspective, society appears as a self-regulating system analogous to an organism, such that the systems analyst should be able to judge whether disequilibria have reached a critical point affecting survival. However, unlike organisms, there is no obvious sense of the limits of social systems. The social scientist can speak of crises only when relevant social groups themselves experience systemically induced structural changes as a threat to their continued existence or identities (LC, 3). In the case of social, rather than natural, evolution, the criteria for what constitutes a good society are open to ongoing determination by social members. Societies must be grasped hermeneutically, from the internal perspectives of participants. As Benhabib points out, Habermas sees the concept of a lived "crisis" as presupposing that individuals interpret their needs, desires, and wishes in the light of norms and values made available by their culture (Benhabib, 232). In contrast, it seems that the currently dominant paradigm of neoclassical economics presupposes an individual's needs, desires, and wishes as being part of an immutable human nature which necessarily responds to the "signals" of the economic system. Against this view, Habermas sees inner nature as the result of linguistic socialisation. We only become individuals within society, not prior to society.

A social theory that sees society totally absorbed into the lifeworld is also problematic. It offers only an internal perspective, whereby the lifeworld that members construct from common cultural traditions is coextensive with society. This perspective regards social actors as completely autonomous and

---

[7] Habermas is critical of Parsons's attempt to reduce the lifeworld to a subsystem by introducing the notion of intersystemic relations regulated by media modelled on the medium of money. He says Parsons is over-generalising the notion of media since only material reproduction can take place via non-linguistic media (*TCA2*, 265–82).

able to rationally orient themselves to criticisable validity claims. Yet the web of interactions extending across social groups throughout history cannot be explained as produced by the conscious intentions of actors who cannot fully control either the possibilities for mutual understanding and conflict, or the consequences and side-effects of their actions. The counter-intuitive aspects of systematically distorted social reproduction are not apparent from the lifeworld perspective (TCA2, 150–1).

By combining lifeworld and systems analyses, Habermas' social theory aims to account for the relations, interchanges, and influences between the two levels. However, only the lifeworld can define the social system as a whole. Habermas insists on the priority of the lifeworld because the very objects that systems theory concepts pick out presuppose their prior identification from a lifeworld perspective (TCA2, 151). Systems remain anchored in the lifeworld on which they depend for legitimation, despite the dependence of lifeworlds on increasingly autonomous systems to relieve the pressures and risks of communication. However, we will see that in late modernity, a crisis is precipitated as systems come adrift from the lifeworld, only to return and colonise and restructure its communicative capacities.

The Rationalisation of Society

An adequate basis for the normative nature of Habermas' diagnosis of modernity cannot be furnished by formal pragmatics and systems theory alone. His social evolutionary theory plays a crucial role by charting the acquisition of communicative competences as progressive learning capacities which respond to internal and external challenges. Communicative competences cannot be assumed as simply given, but are developed with an identifiable ontogenesis which Habermas extends into a developmental history of societies. Much of this account is assumed in TCA, so we will need to backtrack to Habermas's earlier *Communication and the Evolution of Society* (hereafter CES). Borrowing from Piaget's structuralist account of ego development and Kohlberg's work on moral competence, Habermas hypothesises a number of developmental stages of societies, premised on the analogy with children who proceed through an irreversible sequence of learning stages, where later stages incorporate and modify elements of earlier ones. He thus reconstructs the developmental logic of the learning processes of societies by relating forms of societies to developmental sequences adapted from Piaget's and Kohlberg's models of individual development (CES, 156–8).

This staged development involves a process of "decentring", a term borrowed from Piaget for stepping back from the immediacy of a situation to adopt a broader perspective (TCA1, 69). What is taken for granted at one level becomes the object of critical reflection at the next. Having acquired the basic distinction between the lifeworld and the world, the mature individual is thus able to question the taken-for-granted assumptions of the lifeworld (MCCA, 138). This requires the ability to differentiate between cognitive, normative, and expressive speech acts and to treat one's assumptions as hypothetical by suspending the binding force they exert on one's beliefs. Habermas's developmental stages are speculative accounts of social progress which are open to empirical investigation. He remains cautious of any simplistic conflation of individual and social development (CES, 98). While the patterns of individual developmental psychology may turn up in social evolution, they offer only a tentative guide to a social progress that moves from particularistic to universal (CES, 99). Habermas's claim is that societies "learn" in a derivative sense. Only a framework for learning is provided, a "logic" within which social individuals learn to deal with conflicts and solve problems (CES, 121). The actual development of the lifeworld is also influenced by the "dynamics" of development, comprising forces arising from problems with society's material reproduction (TCA2, 148). Habermas's developmental logic establishes a normative standpoint from which deviations become visible.

Applying this social evolutionary theory, Habermas sees the lifeworlds of archaic societies as exhibiting a high degree of congruence in relations between social institutions, cultural worldviews, and persons (TCA2, 156). In "the magical-animistic representational world of paleolithic societies . . . all natural and social phenomena were interwoven and could be transformed into one another" (CES, 104). Categorical distinctions between society and its natural surroundings tend to be blurred. This homogeneous lifeworld becomes rationalised in the process of "linguistification of the sacred", the ongoing development of communicative competence which differentiates the lifeworld's structural components (TCA2, 77). As the legitimating power of metaphysical and religious worldviews declines, the sacred is transformed into the rationally binding force of criticisable claims to validity. The reproduction of cultural traditions, the normative integration of societies, and the formation of individual identities are thus increasingly brought within the medium of linguistic communication. This process represents successive releases of communicative potential by a decentred learning process that allows a critical distance to open interpretations to discursive validation. Ultimately, any validity claim we take for granted can be challenged. When disruptions in background consensus occur, they can only be repaired with argumentative procedures. Tradition is now constantly subject to challenge.

Habermas's account of the progressive unfettering of rationality enables a developmental ordering according to the degree to which aspects of validity have been differentiated from one another. At one end stands ritual practice, at the other is the practice of argumentation by which validity claims can be picked apart. In the course of social evolution, action oriented to mutual understanding gains increasing independence from the normative contexts provided by traditional, religious, or metaphysical worldviews. Institutionally required value orientations become general and formal, law and morality become universal, and individuals become autonomous. Consensus formation takes place according to the strength of arguments governed by abstract rules. Language takes over the functions of achieving understanding, coordinating action, and socialising individuals, thus becoming the medium of cultural reproduction, social integration, and socialisation. As lifeworlds become rationalised, systems arise to relieve pressures on the communicative resources of societies. Systems are initially uncoupled from the lifeworld in a differentiation of coordination by communicative action directed at intentions from coordination based on functional interconnections between action consequences. As lifeworlds increase their rationality through the separation of culture, society, and personality, systems increase their capacity to autonomously regulate material reproduction behind participants' backs. The rationalised lifeworld becomes both uncoupled from, and dependent upon, increasingly complex, formally organised systems of action like the economy and state administration (TCA2, 153–4).

Although Habermas views the development of systems as a necessary complement to rational progress in the lifeworld, he also recognises their complexity and opacity as a threat. As the market economy emerged, money became an intersystemic medium serving to regulate exchanges with the social environment, while also weakening the primacy of the lifeworld (TCA2, 171). In the cooperative social frameworks of traditional societies, success-oriented action is bound to norms embedded in communicative action. Now, with money as a medium, success-oriented action can be steered by egocentric calculations of utility, calling for "an objectivating attitude even in regard to interpersonal relations" (TCA2, 196). Societies become steered by mechanisms such as money and power in a manner largely disconnected from the norms and values of the lifeworld. By the use of highly simplified codes and stereotyped utterances or symbols, such media enable individuals to coordinate behaviour, not by pursuing mutual understanding but by successful performance. Subsystems of purposive-rational economic and administrative action thus become independent of their moral-political foundations. However, the tendency towards ever-expanding bureaucratic and economic subsystems is by no means unavoidable. Although communicative action is partly suspended in formally organised system domains, the steering media of power and money remain anchored in the lifeworld, at least in principle (TCA2, 311–2).

Chapter 3

THE DIAGNOSIS OF MODERNITY

In his acceptance speech for the Theodor Adorno Prize awarded by the city of Frankfurt, Habermas called for commitment to the ongoing project of modernity, enlightened by reflection on its *aporias* and failures (Habermas 1996, 38–58). He characterised modernity as an ongoing historical phase in which complex societies emerge that incorporate diverse conceptions of life. Modernity's rationalised lifeworld is increasingly differentiated, leading to demands for reasons to be given, the intellectualisation of life, and the calculability of things. As well as the growth of systemic mechanisms like economies and bureaucracies, modernity has yielded modern science, positive law, secular principle-governed ethics, and democratic forms of collective will-formation. Although he defends these achievements, Habermas thinks "that something is deeply amiss in the rational society in which [he] grew up and now live[s]" (Habermas 1986, 126). His diagnosis builds on Marx and Weber. Marx recognised the distinction between lifeworld and system but failed to appreciate "the *intrinsic* evolutionary *value* that media-steered subsystems possess".[8] Weber made the mistake of explaining the ills of modernity as the inevitable outcome of "disenchantment", which he equated with the growth of purposive rationality. He failed to discern that critical reflection enables distantiation from the parochial nature of particular views lodged in the lifeworld (TCA1, 72, 236). According to Habermas, the threat to the lifeworld is not, as Weber thought, rationalisation per se, but a form of rationalisation which now enables "the emergence and growth of subsystems whose independent imperatives turn back destructively upon the lifeworld itself" (TCA2, 186).

Habermas's critique is not directed at capitalism per se. He is concerned with the monetisation of the lifeworld in which a form of systemic rationality has become autonomous. If money only circulates within the economic system, its function as mediator between the supply and demand of material goods is not necessarily problematic. However, when the economic system penetrates the communicative practices of everyday life, it reifies social relations. Reification also derives from the political system. Seeking mass loyalty to secure its own continuation, political power employs strategies to prevent and manage crises, raising practical-political questions to the status of special problems to be dealt with by experts, thus short-circuiting genuine public debate (TCA2, 350). Habermas identifies two interrelated tendencies connected with disturbances in lifeworld reproduction, "cultural impoverishment" and "colonisation of the lifeworld". Cultural impoverishment of the

---

[8] Habermas has in mind the rationalisation of economic and legal systems which can increase communicative rationality, whilst liberating actors from its excessive demands (TCA2, 339).

lifeworld results from the "splitting-off of elite subcultures" (TCA1, 355). A culture of experts emerges which is incapable of transmitting the achievements of specialised areas back into the lifeworld. This is clearly true of modern science. Since the nineteenth century, the increasingly professionalised sciences have enforced boundaries between themselves and other scientific and non-scientific disciplines. Communities of scientists develop specialised vocabularies which exclude communicative dimensions extraneous to their narrow purposes. Within science, this enables intersubjective agreement within the scientific community and precision and economy in scientific practice by an exclusive focus on truth within restricted domains. The sociologist of science Robert Merton has argued that science has developed an elaborate system of rewards and status for those who live up to its norms, particularly regarding the originality and priority of significant findings (Merton 1979). However, by encouraging secrecy and exclusion, this arrangement can work against the universalistic ethos of science, a tendency amplified by financial rewards in a market economy. We also see specialised scientific-technical expertise leading to centralisation, hierarchical management, conformity, and routinisation. Through this exclusivity of scientific practice and discourse, citizens become increasingly disengaged from politics and, as passive consumerism takes over, are robbed of their critical capacities. Habermas is correct to highlight the tendency of specialised discourses such as science to actively withdraw and become increasingly inaccessible to the lifeworld. Against this trend, he articulates the ideal of a lifeworld which seeks to draw everything that can be grasped by reason into a comprehensive network.

"Fragmentation of consciousness" marks a threshold beyond which knowledge has been irredeemably "split-off" (TCA2, 355). This notion recasts Weber's thesis of the separation of spheres of value and the resultant loss of meaning and freedom. The breaking apart of value spheres represented for Weber irreconcilable and competing spheres of rationality and an irresolvable value struggle. Tremendous psychic burden is placed on the modern individual, who nonetheless has to create meaning in a fragmented world. Unlike Weber, Habermas endorses value sphere differentiation as a progressive cultural development. Claims within any value sphere are still subject to universal criteria by virtue of the unity of form of the argumentation procedures in the light of which the validity claims to truth, rightness, and truthfulness are redeemed. This is what Habermas has sought to establish in formal pragmatics (TCA1, 249). Accordingly, the ills that Weber identified should not be attributed to conflict between value spheres. Deformations of the lifeworld "ought not be attributed either to the rationalisation of the lifeworld as such or to increasing system complexity as such . . . [but to] the elitist splitting-off of expert cultures from communicative action in daily life", leading to impoverishment of communication and the dying out of vital traditions (TCA2, 330). Habermas wants to drive a wedge into Weber's account of disenchantment by insisting that the problem is the splitting-off of specialised areas of knowledge from the lifeworld, not the differentiation of validity spheres per se. The validity claims of truth, rightness, and truthfulness can still be brought together within the lifeworld's "unity of reason", where these irreducible dimensions of validity can be raised and redeemed by shared communicative norms. However, by becoming the exclusive domains of expert specialists, discourses such as science can no longer be brought within the ambit of the lifeworld's communicative resources.

The resulting fragmented consciousness is incapable of integrating cognitive, normative, and aesthetic understandings of reality in a critical way (TCA2, 355). Pathologies develop across the structural components of the lifeworld—loss of meaning, withdrawal of legitimation, crises in orientation and education (cultural reproduction), insecurity of collective identity, anomie, alienation (social integration), breakdown of tradition, decline of motivation, and psychopathologies (socialisation) (Habermas 1975, 1982). Holistic interpretations can "never attain that level of articulation at which knowledge can be accepted as valid by the standards of cultural modernity". The layperson loses any authority in these areas of knowledge. In effect, capitalism has found "a functional equivalent for ideology" by alienating us from our critical aptitudes (TCA2, 355). To be critical, we must be able to synthesise different aspects of validity (TCA2, 355). Lacking this capacity, we experience a loss because in everyday life, "cognitive interpretations, moral expectations, expressions and valuations have to interpenetrate and form a rational interconnectedness via the transfer of validity that is possible in the performative attitude" (TCA2, 327). To resolve real-life situations in which different value spheres are inextricably compounded, we require the competence to analyse, discuss, and balance different validity claims, rather than handing them over to expert specialists with partial perspectives.

With the splitting-off of expert knowledge, the lifeworld's competencies are weakened. Its fragmented consciousness lacks the holistic resources to recognise and articulate resistance to the systemic imperatives which now turn back to colonise it. Strategic forms of economic and legal action, mediated by money and power, replace communicative forms of action responsible for socialisation, cultural transmission, and social integration. Everyday life is restructured by systemic imperatives which increasingly reduce subjects to objects of processes beyond their grasp. Habermas's powerful metaphor of 'colonisation' suggests that the lifeworld, weakened by fragmentation and loss of internal cohesion, is no match for systems which, barely comprehended, invade it to restructure its resources according to their autonomous imperatives.

When stripped of their ideological veils, imperatives of autonomous subsystems make their way into the lifeworld from the outside—like colonial masters coming into a tribal society—and force a process of assimilation on it. The diffused perspectives of the local culture cannot be sufficiently coordinated to permit the play of the metropolis and the world market to be grasped from the periphery. (TCA2, 355)

While accepting the fact of reification in advanced capitalism, Habermas doesn't accept Horkheimer's and Adorno's totalising critique of instrumental reason, which leaves no position from which reason can speak (Adorno and Horkheimer 1997). Nor is reification inevitable. While colonisation marks a threshold beyond which systems, originally anchored to the lifeworld but now adrift, return to undermine the lifeworld's communicative resources, this process only reflects the dynamics of development, that is, contingent empirical factors. It does not reflect the logic of development, the learning process. Rationalisation could have been balanced, not one-sided as it has turned out to be under advanced capitalism. It could, and still might, allow the democratic normative control of subsystems and their steering mechanisms (Habermas 1990a, 109). Habermas wants to resolve these tensions between capitalism and democracy in ways that establish the primacy of democratically legitimated norms.

Habermas elaborates his thesis in the context of the development of contemporary advanced capitalism, characterised by the welfare state.[9] According to his account, the welfare state largely pacified class conflicts arising from acute economic crises by manipulating conditions around the margins to avoid social impacts while not restricting the free market (TCA2 350–1). Habermas accounts for government interventionism, the welfare state, and mass democracy by analysing multiple interchange relations between lifeworld and economic and administrative subsystems. The economic subsystem exchanges wages for labour and goods and services satisfying consumer demands. The administrative subsystem exchanges organisational achievements for taxes, and political decisions for the mass loyalty of voters. These relations of exchange define new social roles within the lifeworld (TCA2, 343). Freedom is lost in the dependence of these roles on their respective subsystems and the limitations that identification with these roles entails. Meaning is lost by the tendency towards abstraction, which occurs "whenever the lifeworld, in its interchanges with the economic or administrative systems, has to adapt itself to steering media" (TCA2, 322). For example, in the administration of the welfare state, all clients are subjected to the same formal categories and procedures. This one-sided rationalisation manifests in everyday practices in

---

[9] Although the welfare state has been under challenge since the 1980s, it is far from disbanded and remains as a paradigm, although one only partially realised. For this reason, Habermas's analysis of colonisation in terms of the welfare state requires some updating and adjustment, but is not obsolete.

which consumerism, achievement motivation, and competition increasingly predominate over moral-practical orientations.

Habermas discusses colonisation in terms of "juridification", the increasing tendency towards formal law in modern societies. Juridification opens up informally organised areas to bureaucratic interference and judicial control. Juridification is ambiguous. The protections offered in its latest stage—collective bargaining, award wages, social insurance—are experienced as both liberating, enabling citizens to exercise rights, but also as alienating. Just as the economy only processes things as commodities or exchangeable equivalences to which the lifeworld must conform, the law processes the concrete situations of individuals' lives by "administratively digest[ing]" them, forcing them into abstract categories. What was once "private" now becomes subject to public policy and controversy. Entire domains of social relations become politicised. The role and meaning of tradition in these spheres is now challenged, so that even well-meaning reforms instituted to correct the dysfunctions of capitalist growth have the consequence of undermining the lifeworld.

Although barely discussed by Habermas, juridification processes clearly rely on science. Science can be considered as located at the seam between lifeworld and system, where it constitutes, formulates, controls, and maintains systems, both justifying systemic power to the lifeworld and at the same time challenging it. But these roles that science plays cannot readily be comprehended by the lifeworld because science has been subsumed into areas of expertise which exclude the lifeworld's comprehensive perspectives. Fragmented and defenceless, the lifeworld is restructured by systemic imperatives. Legal and economic categories are regimented to align with the sciences which grant them authority. Juridification relies on systems of classifications, social ontologies developed and given authority by human, social, and medical sciences. Exactly what it is to be a refugee, a dependant, a delinquent, or disabled is determined by legal definitions which borrow from these sciences to define and explain, so lending the classification both objectivity and authority. The criteria by which someone qualifies for a certain relationship with the administration can often be traced back to scientific discourses which authorise and justify administrative actions and decisions. Just as the natural sciences subsume multi-faceted nature under commensurable stable categories, so too do the human and social sciences sort their objects. It is not only the case that scientific categories serve instrumental purposes of classification but also that, as we will see Foucault argue, they feed back into socialisation processes in a way that socially constructs human subjects, crafting dependencies and reconfiguring lifeworlds behind our backs.[10]

---

[10] While on one hand science restructures lifeworlds, on the other hand it is itself restructured by systems of power and money which distort its communicative structures. Today the academic research

One effect of colonisation that Habermas raises, but doesn't analyse in any detail, is the development of new potentials for emancipation and resistance (TCA1, 391). As systems cast their nets over the lifeworld, not only do we have fragmented consciousness and its pathologies but also the emergence of social struggles which differ from the usual institutionalised conflicts over economic distribution. These struggles are not about domains of material reproduction and are no longer channelled through political parties. They arise in domains of cultural reproduction, social integration, and socialisation concerned with the restoration of damaged ways of life. The "new politics" of alternative movements turns on quality of life, self-realisation, participation, and human rights. It includes environmental, peace, and squatter movements and struggles for regional, linguistic, and religious independence. This new politics demonstrates the aptness of the colonisation metaphor, suggesting that subjects within a threatened lifeworld can resist, remake their identities, or gain independence.

INSIGHTS AND *APORIAS*

Habermas's ambitious theoretical edifice brings modernity's pathologies into sharp relief. It bears on the dysfunctional state of modern democracies with concepts like colonisation of the lifeworld, fragmentation of consciousness, and systematically distorted communication. It enables analysis and questioning of contemporary society from a standpoint outside everyday knowledge. Its fallibilistic, formal, and procedural character means that it does not aim at a concrete utopia achieved for all time. The pragmatic force of its context-transcending ideal of a society based on agreements arrived at in free and equal exchanges exerts a regulative function, enabling a foothold for criticism and opening reason to self-correction.

Like his Frankfurt School predecessors, Habermas intends his theory to bear critically on contemporary society through its capacity to undermine the pseudo-objectivity of scientific discourses which support unjust social processes. However, unlike his predecessors, he thinks that emancipation does not entail rejection of the cultural legacy of Enlightenment, but rather its completion. He therefore integrates science into his theory more ambitiously than them, to build a comprehensive social theory with a normative dimension buttressed by scientific perspectives. He weaves an account of social evolutionary progress into his formal account of the presuppositions necessary for communication. These counterfactual abstract norms orient

---

career path of "publish or perish" has led to concerns about the relevance and veracity of scientific research papers and the consequent loss of trust in science (see Grimes, Bauch & Ionnidis, 2018)

judgement and action while permitting the freedom demanded by modernity. He defends the threatened lifeworld, which still contains unexhausted normative potential. Although Habermas's earlier paradigm of deep-seated anthropological interests has been transformed by the shift to a communicative paradigm and his incorporation of systems theory, key structural elements have been retained. The role he allocated to "labour" in KHI now appears as systems directed towards action consequences for the material reproduction of society. "Interaction" is subsumed within the richer notion of the lifeworld which provides communicative resources for the differentiated symbolic reproduction of society. "Ideology" is now "systematically distorted communication".

On Habermas's account, the fundamental problem of modernity remains one of balance. The dimensions of rationality available to modernity have been actualised and institutionalised in a lopsided way. However, whilst I agree with this diagnosis, I can't help but think that Habermas's ambitious social theory has not resolved, but added to, the tensions inherent in his earlier theoretical structures. TCA's increased explanatory reach has been bought at the cost of the gap between the empirical and the transcendental becoming more entrenched. The "quasi-transcendental" reconstructive sciences yield an ideal realm of communicative action governed by norms seemingly insulated from the empirical distortions induced by power within systems. This gap will remain a major point of contention for Habermas's critics and something he will continue to struggle to clarify. In order to elucidate how this tension plays out in Habermas' theory, I will first discuss the criticism that he fails to adequately account for power in the lifeworld. Second, I will offer a functional account of the role of science as it enters the lifeworld, in order to illustrate the complexity of relations mediated by science at the boundary between system and lifeworld, a complexity that escapes Habermas's theoretical structures. Third, I will argue that technology is functionally analogous to a systemic medium, further confounding the lifeworld/system distinction. Fourth, I will follow with some criticisms of Habermas's ethnocentric tendencies. I will conclude the chapter by discussing some ways in which Habermas' schema might be recast to narrow the gap between the transcendental and empirical, softening the boundary between lifeworld and system and enabling a less ethnocentric view of rational progress.

Power in the Lifeworld

While the notion of a lopsided modern rationality has both a long tradition and prima face plausibility, TCA's implication of a clean divide between colonised and coloniser and a one-way movement of systems colonising the lifeworld has been much criticised (Fraser 1989; Allen 2008). By making firm institutional divisions, such as between families and the public sphere on one side and business and state bureaucratic organisations on the other, Habermas tends to portray systems as the locus of power which colonises the lifeworld. However, as Fraser points out, families are also responsible for material reproduction of society, just as business and state organisations are also responsible for symbolic reproduction. Determining influences appear to go in both directions. By confining his discussion of power primarily to systems contexts, thus implying a one-way movement of colonisation, Habermas doesn't adequately address the significance of the power which is endemic to the modern lifeworld. His account of colonisation by systemic imperatives fails to adequately capture the reproduction of forms of social subordination, for example, racist or gender stereotypes, which are perpetuated by lifeworld socialisation. Fraser argues that the family perpetuates relations of oppression as much as it reproduces values and cultural norms (Fraser 1989, 121).

In response to such criticisms, Habermas softened the lifeworld/system distinction. The economy and the state, he tells us, are "primarily" integrated systemically. Rather than the distinction between lifeworld and system being based simply on different types of action (instrumental or communicative), Habermas sees the difference in the *dominance* of one action type within a certain domain (Honneth & Joas, 254). The distinction is not an "either/or" but a "more or less". Habermas concedes the possibility of a reverse movement which subjects systems to the normative restrictions of the lifeworld. "The question as to which side imposes limitations has to be treated as an empirical question which cannot beforehand be decided on the analytical level in favour of systems ... [the] colonisation of the lifeworld and the democratic control of the dynamics of systems . . . represent two equally justified analytical perspectives" (Habermas 1990a, 109). However, the possibility of a counter-movement of power from lifeworld to systems does not tell us anything about the domination which Fraser discusses, which appears endemic within the lifeworld. Habermas concedes that the lifeworld is "by no means an innocent image of power-free spheres of communication" (Honneth & Joas, 254). Notwithstanding these concessions, his primary focus has remained on power threatening modern lifeworlds through autonomous systems which have broken free from the communicative realm of social reflection and accountability. Habermas's colonisation thesis tells us little about the racism or homophobia which is generated and reproduced within lifeworld structures. To be truly critical, Fraser argues, critical theory must be capable of foregrounding the domination endemic in the lifeworld (Fraser 1989, 138).

Habermas might respond to these concerns by saying that he sees himself addressing a world-historical problem—the undermining by autonomous systems of the very communicative resources of the lifeworld upon which emancipation relies. The lifeworld, its communicative resources weakened by colonisation, can no longer recognise and challenge the abuses of power that it perpetuates. Abusive power arising in the lifeworld *could* be addressed by the increasingly differentiated communicative resources with which modernity demands justifications, but the problem is that those resources have been weakened by systemic incursions into the lifeworld. The problem that Habermas would now face is to give an account of exactly how the rational resources needed to address abuses of power have been eroded. It is by no means clear that systemic incursions into the lifeworld are responsible.

While TCA can be read as an epic struggle between "good" lifeworld and "bad" systems, I don't think that was Habermas's intention. In any case, Habermas is correct to point to the remoteness from the everyday lifeworld of specialised discourses such as science. He is correct to analyse this situation as a fragmentation of consciousness which weakens the resources necessary to recognise and respond to the sources of subjection. And he is also correct that communicative reason has been truncated by the logic of systems, such as the economy and the bureaucracy, producing regions of the unspeakable and unthinkable. The problem with Habermas's account is not that the lifeworld/system schema is incorrect but that, even admitting the two-way movement of power, it insufficiently penetrates modernity's opacity and complexity. The complexity of relations between lifeworld and system cannot be rendered in such broad strokes as his schema suggests.

The Social Function of Scientific Authority

To see what Habermas's broad strokes miss, we need to consider the functional role played by the sciences as they diffuse into the lifeworld. Habermas's developmental account of the differentiation of validity spheres is a story of our species collectively learning to cope with an unpredictable world leading, in modernity, to the spliting-off from the lifeworld of specialised discourses such as science. This leads to laypeople losing the capacity to integrate cognitive, normative, and aesthetic understandings of reality in a critical way (TCA2, 330, 355). Habermas's account is not so much false as hyperbolic. It requires qualification, without which it glosses over the details of how science is actually taken up by the lifeworld and the role it plays. Science does in fact permeate lay culture—in newspaper articles, popular science journals, documentary broadcasts, podcasts, presentations to decision-making bodies, and so forth. In the light of this penetration of scientific discourses into lay culture, I want to offer an account of the role science plays in the modern lifeworld that complicates Habermas's developmental story.

As scientific discourses are absorbed into the lifeworld, they acquire, by the authority vested in them, an indispensable function in maintaining social stability by generating consensus. A classic study is provided by *Leviathan and the Air Pump*, in which Stephen Shapin and Simon Schaffer examine what gives rise to the credibility of modern science's experimental methods without simply assuming, from our current standpoint, their self-evident truth (Shapin and Schaffer 1985). In their account, Robert Boyle succeeded in constructing scientific facts by enlisting an ideal public of gentlemen to witness experiments. This public stood for the universal and enabled the generation of probable knowledge which we could all reasonably agree upon, while accepting that there were some things on which we might reasonably differ. Boyle was opposed by Thomas Hobbes, who sought certainty from a purely deductive approach. Hobbes thought that the uncertainty of merely probable truths generated by Boyle's approach could only lead to chaos. Epistemology is political. In the wake of the English Civil War, the dispute between Boyle and Hobbes about obtaining reliable knowledge clearly had much to do with the problem of social harmony. Closer to the present, the logical positivists vehemently argued for a universal language of theory observation, the unity of science, and a clear demarcation between science and non-science as a bulwark against fascism. Today, as I write, American "patriots" are rioting at their nation's Capitol, convinced by news-feeds, talk-back radio, and social media that they have been deceived and betrayed regarding the 2020 presidential election. Clearly there is much at stake. Solutions to the problem of knowledge are inseparable from solutions to the problem of social order.

Since the scientific revolution of the seventeenth century, until very recently, few have doubted that it is the role of science to tell us what we ought to really believe. In situations of potential conflict, faced with disconcerting irreconcilability and wanting to steady our uncertainty, we turn to science, our most authoritative knowledge, to provide answers beyond reproach. This role is clearly seen in "trade-offs" which result from the strategic interactions between the advocates of irreconcilable claims, for example, those addressed in an Environmental Impact Statement concerned with the impacts of uranium mining on the natural and social environments, the economy, and Indigenous heritage. Here there are no rationally compelling criteria beyond those of legal proceduralism to guide adjudication between the irreconcilable values of disenchanted modernity. In response, we appeal to scientific objectivity to forge a consensus. Adjudication in such cases often requires adopting an observer's perspective on matters imbued with meanings, significances, and values only available from the participants' first-person perspectives, thus frequently reducing the situation to abstract "resources" to be weighed against other resources. We moderns both lament and endorse this impoverishment, since we have no alternative but to demand that, for example, economic values must *somehow* be "balanced" against aesthetic, traditional, or cultural values in a consistent and accountable manner. For us, only science has the authoritative status to settle such questions. Wanting to avoid open conflict, we demand decisions based on unimpeachable reasons which bear the authoritative stamp of science, even though science itself does not claim that authority.

I am suggesting a view of scientific reason as inextricably mixed with power in a way that doesn't necessarily undermine truth, but which nonetheless retains an ideological potential not easily picked apart by reason. In contrast to this functionalist perspective, Habermas assumes the existence of general interests lying beneath the "impenetrable pluralism of apparently ultimate value orientations" which can play a role in achieving a balance between conflicting validity claims (LC, 108). However, it is far from clear whether any such general principles can produce concrete political, economic, and moral arrangements satisfactory to all. By overestimating the efficacy and universality of rationality, Habermas gives too little weight to the irreconcilability of different social voices. The universalism underlying his work fails to address the conflicts articulated through discourses in which positions cannot be pushed into consensus without distortion or subtle forms of coercion. The function of science to generate consensus and maintain social cohesion doesn't guarantee that this is the *right* consensus. This function is also able to be ideologically employed to close off questions from further deliberation. What I have attempted to show is the intractable vulnerability of the lifeworld to the ideological employment of scientific authority, an intractability not visible from either Habermas's developmental theory of communicative competence or his lifeworld/systems schema. This schema is built on Habermas's conviction that he has discovered an immutable structure underlying our arguments and confusions which, at the same time, points the way beyond them. We shall return to this when we bring him into dialogue with Foucault in chapter 8.

Science and Technology

The distinction between lifeworld and system comes under further pressure when considered in relation to the translation of science into scientific technology.[11] As with the three distinct rational orientations we saw in KHI, the colonisation of the lifeworld is presented as a problem of balance. Systems and the reductive scientific rationality they entail are only threats when they go beyond their proper bounds to attack the communicative resources of the lifeworld. This formulation suggests that "in itself", technology limited to its proper relation to the lifeworld is a "neutral" norm-free manifestation and tool of instrumental action. What this misses is that technology is never "in

---

[11] With the expression "scientific technologies", I am referring not only to the inseparability of science and technology due to a network of feedback mechanisms, but also to disciplines such as economics which tend to be both descriptive and prescriptive.

itself" but, as I have argued in chapter 2, always comes with embedded social norms. Like science, technology mediates the boundary between lifeworld and system. It does so in a non-discursive bodily manner which we will see Foucault, unlike Habermas, grasp.

Following his debate with Marcuse, Habermas retained the view that "there is for this domain of reality [the objective world] only one theoretically fruitful attitude, namely the objectivating attitude of the natural-scientific experimenting observer" (Habermas 1982, 243–244). Habermas does not take account of the social norms that are unavoidably and inextricably embedded in technology because his resolutely abstract account does not adequately address the material actuality of bodies and things in their complex interactions with concepts, institutions, norms, values, and other objects. Actual technologies are absent from his account. Technology only appears as an abstract product of the natural scientific objectivating stance directed towards control and prediction of consequences. However, actual technology cannot avoid being infected by a vast range of the lifeworld's values. Technologies *unavoidably* materially manifest latent social values and hidden presuppositions that seldom become candidates for contestation within the lifeworld. At the same time, they manipulate our communicative possibilities, both enhancing and constraining, moulding us in ways which we are barely conscious of. Just consider social media. In this regard, Feenberg is correct to suggest that the function of modern technology can be characterised in Habermas's system-theoretical terms as a steering mechanism. Modern technology unwittingly, yet unavoidably, embodies particular "norms" and "prescriptions" that steer behaviour non-consensually, in much the same way as systems media like money or power (Feenberg 1999, 166–173). This is clearly seen with the technologies made available by the rapid advance of the internet. Here Habermas's analysis could be broadened by consideration of the concept of "trust", a concept underlying his brief discussion of the condensation of communication (TCA2, 185). As an institution, science does not rely so much on scepticism, as usually claimed, as on a vast background of trust. As a lifeworld condition, trust has a "given" quality, involving risk beyond calculation or justification. Today, trust in institutions which embody the residual norms of the public sphere—democracy, science, legal institutions, the traditional press, and so forth—are being drastically eroded. Trust is being transferred to what can only be considered systemic media—social media, internet search engines, and news-feeds—linked tightly to the neoliberal market economy. At the heart of this transfer of trust are the algorithmic technologies which link "information", not to claims ultimately redeemable by responsible actors, but to consequences that return maximum profits to advertisers and data owners.[12]

---

[12] Smith provides a helpful account of the operation of internet algorithms (Smith 2019).

The systemic nature of technology and the science on which it relies can be seen more clearly if we compare the ideological structures of scientism and late capitalism. In this regard, Paul Giladi draws our attention to

> how scientism and capitalism are *different* yet *logically bound instantiations of formal reason*: scientistic varieties of naturalism are typified by systematic practices of nomothetic reason aimed at subsuming phenomena under the laws of fundamental physics; capitalism is typified by systematic practices of strategic reason aimed at subsuming phenomena under the commodity form. Conceived in this way, one can see the analogous relationship between the colonisation of the lifeworld (neoliberal capitalism) and the colonisation of the normative space of reasons (scientism). (Giladi 2021, 10)

Habermas's analysis could be extended to address this analogy by teasing out the complex interactive web of entwinement between ideology, the discourses of natural science, technology and capitalism. This task however, would not be helped by the fact that Habermas not only lacks recourse to his earlier notion of "technocratic consciousness" but also retains his strategy of divisions that grants neutrality to science and technology.

Scientific technology exhibits the same sort of constraining and enabling functionality as systems like market economies or bureaucracies. However, rather than being non-normative, scientific technology comes with norms already built-in. In chapter 2, I referred to Latour's notion of delegation where humans delegate social norms to technologies which enforce, dictate, or prompt them. Material technologies are best understood as parts of *ensembles* that include other objects such as production lines, media such as power and money, and concepts such as "efficiency". Inserted into such ensembles, technology presupposes, reinforces, and prescribes certain social arrangements while precluding others.[13] However, while technology coordinates action somewhat like a systemic medium, its artefactual materiality ensures that it belongs equally to the lifeworld. As an amalgam uniting particular social values with abstract universal functions, technology not only resists the lifeworld's contestation but is also potentially open to that contestation. Rather than unconsciously incorporating the abstractions and differentiations fetishised by industrial capitalism, technological design could consciously embody other values, purposes, and meanings. As scientific technologies are drawn into the lifeworld the role that they play can be challenged by bringing to bear broader notions of reason. Given this possibility, Habermas'

---

[13] We will see in chapter 6 how Foucault's notion of *dispositif* captures this idea of ensembles. Latour and Woolgar develop this idea as an explicit approach in 'actor network theory' (Latour and Woolgar 1979).

theory that modernity's ills are primarily due to an unbalanced relationship between lifeworld and system is unhelpful. Scientific technology confounds any binary division between lifeworld and system, colonised and coloniser. It is not clear how the complexity of its interactions can be fruitfully accommodated within Habermas's lifeworld/system schema.

Progress and Reason

We have seen how Habermas's lifeworld/system schema is buttressed by a reconstructive scientific account of the lifeworld's progressive differentiation. Drawing on evolutionary and developmental theory, he describes a sequence of *changes* which he then endorses as *progress*, yielding a modern lifeworld accommodating differentiated but connected realms of validity. Critics have questioned the role of science in this account.[14] Critics have also questioned whether Habermas's account actually captures the universal structures upon which it relies.[15] McCarthy notes that the universal and unavoidable presuppositions for reaching understanding in language—which Habermas presents as the capacity to distinguish between different "worlds" and between their corresponding validity claims, attitudes, and value spheres—are found only in certain cultures at certain times (McCarthy 1991, 135).

Habermas responds with a Hegelian move. Communication has a history and evolves. "This does not mean that such cognitive structures appear all at once, whether in ontogenesis or in social evolution" (cited in McCarthy 1991, 135). The universal significance of these structures is shown by their mastery, which represents the developmental-logical unfolding of species-wide competences and reveals unconstrained agreement in language as a fundamental ideal of social cooperation. The problem is that Habermas's characterisations of such progressive rationalisations are inevitably drawn from a Western, modern perspective and hence assume their progressive nature. McCarthy argues that such cognitive-developmental paradigms are open to charges of ethnocentric, scientistic, and rationalistic bias (McCarthy 1991, 134–139). Lacking an adequate grasp of the developmental goals of other cultures, we rely on our own dominant, yet provincial, criteria which we project as universal. Habermas's analogy of premodern cultures as the childhood of

---

[14] Zimmerman appears to have no difficulty with Habermas's utopian orientation, but rather with the attempt to render it scientific. "Do we really have to combine the rather obvious step towards a radical-democratic principle of social organisation . . . with rather daring claims about consensual rationality explicated on the basis of a theory of truth and language?" (Zimmermann cited in Rasmussen, 43).

[15] Gilligan argues that Kohlberg's account of moral development, upon which Habermas relies, was based on research conducted solely on boys and thus provides a stereotypically masculine account of moral development (Gilligan 1982).

humanity reduces cultural complexity to a single universal form. While in some sense we may be able to consider the differentiation of domains of reality and the corresponding validity claims as progressive achievements of cognitive development, we should be wary of any unqualified endorsement of this as the ultimate end-point to which rational development is, or should be, directed. As McCarthy suggests, there is much we could learn from premodern cultures, not only the things we have forgotten or repressed, but "something about how we might put our fragmented world back together again. It is not a matter of regression but of dialogue" (McCarthy 1991, 151).

REINTERPRETING HABERMAS

Over the past two decades, reinterpretations of Habermas's work have emerged which aim to expand critical theory's explanatory and critical reach. Although I wholeheartedly support this aim, it is not the main focus of this book. Nevertheless, these reinterpretations which recast Habermas's quasi-transcendentalism look promising for addressing the concerns I have raised. They render the boundary between lifeworld and system more permeable, leaving room for accounts of the hybrid phenomena I have discussed. They also mitigate the strong universalism of Habermas's developmental-evolutionary account.

As McCarthy points out, Habermas's earlier work attempted to "open up and chart a territory lying between the realms of the empirical and the transcendental" (McCarthy 1978, 91). It is within this territory that an analysis of the mixture of power and reason could be accommodated. However, after KHI, this possibility seems ruled out. KHI identified three fundamental anthropological interests rooted in the structures of human life which yield knowledge in three distinct and irreducible ways. Work, interaction, and power "have a transcendental function but arise from actual structures of human life" (KHI, 194). Initially, these three interests were equally primordial. However, after critics put the status of the emancipatory interest under pressure, Habermas relocated power to the empirical "dynamics" from where it could colonise and distort the "logic" of development he had established in his developmental communicative theory. As Amy Allen points out, this set up the possibility of a form of sociocultural life completely free of the operations of power (Allen 2008, 128). After this move, power and ideology faded into the background of Habermas's work as empirical deviations from the formal development track laid out by the quasi-transcendental reconstructive sciences. The kind of knowledge yielded by rational reconstructions, he tells us, has "a special kind of status; that of pure knowledge" (Habermas 1975, 184). However, as McCarthy notes, if "transcendental reflection appears an exception to the 'interest-ladenness' of cognition ... we are back to something like the traditional notion of disinterested reason" (McCarthy 1978, 102).

TCA effectively opened the gap between the transcendental and empirical that Habermas had previously sought to close. The lifeworld becomes the realm of reason following its logic of increasing differentiation and transparency, guided by the idealising presuppositions of the ideal speech situation revealed by reconstructive science. Systems become a realm of power governed by their own opaque logic and turn back to damage the communicative rationality of the lifeworld. However, the problems I have raised—the power endemic in the lifeworld, the ideological role of science, the normativity of technology operating as a systemic medium, and ethnocentrism—all point to the need to view the system/lifeworld dichotomy as not a strict one. The problem is that, while softening the distinction between the two by admitting the two-way movement of power, Habermas nevertheless seems to retain a dichotomous contrast between the formal-transcendental and empirical levels. The tension between these levels is now located in discourse, specifically in the context-transcendence of validity claims. Truth and normative rightness, Habermas tells us, "have a Janus-face: As claims, they have to be raised here and now and be *de facto* recognised if they are going to bear the agreement of interaction participants that is needed for effective cooperation". However, "the validity claimed for propositions and norms transcends spaces and times, *'blots out' space and time*" (PDM, 223). For Habermas, without this presupposition of transcendental validity, our sociocultural form of life as a community of beings who enter into communicative action would be impossible.

Alert to the apparent insulation of validity from the workings of power in Habermas's project, Amy Allen has called for a more pragmatic and contextualist reading. Such a reading would offer a truly critical theory of racial, gender, and sexual subordination by addressing the role that cultural/symbolic power plays in the formation of subordinated identities (Allen 2008, 124–125). Drawing on Cooke, Allen argues that by adopting "a dynamic idealising projection, rather than a static, concrete or realisable end state" we need not, and should not, consider validity as purified of power (Allen 2008, 137; Cooke, 20). Rather, we should see Habermas's quasi-transcendentalism as "a more principled version of contextualism, one that recognises the force of our normative ideals but also understands that they are inextricably rooted in our practices and forms of life". Such an account "neither holds out hope for the possibility of actually transcending our rootedness in our context . . . nor does it seek to reduce our normative ideals to mere illusions grounded in our power-laden contexts" (Allen 2008, 143). It would allow us to "give up the demand for purity [reason unmixed with power] altogether" (Allen 2008, 125).

Like Allen, Colin Koopman addresses the question of how to bring Habermas and Foucault closer together in order to craft a critical theory which is more critical. I will return to this reconciliation project in chapter 8. At this point, I only want to consider proposals for recasting Habermas's project. While Koopman agrees that Habermas's project calls for a more pragmatic and contextualist reading, he remains cautious of Allen's proposal to scale back Habermas's ambitious claims "regarding the possibility of untangling validity from power, a possibility that he frames in terms of the context transcendence of validity claims" (Allen 2008, 8; Koopman, 222). Koopman thinks that by assimilating Habermas and Foucault at a robust philosophical level in this way, Allen risks reducing each theoretical apparatus to the other. He therefore adopts the strategy of delegation by assigning "methodological tasks to each tradition, according to their strengths, in such a way as to affirm their persisting differences" (Koopman, 224). Koopman articulates these differences in terms of Benhabib's description of the two aspects of critical theory essential for any form of rigorous critique, "the explanatory-diagnostic aspect through which the findings and methods of the social sciences are appropriated in such a way as to develop an empirically fruitful analysis of the crisis potential of the present . . . [and] the anticipatory-utopian [aspect which] constitutes the more properly normative aspect of critique" (Benhabib, 226).

According to Koopman, the strengths of Habermas's critical theory lie in the more normative aspects of critique's "anticipatory-utopian capacities" (Koopman, 264). Koopman thus aims to preserve the distinctive elements of Habermas's normative commitments by arguing for the compatibility of universality with contextuality. He interprets Habermas's standard of universality as context-transcendence through an ongoing "process of universalis*ing*" validity claims across contexts. This process does not require a universalist assertion of validity as already warranted in every context before being interpreted within the constraints of specific contexts (Koopman, 260). By this interpretation, Koopman retains what Habermas sees as "the sense in which statements referring to objects of an objective world that exists independently of its description are true—a sense which *transcends the justification* for these statements" (Habermas, cited in Muller-Doohm, 372). This moment of unconditionality enables "the process of justification [to] be guided by a notion of truth that transcends justification" (OPC, 372). In guiding justificatory discourse, context-transcendent claims also function to expose themselves to criticism, inviting an ongoing critique of dogmatism, prejudice, and self-deception. The fact that validity claims transcend their contexts is crucial if rational reconstruction is to serve as a resource for moral and political criticism of dominant cultural practices. Habermas's "weakly transcendental" approach involves a delicate balance between his "postmetaphysical" defence of context-transcendent validity and his rejection of metaphysical attempts to theorise from a transcendental perspective.

Regarding the ethnocentrism and scientism implicit in Habermas's cognitive-developmental account, McCarthy suggests that *some* type of functional perspective is required in order to construct a critical emancipatory project (McCarthy 1991, 177). I agree. A critical emancipatory project achieves traction by raising the question of how communication, knowledge, and reason function. While the lifeworld concept points at processes of symbolic reproduction available to lifeworld members, a functional account also brings into view social processes that occur "behind our backs". As McCarthy points out, the thought that human consciousness cannot master history, since consciousness is itself a product of history, is quite naturally elucidated by a functional account. The problem that I see with Habermas's mature theory is that it seems to exempt itself from history by employing the quasi-transcendental authority of reconstructive science to occlude its own standpoint within history's flux and indeterminacy. The sort of functionalist perspective that is required was already articulated in Habermas's earlier critiques of systems theory where he argued that functionalist social theory should not be a form of empirical-analytic inquiry but an "interpretive framework". McCarthy cites Habermas's earlier call for a functional analysis "that is hermeneutically enlightened and historically oriented and has as its aim not general theories in the sense of a strict empirical science, but a general interpretation of the kind we examined in the case of psychoanalysis". By retaining a hypothetical moment, this historically oriented functionalism suggests a type of hermeneutic approach that reveals hidden power structures without itself appearing to stand outside power (LSS 187–9).

This notion of a hermeneutically enlightened and historically oriented functionalism dovetails with Koopman's emphasis on universalisation as an ongoing process. Habermas's regulative norms would then be understood in terms of evolving hermeneutical processes, rather than fixed ideals legislating the lawful use of reason. This view is unavoidable from a perspective that sees "unconditional" validity claims as always conditioned by factors beyond the immediate knowledge of claimants. General rules must understate the complexity of their actual contextual application to cover an indefinite range of specific instances. On this basis, rather than authoritarian rules that demand universal adherence across social space and historical time, Habermas's idealised presuppositions of communication could be seen as ongoing appeals to future communities which will interpret them within new contexts. The actual meaning and force of communicative norms is always to some degree contextualised in actual processes of communication. Although Habermas recognises this context-boundedness, we will see in chapter 8 that his strategy precludes an unqualified acknowledgement of it within his own theoretical discourse.

Highlighting the situatedness of knowledge in this way is still compatible with raising claims to situation-transcendent validity, enabling us to view context-transcendence as "marking a normative surplus of meaning that critical theorists can draw upon in seeking to transcend and transform the limits of their situations" (McCarthy in Hoy and McCarthy, 21). McCarthy recasts Habermas's notion of communicative rationality by understanding it pragmatically (never absolute, but always for all practical purposes), temporally (an ongoing accomplishment), and contextually (in ever-changing circumstances) "*without surrendering* transcendence (it turns on validity claims that go beyond the particular contexts in which they are raised) or idealisation (and rests on pragmatic presuppositions that function as regulative ideas)" (Hoy and McCarthy, 72). We could, suggests McCarthy, regard context-transcendence as "a promissory note issued across the full expanse of social space and historical time" (Hoy and McCarthy, 75).

The reinterpretation projects outlined earlier suggest to me that the strengths and possibilities of Habermas's thought can be retained. Specifically, they suggest that the difficulties with the lifeworld/system distinction and Habermas's ethnocentrically Western view of progress can be addressed. Most importantly, the normative basis of Habermas's theory can be retained. After TCA, Habermas began to move towards a more relaxed standard of universality, avoiding ontological formulations while maintaining the moment of context-transcendence as an unavoidable presupposition of communication. Stressing the specifically Kantian sense of "ideal" to designate a regulative function unattainable in actual fact, Habermas insists that the "idealising presuppositions of communicative action must not be hypostatised into the ideal of a future condition" (cited in PMT, xi). It seems to me that the plausible reinterpretations of his quasi-transcendentalism may not cut across Habermas's core commitments so much as merely tone down the unnecessarily ontological language that gives his project such a resolutely transcendental flavour.

Habermas's theory of communicative action captures the sense in which the dynamism of modern rationality is both liberating and oppressive, expansive and truncated. It captures the sense of modern reason as being skewed, as having got out of balance and occluded certain questions by supressing the fundamental *telos* of the ever-open question of the good life. Although adamant that the project of modernity should not be abandoned, Habermas is acutely aware of the irony that modern scientific reason simultaneously enhances and thwarts emancipation. However, it seems to his critics that Habermas's theoretical commitments, despite their purely formal and abstract nature, narrow down possibilities. The lifeworld/system split struggles to accommodate the hybrid phenomena of the actual world. Clearly, the relationship between lifeworld and system requires a more nuanced account than

Habermas's colonisation formulation provides, even with his admission of power into the lifeworld and the softening of his original thesis to accommodate a two-way movement of power. While Habermas harnesses science as an authoritative basis for systematically constructing social theory, his account of science and technology within this construction is not sufficiently developed.

It is beyond the scope of this book to develop a robustly defensible reinterpretation of Habermas's theory of communicative action which addresses these difficulties. However, the various efforts of this reinterpretation project so far suggest that this can be achieved, while retaining the theory's strong normative core. Despite the difficulties, Habermas's colonisation thesis has been vindicated by events which few foresaw at the time he was writing—the explosively rapid penetration of digital technologies into every living and working environment. No region or people, from the educated and wealthy of the metropolitan cities to the poor of the shanty towns, has been exempted. Google searches, dating apps, news-feeds, and social media have now taken over vast areas of social interaction which were previously governed by the counterfactual norms of communicative action. Institutions such as newspapers, which once embodied these norms, are being transformed. Although such institutions were by no means free of coercive power, that power was more vulnerable to being picked apart by communicative action, unlike the digital media now burgeoning, whose apparent ease of use, transparency, and naturalness in fact render power less visible and contestable. In the age of Trump, "fake news", Brexit, rampant climate-change denial, all fed by the internet's algorithmically powered engines of distortion, denial and division and justified by an ideology of unconstrained markets, we see the accelerating erosion of the public sphere in a way unimaginable in the early 1980s. The need is urgent for an articulation of the norms of the public sphere in a form we can all endorse. In the following chapter, we will consider some of Habermas's more recent work on the public sphere, science, and philosophy, and a number of his essays which I will argue are less burdened by the difficulties of his overarching theoretical structures.

Chapter 4

Science and Deliberative Democracy

Over the three decades since TCA's publication, Habermas has remained steadfastly committed to the concerns that defined all his earlier work, specifically democracy, communicative reason, and a form of non-reductive naturalism which accommodates lifeworld intuitions. As a public intellectual, he has continued to contribute to political culture by bringing his philosophical work to bear on particular issues of public concern. Just some of the issues on which he has spoken out—German reunification, post-national Europe, the "historians' debate" about German war guilt, the Gulf War, immigration, multiculturalism, religious fundamentalism, and the GFC—indicate his ongoing commitment to the public sphere (Muller-Doohm, 265, 277, 265).

In Habermas's major work of this period, *Between Facts and Norms* (hereafter BFN) published in 1992, we see the theoretical counterpart to this engagement. BFN reworks Habermas's ideas on the bourgeois public sphere, which he set out thirty years before, by incorporating his theory of communicative reason into a theory of deliberative democracy. After discussing this work in relation to science, I will turn to the role of philosophy and its relationship to science in the contemporary post-metaphysical context. I will then discuss Habermas's more recent essays, in which his analyses of contemporary issues exemplify his understanding of the role of philosophy in relation to science in addressing concrete social questions. I will argue that these essays are less burdened by the difficulties we see in his more systematic theorising. In *The Future of Human Nature* (published in 2003, hereafter FHN), Habermas questions genetic interventions which threaten the lifeworld's categorical distinctions. This leads to a consideration of the place of religion in relation to both philosophy and science in post-secular societies. I will conclude by discussing Habermas's treatment of science in relation to the lifeworld in his discussions of free will.

BETWEEN FACTS AND NORMS

Equipped with substantial theories of modernity, society, and communicative action, BFN returns to the discussions of the public sphere that Habermas had focused on in STPS. Here he had argued that the bourgeois public sphere of the eighteenth century had been taken over by organised capitalism, with decisions increasingly manipulated by vested interests and legitimated by a profit-oriented mass media which trivialised democratic aspirations. TCA's colonisation thesis emphasised the systemic obstacles to the lifeworld's engagement with the political system. BFN marks a distance from this model by providing a more differentiated analysis of power which distinguishes between communicative, administrative, and social power and how these operate within the structure of the public sphere in a democratic society.

By incorporating these complexities, sovereignty can be understood in intersubjective terms as the formation of public opinion by reasoned argument, within both the formal political sphere and the informal public sphere (BFN, 301). To articulate the nature and role of public opinion in relation to decision-making, Habermas introduces a two-track model which distinguishes formal from informal discourses. The informal track revolves around the public sphere, the network of communication flows extending from face-to-face egalitarian discussions to impersonal mass media communications. These streams of communication are "filtered and synthesised in such a way that they coalesce into bundles of topically specified *public* opinions" (BFN, 360). Informal public opinion formation is inclusive and critical and is primarily oriented towards the discovery and definition of problems. Because it is diffuse, unregulated, open, and chaotic, it cannot coalesce into binding decisions.

To be legitimate, binding decisions must be steered by this "communicative power", in other words, the public opinion arising from the periphery and reworked by argumentation in a process of will-formation. It then passes into the formal track in which it is steered by democratic and constitutional procedures (BFN, 356). The formal track comprises discourses within parliaments, government bureaucracies, and courts, in which formal fairness, civility, and equality are ensured by procedural regulations. Discourse in the formal track is oriented towards arguments that issue in decisions. Only the legislature is capable of working through conflicts to reach binding decisions. Only judges and administrators can act decisively on laws. This procedural orderliness comes at a cost. Public opinion is filtered to fit the agendas of political parties or bureaucracies. By inhibiting the spontaneous and inclusive exchanges of opinions that characterise public opinion formation, the formal track reduces the quantity and quality of opinions that get heard (BFN, 354–62).

Nonetheless, a sensitive and active periphery can detect and identify new problems which it can "introduce via the parliamentary or judicial sluices into the political system in a way that *disrupts* the latter's routines". If civil society is structured to allow active and equal participation in political life, it can "ferret out, identify, effectively thematise latent problems of social integration which require political solutions" (BFN, 358). Habermas notes that the nuclear arms race, genetic engineering, ecological threats, and third-world poverty were all issues not initially brought up by representatives of the state apparatus, large organisations, or functional systems (BFN, 381).

Habermas situates his theory of deliberative democracy as a response to the dominant traditions of liberalism and republicanism. He sees modern autonomy expressed politically as the freedom that citizens have to participate in the making of laws which they must also live under. This results in two forms of autonomy being coded as rights, the mutually supporting and interlocking republican and liberal notions of *public autonomy* to participate, and of *private autonomy* as subjects with equal freedom from unjust interference (BFN, 408). However, to avoid pre-empting public deliberation on these rights, Habermas rejects attempts to prescribe *concrete* rights beyond their abstract formulation. He thus derives very general norms that regulate communicative action (BNF, 127). We encountered these general communicative norms of equality, openness, inclusion, and generalisability in STPS (STPS, 36). These are not preconditions for democracy but counterfactual presuppositions which provide ideal standards for deliberation against which actual deliberation can be criticised (BFN 322–3). Guided by these norms, democratic deliberation is "the result of constructive opinion and will-formation" (BFN, 336–7). In the course of deliberations in which the fully prescriptive meanings of rights are worked out, the initial interests of citizens and the definitions of problems may be transformed.

Despite Habermas's well-documented concerns about contemporary democracies, critics have suggested that his account ignores their obvious shortcomings, such as cynical manipulation by media, apathetic citizenries, and abuse of social power. His account is thought by some to have abandoned the critical impulses of his earlier work (Scheuerman in Dews, 153–178; Lyotard 1988, 157; Cook 2001, 95). In response to these substantially unfair criticisms, we should note that Habermas does not deny the internal threats that actual democracies face. Rather, he insists that the norms of deliberative democracy can still be discerned within actual democratic-legal discourses and that these general communicative norms provide a standard against which actual practices can be judged. Habermas is guided by the conviction that law and morality are connected, despite having separated into independent spheres in modernity. Legal norms require moral legitimation which they acquire from their free and rational acceptance by citizens (BFN, 408). People see the law as being just by virtue of the deliberative procedures through which it is instituted. While Habermas has remained conscious of the very real deficiencies of actual democracies, there is no doubt that BFN lacks the bleak outlook of his earlier work. Giladi has argued that if TCA was overly defensive in its portrayal of a lifeworld threatened on all sides by encroaching systemic power, then BFN shifts to a more positive model that shows how the colonisation of the normative space of reason by systemic imperitives can be resisted (Giladi 2021). Within deliberative democracy, operating within a well-functioning public sphere, legal power and the communicative action from which it ultimately derives can progressively challenge, combat, and transform manipulative social power which circulates in constitutional democracies.

Habermas's description of the processes by which public opinion is ideally translated into binding decisions in modern democracies clearly must encompass questions around how to decide contentious scientific and technological proposals which challenge existing laws, rights, or values. Decisions about these proposals—for example, genetic engineering or nuclear energy—require the arguments for and against to be openly considered and eventually converge in binding decisions. The efficacy of Habermas's model of deliberative democracy will depend on how well it captures the relevant social phenomena. To consider this, I will weave together an account of how science fits into his model.

Science and Habermas's Deliberative Democracy

Due to its level of generalisation, BNF (like TCA) does not explicitly theorise the roles that science and technology play in democratic societies. I will therefore consider those roles by drawing out what is implicit in Habermas's account in terms of the framework sketched earlier. Habermas's account of the two-track model bears out the role of science in democracy that I sketched in chapter 3, for example, its justificatory role in the formulation of legal codes and the use of these codes to assess contentious proposals. This justificatory role has particular weight in the "formal track" of decision-making, where science is often granted an authoritative, even determining, role. Presented to judges, politicians, or administrators, scientific discourses are given primary weight in assessing the costs and benefits of state and administrative decisions.[1] As discussed in chapter

---

[1] By "scientific discourses", I mean a range of scientific disciplines which frame social questions in objectivistic language and are collated in the form of cost-benefit analyses, environmental impact statements, social impact statements, and so forth.

3, formal decisions are guided and justified by scientific studies in which meanings, significances, and values are straitjacketed into methodologies digestible by decision-making authorities. Quantitative approaches enable values, where acknowledged, to be weighed against other "social goods" to make decisions that appear robustly defensible.

On the "informal track", everyday commonsense views have their place alongside science. The polemical nature of scientific discourses is frequently mobilised to influence consideration of issues by drawing on decontextualised facts which are treated as sufficient in themselves to "prove" an argument.[2] Certain scientific discourses are drawn upon to oppose other scientific discourses. In this regard, Schnaiberg distinguishes "production science" from "impact science" (Gould et al.). Under "production science", we might include orthodox neoclassical economics or the "applied" aspects of physics and chemistry and their technologically oriented derivatives like engineering, all of which are oriented towards industrial progress. "Impact science" might draw on disciplines such as ecology and areas of the human sciences which ask awkward questions about damage wrought by applications of production science.[3] A typical conflict between these sciences can be seen in the case of the contrasting climate change recommendations arising from the economic modelling of William Nordhaus (winner of the 2018 Nobel Prize for Economics) and from what appears to be the majority of climate scientists and ecologists.[4] These differences arise not only from socialisation into different scientific disciplines and career paths and the accompanying inculcation of values and norms, but also from a barely articulated background of deeply held prescientific attitudes. On the informal track, these prescientific commitments form a background against which particular scientific discourses—either from "impact" or "production" science—are mobilised to play partisan roles, for or against, in deliberations about scientific/technological questions.

Scientific discourses are themselves inextricably linked to these prescientific attitudes by being embedded in, or framed by, metaphor and narrative. Thus, one of science's founding metaphors, arising in the context of the anxieties and turmoil generated by Europe's religious wars, involved the Baconian search for order, control, and mastery over a passive and

---

[2] As science becomes weaponised, decontextualised anomalous facts are cited as sufficient to categorically reject or accept specific scientific claims. As can be seen in climate change denialism, the social deployment of decontextualised facts trades on the authority of science. Habermas is aware of this. "The political consequences of the authority enjoyed by the scientific system are ambivalent . . . short-lived popular syntheses of isolated pieces of information, which have taken the place of global interpretations, secure the authority of science *in abstracto*" (LC, 84). He doesn't develop this thought.

[3] These are not Schnaiberg's examples.

[4] Nordhaus (2017); see also Keen (2020) for criticism of Nordhaus.

purposeless nature. Running in parallel there remain older accounts of nature as animated by the intrinsic connection between all living things and as having purposes that it hides from us. On one hand we have the hubris of conquest and on the other hand a tragic sensibility that, as mere humans, we can only glimpse nature through a glass darkly.[5] In our times, it is the former that dominates. It is also today's anxieties and hopes, themselves structured by interests and power, which shape the narrative and metaphors in which scientific discourse is embedded.[6] Today, the reality of climate change challenges our most cherished commitments, not only the idea that the free pursuit of individual interests always leads to the general good, but also our less obvious commitments underlying our deeply rooted cultural identities. As Amitav Ghosh argues, there is an excess of climate denialism beyond manipulation and money which suggests that "it threatens to unravel something deeper which gives meaning and purpose to people's existence" (Ghosh, 138). Only by bringing these motivating prescientific commitments into view can citizens on the informal track broaden their frame of discussion.

Rather than passively accepting these prescientific commitments as partially rational, awareness and discussion of them can prompt a deepening of rationality. Habermas gives us a clue as to how this might occur in his discussion of legal paradigms. He refers to the judiciary's tacit reliance on two legal paradigms representing conflicting views of society. The *liberal* paradigm is based on the idea of an association of independent entrepreneurs, exercising their negative freedoms and free from government interference. The *welfare* paradigm responds to economic inequality (which the liberal paradigm sees as just and natural) by holding the politically constructed economic system, rather than individuals, responsible for it. Habermas claims that growing social complexity and diversity has resulted in increasing conflict between these paradigms, leading to the development

---

[5] In a similar vein, Mary Midgley (1992) provides an account of myths that inform science. George Lakoff and Mark Johnson argue that metaphors structure our most basic understandings of experiences and shape our actions without our awareness (Lakoff and Johnson 1980). The idea of a hidden nature is explored in Pierre Hadot's *The Veil of Isis* (2006). The notion of the interconnectedness of all things reflects the ancient notion of *anima mundi* and has a modern form in James Lovelock's Gaia theory (Lovelock 2000). Richard Lewontin sees science drawn into ideology by the overextension of metaphors and cautions against the extension of terms, models, and constructions from one field to another (Lewontin 1991). This caution appears justified in the light of the distorting influence of physics on the origins of neoclassical economics as analysed by Mirowski (1989).

[6] Thus, Stephen Dawkins's notion of the 'selfish gene' portrays the drama of individuals pursuing their own self-interest in endless competition (Dawkins 1976). On this account, humans appear as automatons controlled by their genes. By failing to adequately consider cooperation as a principle for the species' survival, Dawkins's book functions as an ideological justification for neoliberalism, which was beginning to emerge at the time of its writing, although it must be emphasised that this wasn't Dawkins's intention.

of a reflective awareness of their limits. He argues that we need to link the development of the concrete meanings of subjective rights to an ongoing process which harnesses a higher-order *proceduralist* paradigm to determine the most appropriate paradigm for particular situations. In this process of democratic deliberation, citizens collectively discuss *on a case-by-case basis* the extent to which situations can be defined by either paradigm (BFN, 393). As an example, Habermas discusses women's rights. Both liberal and welfare paradigms, appealing respectively to gender difference-blind and difference-sensitive principles, are vulnerable to reliance on false stereotypes of what is biologically determined. Both paradigms employ overgeneralised classifications of gender identity which tend to reflect the *status quo* of contingent social power arrangements (BFN, 422–3).

Habermas is wary of biological essentialism (BFN, 425). Rather than appealing to science to find a truth or essence in human nature and to define rights, Habermas's proceduralist paradigm views women's rights as undefined, emerging only through a struggle for recognition in the public sphere. Women themselves, as the affected group, must clarify the relevant aspects and standards that define equality and inequality in public discourses (BFN, 420, 425). My point in raising Habermas's discussion of legal paradigms is to suggest the possibility of developing a similar reflective awareness of our prescientific attitudes. Once articulated, these deep commitments can be teased apart to reveal how they interact and lend support to scientific discourses, leading to a further consideration of the appropriate limits of their application. By drawing these prescientific elements into discussion in such a way that the outcome cannot be determined in advance of that discussion, rationality can be deepened and discussion can be broadened. Such an approach is consistent with Habermas's account of a functioning public sphere in which communicative action is given scope within deliberative democracy to tease apart competing values. While Habermas would certainly emphasise the difficulties of such an approach, he would also insist that we have no alternative but to continue the modern project by harnessing modernity's communicative potentials.

As well as communicative and administrative power, Habermas also discusses the role of the social power which allows highly organised sectors, like business or trade unions, to monopolise access to government, thereby short-circuiting democratic communicative power. Social power harnesses technocracies which operate as functionally separate specialised and self-regulating spheres of action whose parameters are manipulated by experts, for example, legal or economic specialists. The ordinary citizen is thus relieved of responsibility for problem-solving in certain areas (BFN, 342–52). This trend is clearly a challenge to deliberative democracy. Although Habermas accepts this conclusion, he doesn't over-emphasise it. Softening the thesis of the "splitting-off of elites" which we saw in TCA, he points out that functionaries in any specialised subsystem must communicate with the public, since it is the public who first experience any problems of systemic imbalance. Economists must be able to speak to non-economists. The democratic public sphere remains a necessary condition even for a systems-theoretical approach.

Within the democratic public sphere, scientific-technical matters disseminate into public awareness primarily by the mass media. It is worth backtracking over Habermas's views on the mass media since the 1960s. In STPS, he argued that the eighteenth-century bourgeois public sphere developed to enable public opinion to challenge authority and power in a way not previously possible. Informal collective discourses could ground claims of legitimate authority solely in the rational claim of the best argument. However, as it expanded into wider areas of modern life, the principle of a critical public sphere shrank and gave way to monopoly capitalism, in which big businesses, governments, and trade unions advanced their own agendas supported by a profit-oriented mass media which trivialised democratic aspirations. As autonomy was narrowed, the vitality of the public sphere was diminished to the point that, in advanced capitalism, rational debate based on common interests was overwhelmed by the haggling and bargaining of special interest groups (STPS, 181–235).

This analysis was clearly influenced by the critique of the culture industry and other types of critique dominant in the 1960s.[7] Since then, Habermas has shifted from a view of the public as entirely passive. In TCA, he argued that the modern mass media could be a significant tool of social control due to its highly concentrated ownership and its communication of highly selective messages. However, because the reception of these messages requires the hermeneutic skills of the lifeworld, they can be debated, challenged, and subverted, thus encouraging the media to internalise certain ethical constraints and make use of feedback from their audiences (TCA2, 390–91). Further, "the mass media . . . free communication processes from the provinciality of spatiotemporally restricted contexts and permit public spheres to emerge". While these public spheres are structured by those who control the media, they do not entirely control them "and therein lies their ambivalent potential" (TCA, 390). In BNF, Habermas locates the impersonal messages disseminated by mass media on the informal track comprising the network of communications developed from face-to-face meetings. After filtering, condensation, and refinement, positions and arguments are publicly disseminated by journalists, reporters, intellectuals,

---

[7] See, for example, Adorno and Horkheimer (1997, 120).

and lobbyists. It is the unrestricted diversity of the public sphere that enables it to provide a sounding board on an indefinite range of issues. Habermas acknowledges, however, that because of the concentration of their ownership, mass media are at risk of manipulation by social power which obstructs the public sphere's role in detecting and articulating social problems (BFN, 376–9). Today this situation is not so much a risk as an actuality. Not only the concentration of mass media ownership, but also the emergence of new technologies, has resulted in a level of influence able to shape public opinion. This is obvious in the case of climate science, which is constantly undermined by certain media interests thus short-circuiting democratic deliberation and forestalling action.

While the partitioning tendencies we saw in TCA have considerably softened in BFN, Habermas's broad strokes still fail to capture the fact that scientific discourses are shaped by complex webs of relationships with other discourses. Nor does he touch upon how science, through technology, surreptitiously permeates the lifeworld. Ever-resourceful capitalism invents new technologies—the latest being news-feeds, social media, and internet search engines—which now enable a degree of social and personal manipulation only dimly comprehended at the time Habermas was writing. These technologies now have the potential to replace the public sphere with isolated "communities" which reinforce homogeneous sets of narrow views. Technologies, continuously refined and enhanced, are barely noticed unless they cross the threshold of laws, conventional norms, or values. They mould us by the prescriptions, norms, and values materially embedded in them. This moulding of the lifeworld by scientific technology isn't captured by either TCA's colonisation thesis or BFN's procedural democracy thesis, because in neither work does Habermas venture away from the idealised realm of language and communication to lay out, as we will see Foucault do, a genealogy that includes concrete things and bodies. Given the level of generality of his major works, it is difficult for Habermas to theorise the barely noticed penetration of scientific technology into the lifeworld. That was not his aim. What he intended was to articulate a notion of deliberative democracy, not realised as such but standing as a normative resource which can provide a means of critically testing how far democratic deliberation can be realised within the exigencies of social complexity. To the extent that Habermas's social theory claims comprehensivity, the failure to address technologies is a shortcoming.

PHILOSOPHY AND SCIENCE

Given, on one hand, Habermas's earlier critiques of the ideological potentials of science and, on the other hand, his use of science in his philosophical project, we need to consider his understanding of philosophy and its relationship to science. In an essay "Philosophy as Stand in and Interpreter", Habermas lays out his vision of philosophy's role in the scientifically informed modern world (chapter 1, MCCA). Many of the themes in the essay are reworked in a later essay "The Relationship between Theory and Practice Revisited" (Chapter 7, TJ). In both these works Habermas situates philosophy within the historical context of a series of realignments in the relationship between theory and practice, showing that philosophy does not pursue the same eternal questions or fulfil the same continuous role. He traces the transformation of the ancient ideal of contemplation of "objective reason" into Kant's epistemology, which then becomes foundational in the form of an insight into the a priori ahistorical transcendental conditions of experience. By this foundational insight, philosophy became an "usher", able to show the sciences their proper place, and a "judge", setting limits on their claims in relation to other areas of culture (MCCA, 2).

Habermas rejects such roles. Today, philosophy must be critically directed to its subject matter, yet reflexively aware of its own embeddedness in an intersubjective lifeworld. He rejects any form of metaphysics committed to a theoretical attitude that claims freedom from distortions by interests, to describe the universe just as it is, in its law-like order (see also KHI, 303). In a post-metaphysical world, Habermas insists that we can no longer hold to notions of completely detached observers of an independent world. Rather than informed by metaphysical conceptions of the whole of history or nature or the imagined aspirations of a collective subject, theory should be seen as fallibly informing practice while at the same time itself being modified and corrected by this encounter. However, although he accepts naturalism's claim that everything that exists is part of a natural world, Habermas denies that this makes transcendental claims unjustified or unintelligible. Positing the presupposition of a shared objective world as a condition of communicative rationality, he conceives this world as something we unavoidably refer to when advancing knowledge claims. By rejecting the reductive scientific naturalism that aims to reduce all knowledge to scientifically demonstrable knowledge, his soft naturalism demarcates philosophy's role from that of empirical science.

Against the threat of science undermining the lifeworld's communicative resources, Habermas invokes philosophy in the role of interpreter. In this role, it draws science into the broader validity basis of the lifeworld's "unity of reason". However, he doesn't think science is completely distinct from philosophy. He is critical of a tendency within hermeneutics and pragmatism which contrasts a "narrow objectivistic conception of science" to philosophy which is "illuminating or awakening instead of being objective" (MCCA, 12–13). Rorty varies this contrast by his distinction between "normal discourses", which have reliable criteria for settling disputes and making progress, and "abnormal discourses", which are incommensurable when their basic orientations are contested. Rather than dying out or morphing into normal discourse, some abnormal discourses persist as "interesting and fruitful disagreement" sufficient to themselves. According to Rorty, philosophy verges on this "edifying" condition once it has got beyond any pretensions to problem-solving. While sympathetic to Rorty's view that "philosophy has no business playing the part of highest arbiter in matters of science or culture", Habermas rejects the idea that philosophy can be "*outside* the sciences, without being immediately drawn back into argumentation, that is, justificatory discourse" (MCCA, 14). He is uncomfortable with Rorty's claim that philosophy, unlike science, is merely edifying. I share this discomfort. Philosophy entails holding to rational commitments that connect practically to the social and objective worlds which science has in view.

Habermas rejects naturalistic reductions like the reduction of lifeworld to system, reasons to causes or, as we will see, free will to brain states. By keeping both poles of such dichotomies in play, philosophy can criticise both reductive naturalism and metaphysics. In taking a middle path between reductive objectivism and radical contextualism, philosophy must respond to "the particular need of modernity, which is, after all, bereft of any guidance by models of the past" (TJ, 284). Modernity has broken with the continuity of traditional models and must develop its own normative understandings. Against the notion of philosophy as a remote and highly specialised discipline, Habermas champions approaches which "have public impact because they face up to the problems that confront philosophy from both private and public life" (TJ, 284). With the modern fallibilist sensibility of a discipline aware of its situatedness in the world, "philosophy is forced to drop the claim of holding the key to Truth" and finds its place in functionally differentiated roles within the modern world (TJ, 285).

However, "philosophy cannot completely immerse itself in any one of its social roles". It can only fulfill these roles by "at the same time transcending [them]". To fully immerse itself would "rob it of its best . . . a kind of untamed thinking that is neither channelled nor fixed by any particular method" (TJ, 286). For philosophy to completely immerse itself in, say, the role of a science, it would have to conform to science's strictures—the idealisations, abstractions, and limitations that scientific method necessarily requires. While science must be guided by methods which rule out *in advance* certain types of objects and claims as unscientific, philosophy is an interpretive practice and cannot be overly constrained by method. However, philosophy can keep the strictures of method in view, while retaining its connection to the lifeworld.

Habermas notes that, since the seventeenth century, science has become increasingly independent of philosophy which, stripped of its foundationalist claims, has become fallibilist (TJ, 286). But rather than a clear-cut division of labour between them, philosophy can enter into cooperation with the sciences. Habermas refers to the blend found in Marxism and psychoanalysis, which is typical of many social sciences by virtue of containing "a genuine philosophical idea". He cites Freud (symptom formation through repression), Durkheim (creation of solidarity through the sacred), Mead (the identity-formation function of role-taking), Weber (modernisation as the rationalisation of society), and Chomsky (language acquisition as hypothesis testing) as paradigmatic examples of theories which contain embryonic philosophical ideas and pose both empirical and universal questions at the same time (MCCA, 15). These sciences are not "immature", eventually converging as a unified science in "the triumphal march toward objectivist approaches such as neurophysiology", but rather stages "on the road to the philosophisation of the sciences of man" (MCCA,15). Here philosophy operates as a "stand-in" for "empirical theories with strong universal claims, the primary candidates being 'reconstructive sciences' which presume to explain the universal bases of rational experience, judgement, action, and linguistic communication by anonymous rule systems that competent subjects follow" (TJ, 287). Rather than philosophy pretending to the roles of usher or judge, it collaborates with science "by an auspicious matching of different theoretical fragments". Habermas seems to suggest that philosophy is more quasi-empirical than quasi-transcendental. It is now a reconstructive science based on, and falsifiable by, empirical evidence.

While endorsing the rationalisation process through which different validity spheres became autonomous, Habermas wants to establish a balance between the separated moments of reason in the lifeworld. As we saw, increasing specialisation has led to science's remoteness from the lifeworld. Split-off elitist specialisations fragment consciousness and weaken the lifeworld's communicative resources, which offer the only real alternative to exerting influence in more or less coercive ways (MCCA, 19). To coordinate actions without resort to coercion, we must draw on "a cultural tradition that ranges *across the whole spectrum*, not just the fruits of science and technology" (MCCA, 18). Here philosophy takes on "the role of interpreter on behalf of the lifeworld . . . refurbish[ing] its link with the totality" (MCCA, 18–19). Philosophy is able to "preserve unity across all the disparate aspects of validity" without reducing or levelling them (TJ, 287). It can initiate an interplay between various dimensions of validity that have "come to a standstill like a tangled mobile" (MCCA, 19). Philosophy should drop its role of arbiter, inspecting and judging aspects of culture, and act as interpreter, unravelling these tangled parts and connecting them into the comprehensive whole of the lifeworld. By thinking in terms of the whole, philosophy is able to develop interpretations that are normatively charged with practical intent. Through its close relation to both sciences and to common sense, it "can criticise the colonisation of a lifeworld that has been gutted by trends of commercialisation, bureaucratisation, and legalism as well as scientism" (TJ, 290). A dialectic between scientific and technological potential can only be established within the shared lifeworld in which reasons can connect, back and forth across validity spheres, ranging from descriptions of what is or what could be through to prescriptions of what should be.

Philosophy must maintain its interface with science or "it would lose the very insights of its own that it needs in order to fulfil its exoteric roles" (TJ, 287). By doing so, it can undertake a "diagnosis of our time in terms of which modern societies come to understand themselves" conducted "primarily in the philosophical form of an auto-critique of reason" (TJ, 290). Habermas is referring to modernity's reflexive attitude and its capacity for distantiation by which we moderns come to understand ourselves from an "external" perspective. Philosophy as critical theory enlists science as a partner to expose distortions of the lifeworld. While science provides a counter-intuitive perspective unavailable to the lifeworld, it cannot necessarily claim to be the ultimate authority, but rather a valuable perspective, albeit one in need of translation and contextualisation within lifeworld commonsense. This is because macrosocial problems "are no longer visible from the perspective of closed self-referential functional systems such as institutionalised science and its discourses" (TJ, 289). Questions about the overall direction of science and technology should be considered within the "diffuse network of a public sphere anchored in civil society . . . [in which] highly complex societies become aware of significant failures and risks and can politically deal with [them]" (TJ, 289).

It follows from Habermas's account that science can challenge philosophical commitments by demanding that philosophy make its positions perspicuous, as it must do in the light of commonsense intuitions which like science must be taken seriously and cannot easily be abandoned. As we will see in the following sections, this dialogue between philosophy and science can also serve as a corrective to incautious claims by scientists and their uptake by a naïve laity. In the remainder of this chapter, we will see how Habermas employs philosophy as an interpreter between science and the lifeworld in order to contribute to public discussion of pressing contemporary concerns. We will now turn to Habermas's discussion of genetic engineering, the place of religion within modernity, and finally, free will and neuroscience. By focusing on such currently pressing issues while not claiming the status of comprehensive theory, these discussions can be read in a manner unencumbered by the difficulties of Habermas's theoretical system.

THE FUTURE OF HUMAN NATURE

In FHN, Habermas approaches the issue of genetic engineering by opening up the rational core of a prima facie intuition—our deep-seated fears about cloning human beings—as an impetus for further reflection. He argues that the increasing technical possibilities for intervening in human nature challenge our identity as autonomous moral agents. Technologies such as genetic engineering thus call for the clarification of central aspects of the normative orientation of the human species. What is at stake for Habermas is the *"uncontrollability* of the contingent process of human fertilisation that results from what is now an *unforeseeable* combination of two different sets of chromosomes" (FHN, 13). Habermas's premise is that the randomness of human genetic makeup is a precondition for the formation of individual identities as well as for the fundamental equality between human beings. Genetic intervention potentially replaces the individual uniqueness and autonomy of a newborn with a new kind of inequality between humans. Manipulation of genetic material threatens the inviolable status of persons by treating them as mere objects.

Habermas's critique is directed specifically against the idea of "liberal eugenics", genetic interventions, freely accessible in unregulated markets, that manipulate traits in accordance with parental wishes. He doesn't oppose biotechnological interventions per se. What he is concerned about is the pre-implantation genetic diagnosis which enables screening of embryos to identify certain traits prior to implantation and opens the possibility of future genetic interventions for enhancement purposes. Habermas's worry is that a future biotechnology which enables characteristics chosen by parents would sever us from what is most human by a slippery slope passing from pre-implantation genetic diagnosis to selective implantation, trait selection, and cloning. By reducing persons to mere means, genetic enhancements have the potential to shape self-understanding in a way that threatens freedom and equality.

What human beings are "by nature" is increasingly coming within the reach of biotechnical intervention. From the perspective of science, this is just another frontier passed in the ongoing extension of the ability to control and manipulate nature. Habermas sees normative regulation as generally adapting to technological development. Given the promises of gains in the scope of individual choice and prosperity, scientific research has generally been aligned with the basic commitments of liberalism which grant it its freedom. In fact, the history of medicine suggests a sceptical attitude to "moralising human nature". Blood transfusions, vaccination, heart transplants, brain surgery, artificial insemination, artificial organs, have all been challenged but are now unquestioned life-enhancing measures. Libertarians argue that parents have the right to an enhanced child and see concerns as just further dubious re-enchantments of inner nature opposing developments that will soon be accepted as normal. Against this, Habermas argues that the extension of technological control over our "inner" nature without our consent is distinguished by the fact that it "changes the overall structure of our moral experience". Shifting the line between choice and chance affects our capacities to see ourselves as authors of our own lives and to recognise others as autonomous persons (FHN, 29).

Given the systemic imperatives of economics linked to rapidly developing technologies, Habermas thinks we risk unthinkingly endorsing a right to scientific research, and only later realising that technologies have been normalised prior to any informed consideration. We are, Habermas argues, being overwhelmed by a lack of perspectives (FHN, 18). "The development of biotechnology generates a dynamic which again and again overtakes the time-consuming processes through which society reaches a self-understanding about its moral aims" (Habermas cited in Muller-Doohm, 321). According to Habermas's social evolutionary account, the archaic lifeworld adopted a number of fundamental dichotomies—the made/the grown, the technical attitude/the clinical attitude, having/being a body, fate/autonomy—which he considers essential to the self-understanding of human agents. He contends that these fundamental distinctions and categories risk being violated by certain genetic technologies. It is the task of philosophy to reassert them in a form capable of consolidating the normative and ontological boundary between "the nature that we are and the organic endowments we give to ourselves" (FHN, 12). At stake are "the boundaries between persons and things" (FHN, 13).

Grown or Made

Liberal eugenicists argue that there isn't much difference between the genetic enhancement of future individuals and socialisation. Just as a person's dispositions and consequent life possibilities are moulded by socialisation, so they can be moulded by genetic interventions. Both occur without express agreement. However, Habermas argues that socialisation proceeds by communicative action which, for the parents, is connected to reasons. Even if this "space of reasons" is not yet open to the child, she still has the role of the second person in relation to expectations underlying her parents' efforts. Adolescents still have the opportunity to critically reappraise restrictive socialisation processes. Genetic programming allows no such opportunity when it commits a person to a specific life-project according to another's intentions (FHN, 63). Habermas concludes that we may only have good reasons to assume the programmed person's consent in the case of the prevention of clearly unacceptable diseases and disabilities.

Liberal eugenics affects the capacity for being oneself by establishing an unprecedented human relationship. The genetic programmer's irreversible choice jeopardises a precondition of the moral self-understanding of autonomous actors, which rests on the assumption that there is no definite obstacle to egalitarian interpersonal relations. We assume, as a pragmatic ideal, that we could be in anyone else's shoes. While any genetic inheritance, programmed or natural, is a form of irreversible dependence, Habermas argues that, in the case of a natural inheritance, this only relates to a person's existence, not their essence. Unlike couples who simply decide to have a child, the genetic programmer irrevocably breaches the reciprocity between persons by determining the essence of a future person. The person whose programme has been deliberately fixed by another is, in principle, barred from exchanging roles with her designer (FHN, 65). This relationship is inimical to the reciprocal and symmetrical relations of mutual recognition required for a moral and legal community of persons born free and equal.

The problem with liberal eugenics is that it presupposes the levelling-out of a fundamental distinction constitutive of self-understanding, the distinction between *what is manufactured* and *what has come to be by nature* (FHN, 50). The self-emergent grown life that requires nurturing is distinguished from what a technician makes from inert matter by its capacity for individuation and self-determination that exceeds causal intervention (FHN, 47). This distinction is reflected in fundamental attitudes of engagement with the world. Attitudes of cultivating, healing, or breeding "share a respect for the inherent dynamics of self-regulated nature" (FHN, 45). Practice adjusts itself to the object and is responsive to its inherent purposes, potentials, and meanings (FHN, 45). This is altogether different from a scientific agent who combines the *theoretical* objectivating attitude of the disinterested observer with the *technical* attitude of an intervening actor working instrumentally to impose purposes upon an object presupposed to be passive.

Liberal eugenicists may attempt to assimilate enhancement to therapeutic intervention. But in the case of therapeutic interventions, what matters is the clinical attitude of a therapist who, with the performative attitude of a participant, anticipates the future consent, however virtual, of the patient (FHN, 52). Where a person is genetically "healed" without his actual permission, virtual consent is assumed by the doctor, who is not acting on an object but relating to a future person. This differs from the case of someone who learns that his genetic makeup was programmed solely according to the preferences of a third person. This person may find it difficult to "own" the prenatal intervention as being his. This reflects the further distinction between *having* a body and *being* a body. The primary mode of experience is that of *being* a body (FHN, 49). Having a body is the result of a capacity to assume an objectivating attitude to the *prior* fact of being a body. Habermas fears that upon becoming aware of being "made" by genetic manipulation, an adolescent's participant perspective of being a body would "collide with the reifying perspective of a producer or a bricoleur" (FHN, 51). This raises the question of whether we can actually function as agents if we cannot experience some sense of ourselves as embodied.

Natality, Freedom, and the Limits of Eugenics

Habermas suggests that the shift from the participant stance of living in one's own body to the observer perspective, which governed the intervention one's body was subjected to before birth, is categorically different from the decentring of our geocentric and anthropocentric worldviews by Copernicus and Darwin. The subjection of our body to biotechnology involves *the intention of another person* intruding into our life history before our birth. This intrusion might erode the "capacity to be ourselves", to speak with our own voice as "the person *herself* who is behind her intentions, initiatives, and aspirations" (FHN, 57). To be able to identify with one's body requires that the body can be experienced as natural, "a continuation of the organic, self-generative life from which the person was born". What is important is that "we experience our own freedom with reference to something which, by its very nature, is not at our disposal" (FHN, 58). Since human autonomy can only be understood in contrast to the "fate" that limits freedom, the lifeworld makes the distinction between what we are and what happens to us (FHN, 60). Fate is entwined in an origin whose beginning is beyond human disposal. To know oneself as the author of one's actions and thoughts, one must be able to ascribe one's origin to something beyond human manipulation, to a beginning, such as nature or God, which cannot be compromised by another person's intentions.

Habermas turns to Arendt's discussion of "natality" to elucidate this thought. Arendt argues that in acting, humans feel free to begin something new because birth itself, as a divide between nature and culture, marks a new beginning (Arendt cited in FHN, 59). Habermas interprets this to suggest that the significance of birth is that it marks the divide between the natural fate of the body and the socialisation that follows birth. It is only by maintaining the distinction between nature, which is not at our disposal, and culture, which is subject to historical contingency, "that the acting subject may proceed to the self-ascriptions without which he could not perceive himself as the initiator of his actions and aspirations" (FHN, 59). It seems that the very notion of "authentically being oneself" as the source of one's actions and thoughts involves at minimum a natural history free from others' intentions prior to birth and entry into socialisation. Birth plays the role of a marker in the self's history that gives the reference point for what one is prior to any imposition of others' intentions, and it is this reference point that allows one to assume a reflexive attitude towards one's socialised fate and to revise one's self-understanding (FHN, 60). Habermas suggests that our minimal self-understanding of ourselves as members of the species presupposes that we are all authors of our own lives and that we deal with others as authors of their own lives. By depriving the fusion of two sets of chromosomes of its contingency, intergenerational relations lose the naturalness which has so far been part of the taken-for-granted background of our self-understanding (FHN, 62). Without maintaining the distinction between the grown and the made, we risk a series of cumulative intergenerational violations.

There are a number of objections to Habermas's argument which suggest that questions around genetic interventions cannot be easily settled. Bostrom comments that "had Mother Nature been a real parent, she would have been in jail for child abuse and murder" (Bostrom 2005).[8] Clearly a vast pool of facts supports the view that we can do better than nature, just as a vast pool of facts supports the contrary view. The set of facts that one draws upon perhaps depends on one's prior unarticulated deep-seated prescientific convictions. Bostrom comments that "bioconservatives draw attention to the possibility that subtle human values could get eroded by technological advances" (Bostrom 2005, 1). This trivialises Habermas's argument. It is not "subtle human values" at stake but, according to Habermas, the set of fundamental distinctions that form the basis of autonomy, morality, and sociality.

Habermas's difficulty is in articulating these distinctions as criteria for acceptability of particular interventions. He distinguishes between acceptable and unacceptable interventions on the basis of "attitudes", rather than on what is actually done. Habermas cannot accept the instrumental attitude of parents fulfilling their egotistical dreams by designing their children, but can accept the defeasible anticipation of consent typical of an emergency doctor treating an unconscious patient (FHN, 52). Further, it seems that the dire consequences that Habermas predicts result from a genetically modified person's *belief*, regardless of its truth, about the nature of the intervention and their programmers' attitude. One can imagine an instance of genetic intervention remaining unknown to the mature person, who would then suffer none of these consequences. It is therefore tempting to say we should simply find a way to adjust our attitudes and find ways to accommodate our

---

[8] FHN should be seen within the larger context of the debate about "transhumanism", in which Habermas is typically cast as a "bioconservative", a label he happily accepts (Muller-Doohm, 411). Transhumanism seeks to increase human health and lifespan, extend intellectual and physical capacities, or give increased control over mental states and moods with technologies like genetic enhancement, virtual reality, nanotechnology, and artificial intelligence. Transhumanists promote the view that human enhancement technologies should be widely available, and that individuals should have broad discretion in their use (Bostrom 2005).

beliefs. Habermas himself is aware of the paradoxical nature of the challenge in that "in the very dimensions where boundaries are fluid, we are supposed to draw and enforce particularly clear-cut lines" (FHN, 19).

Habermas's concerns are not unique. Many social theorists worry about the dangers posed by advances in biomedical knowledge to distinctions between the natural and the artificial, and the implications for our unproblematically assumed notions of free will and human dignity. There is also widespread enthusiasm for transhumanism, the idea that we can and should augment so-called natural human capacities with a range of new technologies.[9] Against these opposed positions, Rose stakes out a middle ground (Rose, chapter 3). He sees both anxieties about, and enthusiastic endorsements of, genetic enhancements as not based on what actually happens in contemporary biomedicine, but rather on popular science predictions and speculation. However, in Rose's view, a new way of thinking is emerging from a constellation of discourses and practices—medical, economic, legal, and ethical. He sees genetic enhancement as driven by a notion of optimisation within which knowledge, authority, technology, and subjectivity are reconfigured. Habermas may agree, but where they differ is that Rose does not think that genetic intervention directed at transforming life has undermined, or presumably could in the future undermine, the distinction between nature and culture (Rose, 83).[10] Rose's assurance doesn't get to the heart of Habermas's concerns. The tentative nature of Habermas's analysis, attempting "to attain more transparency for a rather mixed-up set of intuitions", is not directed towards what he thinks has happened or will happen if liberal eugenics proceeds (FHN, 22). No one knows what will happen. However, it is plausible that much like the barely perceived yet profound changes being wrought by the digital revolution—revising notions of trust, truth, society, and self—liberal eugenics, if it were to proceed, could precipitate the consequences Habermas suggests. It seems to me that Habermas is right to raise the possibility, even a remote one, of a slippery slope that leads to the undermining of fundamental distinctions required for autonomy, morality, and sociality. His aim is to raise these commonly held lifeworld concerns to the level of public articulation before they are ridden over roughshod by scientific, technological, and economic imperatives.

In FHN, we see Habermas articulating a rational basis for concerns that are difficult to articulate, in order to bring them to consciousness in a way that enables public discussion. He wants to show what is at stake collectively when one particular conception of the human becomes the basis upon which

---

[9] See for example, https://humanityplus.org/
[10] Rose is not addressing Habermas's argument directly but Rheinberger's view that the shift to intervening in life in order to transform it has overridden the distinction between nature and culture.

a profound and irreversible transformation is enacted.[11] However, he is not attempting to settle the question definitively. Faced with the modern lifeworld struggling to articulate a response, he mobilises its rational resources to resist attempts to make a particular conception, driven by commercial and technological imperatives, the dominant conception. In this endeavour, he remains abstemious regarding strong ontological commitments. He is not concerned with the reality or non-reality of "nature", "fate", "causality", "contingency", or "autonomy". Rather his concern is to defend the lifeworld as a grammar of intersubjective relations conducive to the good life against the imperatives of technocracy and economism and their reductive ontologising. By raising the contingency and naturalness of human life against free-market eugenics, Habermas's analysis can be seen as a self-reflection of modernity on its own limits, a reflection which now urgently needs to be extended to consideration of the fate of the planet.

Faith and Knowledge

The difficulty surrounding attitudes towards the unborn is only one difficulty in resolving questions about genetic intervention. Given the indeterminacy of reasons which can motivate a compelling argument, it is not surprising that Habermas draws on religious language to tease out his intuitions. To grapple with the human dimensions of new technologies like genetic intervention, we need to find evaluative languages which incorporate conceptions of the human good not already structured by the presuppositions of scientific and technological progress. We also need to find a way to discuss our relation to the non-human world other than as users or exploiters of resources, but in a way that would allow us to be critical of such "progress" without immediately being seen as opposed to science or reason.

Habermas's interest in religion is not new. In the 1970s, he analysed religion primarily from a sociological perspective. In TCA, he adopted the Weberian account of religion as withering away in an enlightened world. By the end of the 1980s, he acknowledged that religious images help in coping with the experience of contingency and provide comfort in times of need (Habermas 2010, 15; Muller-Dohm, 384). More recently, he has noted a "growing political influence of religious orthodoxies" in the form of aggressive Islamism, Hindu nationalism, or Protestant fundamentalism (BNR, 1). The potential for conflict raises the issue of post-secular societies which, though largely secularised, must nevertheless reckon with the continued relevance of different religious traditions. Habermas doesn't consider the existence of religious

---

[11] Kompridis cautions that resistance to this conception should not be at the expense of erasing the plurality of conceptions that reflect and preserve human plurality (Kompridis 2009).

traditions within secular modern society as a zero-sum game, a battle between the unbridled capitalistic productivity of science and technology on the one hand and the conservative forces of religion on the other. Rather than one side gaining only at the expense of the other, Habermas invokes a third force, the civilising voice of a democratically shaped and enlightened common sense. Common sense is autonomous and can resist both reductive scientific naturalism and religious tradition, from which it can nevertheless learn.

As we have seen, scientific descriptions become highly disconcerting for common sense as they come closer to our bodily and mental existence. Habermas suggests that usually "scientific theories change the *content* of our self-understanding", but they don't "touch on the *framework* of our everyday knowledge, which is linked to the self-understanding of actors and speakers" (FHN, 105). However, what happens as we progressively subsume ourselves under scientific descriptions? To describe actions in terms of the generation of mental states by biological machinery misses what is relevant to what we recognise as agents. "The scientistic belief that science will one day not only supplement but replace the self-description of actors as persons by an objectivating self-description is not science, but bad philosophy" (FHN, 108). In its insistence on rational justification, science is certainly supported by a scientifically informed common sense which has found its place in modernity. However, "no science will relieve common sense . . . of the task of forming a judgment, for instance, on how we should deal with pre-personal human life under descriptions of molecular biology that make genetic interventions possible" (FHN, 108).

Citizens of the modern liberal state must, according to Habermas, accept that "only secular reasons count beyond the institutional threshold separating the informal public sphere from parliaments, courts, ministries, and administrations" (BNR, 130). Furthermore, religious citizens "must develop an epistemic stance towards . . . the institutionalised monopoly on knowledge of modern scientific experts" (BNR, 137). However, this does not imply a reduction of our knowledge to only that which can be authorised by science. Such radical naturalism "devalues all types of statements that cannot be traced back to empirical observations, statements of laws, or causal explanations, hence moral, legal, and evaluative statements no less than religious ones" (BNR, 141). In any case, scientific knowledge is itself communicatively structured by participants in processes of argumentation and justification. To imagine scientific research proceeding entirely from an observer's perspective only leads to the metaphysician's "Gods-eye view". Habermas's post-metaphysical philosophy doesn't dispute theological affirmations but does assert their lack of determinate meaning. However, he doesn't deny that "religious language is the bearer of a semantic content that is inspiring and even indispensable, for this content eludes (for the time being?) the explanatory force of philosophical language and continues to resist translation into reasoning discourses" (PMT, 51).

Habermas rejects the sharp liberal distinction between reason and religion. Non-believers should shoulder equal responsibility for undertaking "cooperative" translation of religious to secular language (BNR, 131). While not a straightforward task, translation can enable divergent traditions to find a common language other than that of the market or science. For example, in the controversy of how to deal with human embryos, believers might invoke *Genesis*: "So God created man in his own image" (FHN, 115). Here the notion of God's creature expresses the equality of all humans. Habermas glosses this passage to suggest that God remains a "God of free men" only as long as we do not level out the *absolute difference* that exists between the *creator* and the *creature*. He interprets the concept of God as not needing to abide by the laws of nature like a technician or by the rules of a code like a biologist or computer scientist. Rather, God determines man by enabling and obliging him to be free. This contrasts with the entirely different causal dependence involved if the absolute difference assumed as inherent in the concept of God's creation were replaced by a human who intervened solely according to her own preferences.

Habermas's philosophical interpretation brings religious language into "the universe of argumentative discourse . . . uncoupled from the event of revelation". However, translation does not always succeed. "In these moments of its powerlessness, argumentative speech passes over beyond religion and science into literature, into a mode of presentation that is no longer measured by truth claims" (RR, 75). If it does succeed only by eliminating the substance once intended, it leaves irritations (FHN 110). However, "the unbelieving sons and daughters of modernity seem to believe that they owe more to one another, and need more for themselves, than what is accessible to them, in translation, of religious tradition—as if the semantic potential of the latter was still not exhausted" (FHN, 111). I think Habermas would agree that our modern common sense, when beholden exclusively to the authority of science, belittles the authority which we sometimes rightly feel belongs to religious language.

FREE WILL AND DETERMINISM

We have seen the epistemic standpoints of participant and observer used to characterise Habermas's lifeworld /system distinction and employed in his discussion of liberal eugenics. This same distinction is brought to bear on the longstanding debate about the possibility and nature of free will, a debate radicalised in recent decades by claims purportedly supported by neuroscience.

Here Habermas rejects reductionist interpretations of science while leaving intact both the findings of science and the lifeworld intuitions of free will. It is not surprising that he defends the concept of free will, which is available only from the participant's perspective. He wants to embrace a version of critical theory that insists we are free to choose the good life, however circumscribed that freedom actually is. However, Habermas doesn't neglect the emancipatory possibilities made available by the "external" perspective of the scientific observer. The importance of this perspective is also one of the crucial insights of critical theory. He notes that scientific progress has enabled an ongoing decentring which relativises the place of humankind in the world in enlightening ways, for example, the objectivating self-descriptions arising from Freudian or sociological reflection (Habermas 2007b, 23–24).

Similarly, neurological descriptions can help us recognise, and take account of, the ways in which the brain, as the physical substrate of rationality, can be causally impacted. Objectivating neuroscientific accounts can enable a critical engagement with causal biographies, challenging views about responsibility for certain behaviours which we come to understand as consequences of neurological disorders rather than moral failings. What is at stake for Habermas is that the naturalisation of the mind is coextensive with the desocialisation of the person. By subsuming the description of a person into the "extensional concepts of physics, neurophysiology, or evolutionary psychology", we effectively desocialise the person. We thus remove her from that very context of a shared form of life in which the concept of a person can be meaningfully applied and from which it gets its meaning (FHN, 106). Also in play is the fact that claims about free will raise questions of responsibility. "The awareness of authorship implying accountability is the core of our self-understanding, disclosed to the perspective of a participant, but excluding revisionary scientific description" (FHN, 108). In fact, any deliberation about authorship or accountability is short-circuited by the reductive scientific naturalist's insistence on assimilating the perspectives of participants in social practices to causal accounts of the scientific observer. This insistence is grounded by the metaphysical conviction that there can be nothing beyond the one true world described by natural science, and that anything appearing beyond it is an illusion which demands elimination or reduction.

Habermas declares his target of criticism to be "the tendency to jump to philosophical conclusions from a successful and undisputed scientific enterprise" (Habermas 2007b, 84). This tendency is seen in attempts to explain free will as an illusion by drawing on science.[12] Habermas suggests that

---

[12] This tendency is strongly implied, if not argued for, in the work of neuroscientists such as V. S. Ramachandran, philosophers such as Paul Churchland, and evolutionary biologists such as Steven Pinker (Ramachandran 2004; Churchland 1984; Pinker 2002).

we can understand the temptation to "explain" free will in terms of three competing intuitions. First, as knowing subjects, our experience, backed by the systematic authority of the natural sciences, shows us that everything that happens is caused according to law-like regularities. Second, as agents, we are convinced of the irreducible distinctiveness of the causal effectiveness of our minds. We can decide to intervene in the world in a way that is entirely up to us. However, third, as scientifically enlightened persons, we are convinced that the universe is one and includes us as part of nature. In the light of this third intuition, the first two intuitions seem contradictory. This seeming contradiction has become acute in modernity, radicalised by the ever more thorough objectifications of the human mind. This is evidenced in the call to *replace* the lifeworld's "manifest image" with the "scientific image", to use Sellars's terms (Sellars 1963). Against this call, Habermas suggests that two perspectives, that of a *participant* in intersubjective practices of reasoning and that of an *observer* of an external world of causes, are inextricably interwoven and run in parallel (Habermas 2007b, 35–36). The problem is the claim that the objectifying stance characteristic of science should take precedence over the stance of the participant in the lifeworld.

Free will is typically framed as a matter of an agent having a certain sort of causal efficacy, that is, mental causation. However, what Habermas sees as essential to the concept of free will is not mental causation, but the internal link to reasons bound to social practices. As an analogy, we can consider the pawns in a chess game which are, according to natural science, only bits of cellulose or other material.[13] But the chess player is not pretending that they are pawns. The concept "pawn" is bound into our practices. Similarly, reasons for action are real, although they only exist in relation to the social practice of giving and asking for reasons. Free will, Habermas claims, is a natural part of the social world, contrasting with determinism not as mental causation, but as the ability to respond to reasons. As participants, we have reasons for our actions and tacitly assume that *we could do otherwise* and that it is *up to us* how to act. Free will is a presupposition of the language game of responsible agency which reveals itself only to *participants* who take up performative attitudes *vis-a-vis* second persons (Habermas 2007b, 15). Actors are aware that, as responsible agents, they are *always already* operating within a space of reasons which they should be responsive to. The intuition of having free will is reflected in the presuppositions of this language game accompanying our actions. These presuppositions remain inaccessible for the observer, whose viewpoint is that of the uninvolved third person.

---

[13] This example is borrowed from Anderson (2005).

Habermas contrasts causal and rational explanations, arguing that reasons which render actions explicable are different in kind from events linked by laws of nature. To the extent that we are guided by reasons, we submit to intersubjectively shared norms (2007b, 17). Unlike causes, reasons have a degree of indeterminacy. They are not absolutely decisive, but either better or worse reasons. Reasons are reasons *for us*. It does not follow that any given person would reach the same decision in identical antecedent conditions from identical rational explanations. To be a responsible agent, one must be motivated by letting one's judgement be determined by reasons. As an "author" of one's actions, one must take the initiative and attribute it to oneself. Thus an actor is free when she acts on reasons which she has made her own, but she may also have acted against her own better judgement (BNR, 159–60). In the process of deliberation, we must assume that the outcome of action is not predetermined from the outset. This assumption is the very meaning of deliberation because without it deliberation would be purely epiphenomenal, performing no function. Since a will is formed, however imperceptibly, *in the course of* deliberations, we experience ourselves as free only in the actions that we perform more or less consciously (BNR, 155).

The open-endedness of a free decision doesn't preclude it being *rationally* conditioned. The actor is free when she wills for reasons that, on deliberation, she accepts. We only experience as a lack of freedom those external constraints that force us to act otherwise than we would on reflection (BNR, 157). Freedom involves an implicit endorsement of our conditioning. In the course of her deliberations, the actor reaches a rationally motivated position that is neither arbitrary nor a causal process but is the result of rule-governed inferences. Habermas draws on the internal connection between reflection and freedom found in Kant's sense of autonomy. However, rather than assuming an unconditioned noumenal causation intervening in the world while cut off from all empirical contexts, he sees freedom as conditioned by being embedded in the context of reasons as they arise in the lifeworld (Habermas 2007b, 19). To be free is to bind one's will by sensitivity to culturally transmitted and socially institutionalised reasons. This can be contrasted to unambiguously unfree actions like compulsive, habitual, accidental, or neurotic actions which are typically described in terms of causes. I am unfree "if my decisions were determined as a neural event in which I was no longer involved as a person who takes a position" (BNR, 158). It is only by ignoring the slide from the participant's to the observer's perspective that it can seem that the rational motivation of an action forms a bridge to the determination of action by observable causes.

The contrast between reasons and causes is clearly seen in the fact that the causal explanations of natural science are appealed to in pleading that a person has acted unfreely and therefore cannot be held responsible. Habermas notes that in legal discourses, naturalistic explanations are appealed to if actions are unintelligible on the basis of comprehensible motives (Habermas 2007b, 19). As long as we are only talking about *limitations* to free will, that which is limited remains presupposed, but free will entirely disappears when action naturalistically bypasses the propositional attitudes of the actor and can be traced solely to nomologically determined events. Behaviour is then seen to be not decided by persons but rather fixed by their brains. From the neuroscientist's perspective, "decisive arguments within the hierarchy of reasons can only ratify . . . what has already been long decided in regions of the brain far from consciousness" (Habermas 2007b, 20). This form of explanation "requires switching perspectives from being the participant who accuses or justifies to being the observing analyst who . . . explains the behaviour" (Habermas 2007b, 21). If the distinction between what were once complementary patterns of explanation is lost, the naturalistic explanation of behaviour loses any link to the norms of responsible agency (Habermas 2007b, 22). Free will is eliminated and the language game of responsible agency collapses under the assumption that unconscious brain states completely determine all mental states.

John Searle objects to this account by arguing that the language games of neuroscience and of responsible agency are not in conflict, but are rather "different levels of description of the same system" (Habermas 2007b, 71). The problem is that Searle's claim simply begs the question of what this "same system" consists. The objects of neuroscience have colour, weight, and texture; occupy space within the skull; and are connected causally. The objects of mental states have no colour, weight, or texture; don't occupy space but refer to things in the world; are connected inferentially; and are governed by social norms. It is only as a participant that we can pick out these states, after which the neuroscientist switches to the observer's stance to note the neural correlates. Asking if this is necessarily an account of the same processes or "same system", Habermas responds that "we could be sure that we are dealing with descriptions of the *same* processes only if we could translate equivalent statements from the one language into the other". However, "The language we employ for psychological processes and semantic matters cannot be reduced to physicalist or behaviourist language" (Habermas 2007b, 89).

Habermas asks whether, if we assume the neuroscientific debunking of free will to be correct, it is even possible to adapt our normatively moulded consciousness to an objectivating self-description, "according to which one's thoughts, intentions, and actions are not just instantiated by brain processes, but completely determined by them" (Habermas 2007b, 23). Although we may think that we are deliberating freely about reasons and that we could choose to act otherwise, can we persuade ourselves that our actions and thoughts are "really" caused by a series of electrochemical reactions in our neurons in which we don't recognise ourselves? Or is the participant's perspective in the game of giving and asking for reasons unavoidable? In short, are there limits to self-objectification?

In response, Habermas suggests that the conception of ourselves as persons depends on a distinction between *doing* and *occurring*. Neurological description abolishes this distinction by dropping any reference to normative success or failure in both action and reasoning and replacing it with a language of events that simply occur (Habermas 2007b, 26). While an objectivating description of a one-sided determination of the mind by the brain might undermine an illusion, it also "dissolves this perspective from which alone an increase in knowledge could be experienced as emancipation from constraints" (Habermas 2007b, 24). By insisting on a strictly neurological account of the mind, researchers lack the resources to understand what it means to confirm or refute their own theories in the light of reasons. Such a stance amounts to a "performative contradiction" (Habermas 2007b, 24). The limit of self-objectification is encountered when persons describe their actions as spatiotemporally identifiable events that can be explained nomologically. Persons can then no longer recognise themselves as persons.

Habermas regards the explanatory models, terminology, or language based on the two perspectives of participant and observer as being irreducible. "Descriptions of persons and their thoughts or practices cannot be translated into behaviourist or physical terms without losing or changing their meaning" (Habermas 2007b, 25). This is glossed over in neuroscientific accounts by the metaphorical assimilation of reasons to causes. Knowledge and its acquisition are irredeemably normative in a way that resists all attempts at empiricist redescription. The participant perspective cannot

> "be aligned with, and subordinated under, the observer perspective in such a way that we can capture ourselves, in an objectivating manner and observe ourselves from a fictitious view from nowhere—not just as acting and speaking subjects, but also as epistemic subjects engaged in the act of investigating 'ourselves'" (Habermas 2007b, 26–27).

Failure to recognise how deep-seated the participant perspective is can be traced back to a basic scientific assumption that the objectivating perspective of the natural sciences has priority over the participant's perspective. Habermas sees neither epistemic perspective as having priority, since "the perspective of an observer who adopts an objectivating stance towards something in the world, is *a fortiori* interwoven with the perspective of participants in discourse who, in presenting arguments, adopt a performative stance toward their critics" (BNR, 169). This means that the observer's objective reality can only be constituted *together with* the intersubjectivity of possible communication. Our participant's perspective is not something we can step out of and see from the outside. Resistance to the naturalistic redescription of our self-understanding as persons does not reflect a mere illusion on our part but is explained by the fact that "there is *no getting around a dualism of epistemic perspectives* that must interlock in order to make it possible for the mind, situated as it is within the world, to get an orientating overview of its own situation" (Habermas 2007b, 35). "Objective" descriptions do not issue from isolated minds reflecting the world "as it is" but constitute the claims of participants in dialogue with others, even if those others are an imagined ideal audience.

According to Habermas, the participant and observer perspectives arose from adaptations as our species acquired communicative competency.[14] Culture evolves naturally, with our sociocultural forms of life having evolved from prior forms by learning processes. On this account, archaic humans included the objective world within the social relations of the intersubjectively shared lifeworld. As they sought to master unpredictable nature, humans developed a language of objects and events that enabled increasingly desocialised accounts of objective nature (Habermas 2007b, 35). Two distinct patterns of explanation developed, one reflecting the participant perspective, the other the observer perspective. Scientific disciplines developed reflecting this bifurcation, with the natural sciences adopting "explanation" and the social sciences and humanities adopting "understanding" as their respective aims (Habermas 2007b, 36).

In the seventeenth century, philosophy was confronted with the question of "what it means for humankind to understand *itself* in the context of scientifically objectivated nature" (Habermas 2007b, 36). The objectivating stance of the observer turns back to focus on human nature, subjectivity itself, as an object of science. With the development of sciences like neuroscience, which lead the call to *replace* the manifest image with the scientific image of man, the interwoven perspectives which once ran in parallel came to be seen as conflicting. If we see our natural character as contrasting with our location in a normative space of reasons, we can no longer accept the idea of knowledge as a natural phenomenon. Knowing seems mysterious and the knowing subject retreats from the natural world. While much contemporary philosophy of mind sees itself as under an obligation to reintegrate the thinking subject into a natural world by reduction, Habermas's evolutionary account affirms that thinking and knowing are already part of our way of being natural animals.

---

[14] Habermas borrows this line of reasoning from Sellars (1963).

Although recognising the emancipatory possibilities of the observer perspective, Habermas rejects the scientistic claim for its priority. He wants a more liberal form of naturalism that is not restricted to the objects of natural science but also embraces minds, beliefs, reasons, goals, meanings, and morals. "Reality is not exhausted by the totality of scientific statements that count as true according to current empirical scientific standards" (BNR, 153). Habermas's soft naturalism does not compete with attempts to scientifically understand neurobiology by material and efficient causes, but it refuses to reduce our everyday understanding of "acting on reasons" to these causes. In our everyday practices, there is an allocation where naturalistic objectivating explanations are employed for actions which cannot be made sense of from the participant's perspective. It is only with the extension of the logic of natural science to "explain" action and reasoning per se that a metaphysical leap is made. In claiming that reality is *restricted* to the scientific image, the reductive scientific naturalist forgets that her objective perspective is itself interwoven with the normative space of reasons, inaccessible to natural science. The natural sciences cannot embrace all we encounter.

CONCLUDING THOUGHTS

Habermas wouldn't hesitate to acknowledge science's emancipatory potentials in the treatment of genetic disorders and in the neuroscientific discoveries of the biological basis of behaviours which we could otherwise only moralise about. What he criticises is the slide from science to bad philosophy seen in arguments for liberal eugenics or the reductive neuroscientific denial of free will. These scientific arguments, backed by powerful economic imperatives, run roughshod over the lifeworld's intuitions of freedom and responsibility. Eschewing simplistic reconciliations, Habermas draws on the semantic potential of religion to give voice to lifeworld intuitions. His concerns are less directed towards the ontological status of objects recognised by the natural sciences and more towards safeguarding the forms of communicative reason conducive to the good life. In addressing concrete issues such as genetic manipulation, neuroscience, or the place of religion in modern societies, Habermas draws on the theoretical structures provided by his major works, without necessarily being entrapped by the full weight of their commitments. This mode of analysis can be extended to a range of contemporary questions, for example, our responsibilities to other species, future generations, or the care of the planet. While TCA's stark image of colonisation paints an epochal crisis in broad strokes, the fine-grained analysis of concrete issues challenges the efficacy of Habermas's metaphor. It is as though we need to place his system aside to engage with actual concrete issues.

Habermas's concerns can be traced back to his fundamental commitment to the democratic public sphere, which he believes increasingly lacks cohesive perspectives. In the face of its fragmentation and colonisation, he mobilises the lifeworld's resources to bolster the public sphere against the impoverishment of its communicative resources at the hands of reductive scientific naturalism and economism. It is not hard to see the erosion of meaning caused by the reductive tendencies of both. Scientific naturalism aims to reduce human agency, mindedness, reason, and the whole realm of the normative to causal laws, while neoliberal economism aims to reduce evaluative questions of the good life to matters of efficiency and individual preference.

It is ironic that while Habermas criticised Foucault's work—I discuss these criticisms in chapters 6 and 8—his underlying concerns and commitments are similar to those of Foucault's mentor, Georges Canguilhem. In chapter 1, I discussed Canguilhem's view of biological norms as being social and referring to values which cannot be reduced to objective scientific concepts (Canguilhem 1991). Following from this fact, Canguilhem advocated a type of medicine which gives value to the patient's subjective experience. He lamented the loss of the general practitioner's prestige and authority "in favour of specialist physicians, engineers who take apart the organism like machinery" (Canguilhem 2012, 38). In a formulation reminiscent of both Habermas's colonisation thesis and his reflections in FHN, Canguilhem argues that the living conditions of patients are the effect of "the colonization of medicine by the general and applied sciences" due to "the interest (in every sense of the term) of industrial society . . . in the health of working populations, or, as some put it, in the human component of productive forces" (Canguilhem 2012, 37–38). Like Habermas, Canguilhem is acutely aware of the powerful forces that shape and direct science and wants to further a form of democratic deliberation by which the lifeworld can enlighten itself about them. He criticises medicine as having become "a phenomenon at the scale of industrial societies", concluding that "choices of a political character are implied in all debates concerning the relations of man and medicine" (cited in Talcott 2019, 238). These themes are developed and extended by Foucault. I suspect that if Habermas had approached Foucault's work through Canguilhem he might have understood it better. We will come to this later.

We have seen how Habermas's "weakly transcendental" project adopts an empirical starting point, the rational reconstruction of fundamental presuppositions implicit in communication. Philosophy becomes interdisciplinary, drawing on the sciences, while retaining its distinct role by its connection to the lifeworld. By drawing on this connection, philosophy challenges the denial of genuine phenomena beyond the limits of natural science. Certain social problems "are no longer visible from the perspective of . . . institutionalised science and its discourses" (TJ, 289). However, at the same time,

Habermas's project also gains traction by enlisting the "external" perspectives of science to challenge the lifeworld's conservatism and opacity. The danger for Habermas then is the temptation to grant these sciences an authoritative role within an overall system which exempts itself from the contingency and plurality revealed in the fine-grained detail of the world. In his essays dealing with concrete issues, we see philosophy collaborate with science and adopt the role of interpreter between science and the lifeworld. However, we can also see that science cannot insist on its exclusive authority but must ultimately be accommodated within the lifeworld's common sense. It is philosophy that brokers this accommodation. In the following chapter, we will turn to Foucault. In the first section, I will draw out what I believe lies at the heart of the differences between Habermas and Foucault, specifically their taking up of Kant's Enlightenment project. I will then turn to Foucault's "archaeological" period of work.

Chapter 5

Foucault's Archaeology of Scientific Knowledge

FOUCAULT'S RADICALISATION OF CRITIQUE

While there are stark contrasts between the works of Habermas and Foucault, there is also a common orientation best characterised as "critique". Both see themselves as being within the Enlightenment tradition of critique, which seeks to change society by understanding it, freeing human beings from entrapment in systems of dependence or domination, both internal and external. Like Habermas, Foucault thinks that philosophy must understand its position within the particular social and historical formation of modernity, which it attempts to diagnose and change. This diagnosis includes a critique of modernity's paradigmatic form of reason, science. Both thinkers recognise the dangers of reductive understandings of science which grant excessive authority to forms of reason that narrow human possibilities. Habermas wants to defend the lifeworld and the hermeneutic-historical human sciences from their reduction by scientific naturalism. Foucault sees the human sciences as already linked into circuits of power that construct social realities as merely natural. He wants to highlight this constructive activity to reveal its contingency. I contend that Foucault's approach can be understood as a distinctive radicalisation of the way in which Habermas undertakes critique.

To grasp the nature of this radicalisation, we should recall Kant's transcendental critique. Faced with the scandal of metaphysics, Kant wanted to lay down a sound basis for knowledge by marking out its limits. Habermas also establishes limits, invoking science to underscore their necessity and universality. We saw him articulate these limits, firstly, by the three fundamental cognitive interests which necessarily and universally determine our access to reality and then, after his linguistic turn, with the quasi-transcendental notion of universal and necessary presuppositions which underlie communication. He identifies the ills of modernity as consequences of breaching these limits, specifically because one form of reason is now taken as reason per se. Rather than establishing limits, Foucault wants to break limits and open up possibilities beyond them, possibilities of what we are, think, do, and say. If Habermas's approach is critique, then Foucault's is a critique of critique. Foucault sees limits as not restrictions, but possibilities beyond the known. In his final period of work, Foucault articulated his relationship to Enlightenment, Kant, and critique in those terms. Finding valuable resources in Kant's minor works, he takes up this theme in his 1978 lecture *What Is Critique* and the 1983 essay *What Is Enlightenment?*, a commentary on Kant's essay of the same name written two hundred years before. In this essay, Foucault notes that what makes Kant's text different is its engagement with the specific moment at which Kant is writing and because of which he is writing (PT, 104–5). While philosophy paradigmatically deals with eternal and universal truths, Kant's essay asks about the present situation of the Enlightenment (PT, 98–9). What is philosophically distinctive about the present is that it enquires into itself. This historical self-reflection is a "distinctive feature of philosophy as a discourse of modernity and on modernity" (GSO, 13).

Foucault links Kant's account of the Enlightenment to "the art of not being governed or better, the art of not being governed like that and at that cost . . . not being governed too much", a formulation suggesting not ungovernability, but movement within a context imbued with governmentality.[1] Critique seeks to expand the field of possible action and thought in relation to specific forms of government. Foucault's understanding of Kant's account of the Enlightenment and critique are thus both linked to resistance to power (PT, 45). On this basis, Foucault places himself within the Enlightenment critical tradition ensuing after Kant. However, apart from some of his minor works, Kant's critiques are concerned with knowledge and its necessary and universal limits, in contrast to Foucault, who is concerned with "problematising" limits by revealing them to be contingent and local.[2] Foucault came to articulate his project as analysing "the *problematisations* through which being offers itself to be, necessarily, thought—and the *practices* on the basis of which these problematisations are formed" (HS2, 11).

Foucault's critique should be seen as part of a broader tradition of the critique of reason extending from Kant and Hegel through Nietzsche and Weber to the Frankfurt School and Habermas. In this *general* sense, critique

---

[1] By "governmentality", I mean Foucault's notion of the organised practices, mentalities, rationalities, and techniques through which a society is rendered governable (see STP, 108–9).

[2] "Problematisations" are "the way an unproblematic field of experience, or a set of practices, which were accepted without question, which were familiar and 'silent', out of discussion, becomes a problem, raises discussion and debate, incites new reactions, and induces a crisis in previously silent behaviour, habits, practices, and institutions" (Foucault 2001, 74).

is any procedure which makes clear the conceptual preconditions of our thoughts, words, experiences, and actions. This sense can be distinguished from more specific senses such as Kant's *transcendental* and Foucault's *historical* critiques.[3] Kant's three major critiques are transcendental in the sense that they do not turn on any ultimate appeal to sense experience. Experience occurs by virtue of the structure of our mental faculties which necessarily and universally constitute a priori conditions of objects in general. However, in Foucault's historical critiques, the conditions of possibility are historical actualities. What has actually happened in history makes possible what actually happens in the present. Unlike Kant's transcendental conditions, purportedly shared universally and necessitating particular forms of experience, Foucault's conditions of possibility emerge as contingent actualities in history which enable and constrain new frameworks of thought and action in the present.

If Kant's transcendental critiques can be located within a broader tradition of critique, Foucault can also claim to be within this tradition while also rejecting its particular manifestation as transcendental critique. Foucault identifies his work with Kant's essays on the Enlightenment, revolution, history, and anthropology as types of critique quite distinct from transcendental critique. "This other critical tradition does not pose the question of the conditions of possibility of a true knowledge . . . but involves what could be called an ontology of the present, or present reality, an ontology of modernity, an ontology of ourselves" (GSO, 20–1). By "ontology", he doesn't just mean being, but a *way* of being, a way that attends to the relations between power and knowledge by asking how to be governed and how not to be governed too much. Foucault wants to draw a distinction between his more radical critique of reason, and philosophy more generally. While philosophy might consider itself the voice of reason, Foucault's critique reveals reason's underlying conditioned nature. His ambivalence towards the modern philosophical tradition is seen in his location of critique at "the outer limits of philosophy, very close to it, up against it, at its expense" (PT, 42). Critique only exists in relation to something other than itself—some institution, practice, or discourse. Although close to philosophy, to be critical, it must also maintain its distance. It is a means towards an undefined future, "an instrument, a means for a future or a truth that it will not know" (PT, 42).

In contrast to Habermas's explicitly normative stance, Foucault's critique explicitly suspends normative commitments and eschews prescription. "Critique doesn't have to be the premise of a deduction that concludes, 'this then, is what needs to be done.' It should be an instrument for those who fight, those who resist and refuse what is" (EW3, 236). But this is not negation, "a

---

[3] This useful distinction is from Koopman (2013, 109).

work of destruction, of refusal and denial, but rather an investigative work that consists in suspending as far as possible the normative system which one refers to in order to test and evaluate it" (Foucault, cited in Lemke 2012, 61). While judgements subsume particulars under already constituted categories, Foucault's critique suspends judgements in order to enquire into the hidden constitution of the field of categories themselves. Foucault offers no normative grounding, rules, or criteria to guide moral judgement. Rather than telling us what to think and do, we must craft our own ethics. By severing critique from prescription, Foucault's critique of critique challenges the idea that normative grounding is required for criticism.

Foucault's critique does more than the cognitive task of eradicating error or addressing lack of knowledge. It examines the limits that various "truth regimes" impose on autonomy and brings to light aspects of the present categorial order which determine us as being a certain way. The dominant Enlightenment tradition enquires into formal conditions of truth by looking for universal norms to separate rationality from irrationality (GSO, 20). In contrast, Foucault's "history of truth" inquires into the historical conditions and limits of singular rationalities, relativising scientific rationality as one kind of rationality among a plurality. His critiques resist assimilation into prevailing conceptual orders by offering perspectives that reveal the ordering process itself, thus problematising the universality of norms assumed by the Enlightenment tradition (EW3, 238).

Foucault's Enlightenment *ethos* of ongoing critique entails opposition to Habermas's humanism. Foucault thinks that humanism is uncritical, drawing on a stable human essence borrowing images from religion, science, or politics to substitute for the open-ended and undefined work of freedom. It should be opposed "by the principle of a permanent critique and a permanent creation of ourselves in our autonomy: that is, a principle that is at the heart of the historical consciousness that the Enlightenment has of itself" (PT, 112). By rejecting Kant's transcendental critique but retaining the Enlightenment *ethos*, critique no longer has a determinate prescriptive relation to questions of knowledge, truth, and rightness. It is now a matter of perpetual self-transformation, a process carried on within the understanding of the ever-changing present, not from a universal understanding beyond time (PT, 108–9). This also entails a shift from the injunction which has informed critical theory since Marx that the point of philosophy is to change the world. For Foucault, the point is to change oneself. A critical attitude towards the self reflects both an awareness of the contingency of one's situation and a willingness to transform it.

By retaining its *ethos*, Foucault's critique challenges the Enlightenment's determinate outcomes. He consistently criticised the modern myths of liberation—the progress of the human sciences, the liberation of the mad, prison reform, sexual liberation, and so on. In all of these, he found, masked by Enlightenment reason, the forces of normalisation impoverishing and narrowing possibilities. The autonomy that underpinned Kant's critique was legislative. One must conform to the moral law. However, for Foucault, autonomy doesn't involve freely binding oneself to a necessary and universal law. Rather, it involves the freedom to call into question all that is *presented as* necessary and universal. Foucault inverts the *telos* of Kantian critique. "If the Kantian question was that of knowing (*savoir*) what limits knowledge (*connaissance*) must renounce exceeding, it seems to me that the critical question today must be turned back into a positive one: In what is given to us as universal, necessary, obligatory, what place is occupied by whatever is singular, contingent, and the product of arbitrary constraints?" (PT, 113).

What Is Enlightenment? not only sets out Foucault's approach in relation to Kant but also, by extension, in relation to Habermas. Like Kant, Habermas's critique identifies the limits of reason in grounding claims to truth and normative rightness. Although substituting his fallibilistically conceived philosophy of intersubjectivity for Kant's transcendental philosophy of the subject, limits still play the same role of anchoring a theoretical system. While Habermas wants to discover and respect limits and Foucault wants to go beyond limits, they are not the same limits. Habermas's limits have to do with the correct balance between the elements of modern differentiated reason. He sees the natural sciences, with their objectivating stance directed at prediction and control of the objective world, as being incorporated into systems. These systems have breached their rightful limits and colonised and weakened the intersubjective communicative potentials of the lifeworld and, more specifically, the historical-hermeneutic sciences. The limits that Foucault wants to surpass are entirely different. He is concerned with the limits imposed by the intensification of an inexorable modern power that incorporates all resistance into itself. According to Foucault, this power is linked to the human sciences, not the natural sciences, which have to a large extent separated themselves from the power effects of their origins. While Habermas's primary target is the natural sciences that have breached their limits, Foucault's is the human sciences, the limits of which have come to define and dominate us. From the foregoing analysis, it is not surprising that there are areas of convergence between Foucault's and Habermas's work. There are also stark differences. These differences led to their so-called debate in the 1980s and 1990s, which was carried on primarily by commentators. In chapter 6, I will touch on some of the more significant differences.

Foucault wrote "What Is Enlightenment?" six months before his death in June 1984. The essay offers a retrospective view of his own intellectual project, consistent with his frequent claims in the years before his death of its fundamental unity. In this regard, he presented a number of systematic classifications of his works which he saw as revealing the contingent historical construction of the subject along three axes—knowledge, power, and ethics (EW1, 262–3, 318; EW3, 326–7; HS2, 6; FR, 336–8). This chapter examines archaeology, or the knowledge axis, specifically the knowledge encompassed by the human sciences. Foucault's archaeology aims to chart an unconscious and anonymous structure of constraints and possibilities underlying what can count as scientific knowledge. This fundamental structure, only accessible by the archaeological method, reveals the formation and transformation of objects, concepts, forms of cognitive authority, and social functions.

Foucault's doctoral thesis was published in 1961, with the full English translation appearing as *The History of Madness* (hereafter HM) in 2006.[4] Seeking to understand the pervasive power of psychiatry, HM charts the history of the experience of the different ways madness has been socially constructed. *The Birth of the Clinic* (hereafter BC) was published in 1963. Attacking positivist understandings of scientific method as involving mere observation of what is immediately apparent, BC reveals the interweaving of the perceptual and the discursive in the emergence of modern medicine. *The Order of Things* (hereafter OT), published in 1966, is an investigation of the human sciences, charting their emergence and placing them in relation to empirical sciences, philosophy, and what Foucault calls the "counter-sciences".[5] In 1969, *The Archaeology of Knowledge* (hereafter AK) appeared as a retrospective overview in which Foucault intended to order and clarify his archaeological method. In this chapter, I will briefly discuss HM, an ambitious, vast, rambling work written before Foucault had formulated the archaeological method yet anticipating all his later themes and preoccupations—discourses, power, practices, institutions, genealogy, and subjectivity. I will then situate archaeology in relation to more orthodox approaches to the history of science by making use of the framework provided by AK and briefly commenting on BC. I then will turn to OT to consider the place of the sciences within modernity before concluding the chapter by examining Foucault's critique of human sciences and addressing some of the concerns that Foucault's archaeology has raised.

---

[4] HM first appeared in English in abridged form in 1964 as *Madness and Civilisation*.
[5] For Foucault, "empirical sciences" treat human beings as part of nature since their representations are *products* of an external world. They include biology, economics, and philology. The "human sciences" are concerned with human beings as subjects, with their representations *constituting* the world. They include sociology, psychology, and literary analysis (OT, 384). With "counter-sciences", Foucault has in mind something like structuralist Lacanian psychoanalysis, Levi-Straussian ethnology, or structuralist linguistics.

MADNESS

If we recall Foucault's biography—qualified in psychology, working in mental asylums, homosexual at a time when homosexuality was considered both a crime and a disease, suicidal thoughts, depression—it is not surprising that his first book problematised psychiatry and psychology. Nor is it surprising that Foucault remained engaged with the question of marginality, seeking to understand and call to account the pervasive power of the human sciences. HM charts the history of the experience of the different ways madness has been socially constructed. Foucault's account generally aligns with Bachelard's view of the history of science as characterised by sharp breaks or epistemological ruptures.

In the Renaissance, Foucault tells us, madness remained in dialogue with, and was acknowledged by, reason. The pauper was related to the suffering Christ, the madman to the madness of the Cross. But then quite suddenly, in the middle of the seventeenth century, "a decisive event", the Great Confinement, dramatically changed the experience of madness, transforming its meanings and silencing the earlier dialogue (HM, 77). As moral categories were restructured, poverty became an object of condemnation and idleness was regarded as rebellion. This amounted to a new way of seeing madness, as a social problem in which the central danger had become idleness, through the theological promotion of idleness over vanity as a cardinal sin. The characteristics of what had been discrete groups became homogenised within the broad social category of Unreason (HM, 82). Here we see the *constitution* of a new experience of madness, by which "something inside man was placed outside of himself, and pushed over the edge of our horizon . . . creat[ing] alienation" (HM, 80). This gesture was decisive in initiating the objectification of an aspect of humanity, previously an experience of inalienable interiority, and opening the way to its scientific study.

Mental illness became an object of scientific inquiry in the late eighteenth century, as confinement gave way to the asylum, psychiatry, and psychology.[6] According to Foucault, the doctor's entry into the asylum was not due to his scientific knowledge but his moral authority. While asylums granted some freedom, they imposed a crushing psychological confinement by harnessing bourgeois morality to manipulate their inmates to feel guilt. Again, madness did not pre-exist the practices that developed to designate and deal with it. As it "became an object of investigation, a thing invested with language, a known reality . . . madness was alienated from itself through its promotion to

---

[6] Reflecting distrust in the categories of scientific psychology, Foucault doesn't refer to "mental illness" as a fact that needs to be explained. He criticises historians who suppose "an immutable continuity in madness" (HM 79).

a new status as object" (HM, 443). It is the subject itself, what is most interior, that becomes an object, both for others and for themself, an object seen through the lens of internalised critical judgement (HM, 449).

The doctor was seen as a worker of miracles. Foucault notes the paradox of "medical practice enter[ing] the uncertain domain of the quasi-miraculous just as the science of mental illness was trying to assume a sense of positivity". On one hand, "madness is placed at a distance in an objective field where the threats of unreason disappear". On the other hand, "the madman and the doctor begin to form a strange sort of couple, an undivided unity where complicity is forged along very ancient lines" (HM, 507). This complicity "along very ancient lines" suggests the development of an epistemic stance towards scientific authority, perhaps somewhat akin to shamanism, and a slide into the suspension of reasonable doubt about psychiatric knowledge (HM, 509). The resultant form of objectivity was a "magical reification" in which both doctor and patient were complicit, a myth of scientific objectivity that served to disguise moral domination in the name of bourgeois values. Doctors, refusing to recognise "the ancient powers that lent their status full strength" and having no other way to explain their power to heal, could only see themselves as searching for objective truths (HM, 509). Although nineteenth-century psychiatry listened to the mad person, this was not an engagement, but a monologue "that exhausted itself in the silence of others . . . the patient was trapped in a relation to the self that was of the order of guilt" (HM, 496–7). It was only with Freud that the doctor-patient relationship was itself highlighted as a scientific object. Yet while the magical effects of the relationship were granted their true importance, Freud's explanation was still covered over by further myths of scientism, leaving a structure of moral judgement and coercion (HM, 510–11).

From the time the mad person was taken into the asylum, she was alienated, becoming an object in the eyes of others and in her own eyes, a deep, inaccessible, and problematic object. This account is not only about the mad person, since "to recognise the mad was to recognise oneself, feel the same forces, hear the same voices and see the same strange lights rise up within". From within this particular historical stance, the scientific gaze that objectified "could no longer see without seeing itself" (HM, 519). It is by another whom we recognise as a subject, but a subject totally objectified, that not only the mad person but also anyone becomes an object to himself or herself which is fully open to scientific investigation like any natural object (HM, 525).

HM is a vast ambitious book that is not easily categorised into any particular intellectual tradition. Perhaps for this reason, it met with a muted and often hostile reception (Macey, 211–14; PPC, 99). Among psychiatrists, it was immediately labelled a work of romantic anti-psychiatry which denied the reality of mental illness and represented psychiatric knowledge as pseudoscientific. Foucault was associated with figures like R. D. Laing, David Cooper, and Thomas Szasz as part of an anti-psychiatric movement that portrayed Enlightenment rationality as an oppressive power. These various anti-psychiatry discourses shared the notion of psychiatric power as *intrinsically* oppressive and the truth about the mad as distorted by a pseudoscientific body of knowledge. Foucault denied such claims: "Sometimes people have read my book about madness as if I had written that madness does not exist, or that madness was either a myth in medical or psychiatric discourse, or that it was a consequence of mental institutions. I have never said that madness does not exist or that it is only a consequence of these institutions" (FL, 418).

Iliopoulis argues that HM cuts deeper than anti-psychiatry. Rather than the anti-psychiatrists' critique of the abuses of power within institutions, its aim is to dispute the notion that, beginning in the Enlightenment, a sophisticated medical knowledge is gradually converging upon the truth of madness. For Foucault, the truth of a science does not simply equate with the progress it is thought to have made in its history or by its meeting certain epistemological standards. Truth is produced in the form of an ongoing series of crises and events, the result of political, scientific, and ethical battles which are continually fought. Foucault charts how, as the political stakes which these conflicts expressed shifted, so did the truth of madness. Rather than disputing the validity of psychiatric discourse or the therapeutic role of the asylum, he analyses how various practices contributed to the rise of psychiatry's medical status. His critics failed to grasp the historical breadth of Foucault's analyses which, rather than denying the existence of mental illness, explored the conditions by which it emerged as a domain of scientific knowledge (Iliopoulis, chapter 4).

This is quite different from the claims of the anti-psychiatrists, such as Laing, Cooper, or Szasz, who dispute the validity of psychiatry, generally on the basis of the doctor's power and the effects that flow from it. These claims presuppose certain a priori conceptions of truth and power internal to the functioning of the institution, thus the truth that the doctor holds is always seen simply as a scientific justification of his oppression of patients. In contrast, Foucault reveals the historical transformations that power undergoes in the psychiatric institution by contesting the universality of any single type of power relation between psychiatrists and the mentally ill. He can thus reveal how particular types of power relations produce different truth regimes which determine the position of the mad as objects of knowledge without *necessarily* operating as instruments of control and oppression. Although Foucault might agree that power and truth should be targets of critique, he would insist that both terms need to be redefined before criticism of them is possible. As becomes clearer in his genealogies, power rests on an economy of truth. For power to function, a true "scientific" discourse needs to be produced, authorised, and circulated.

Whitebrook's reading of HM places Foucault together with R. D. Laing as an idealiser of madness.[7] Elsewhere, Gutting points to certain puzzling passages in which Foucault refers to the evocation of madness by poets such as Hölderlin, Nerval, and Artaud. He sees Foucault drawn to the idea that "the voice of madness" itself can take us beyond the confines of rational categories (Gutting 1989, 96–99). I think that Foucault is better understood as attempting to reinstate the forgotten dialogue between madness and reason (HM, 518). In the lyricism of poets, opposites remain in play. When we vacillate between seeing the mad person as object and recognising her as subject, or as mad or non-mad, we see how one perspective relies on the other. As Rajchman points out, such a reading is consistent with a 1960s tendency that saw revolution emerging from avant-garde writing. At this time, Foucault saw literature as a counter-discourse which transgresses limits, making them visible and contestable (Rajchman 1985, 111). While archaeology reveals the limits of what can be said, thought or experienced in an age, it also gestures beyond those limits. In avant-garde literature, a domain of freedom opens up, revealing different experiences based on perceptual and practical grids deviating from conventional science and rationality. These avant-garde writers disrupt the sharp conceptual distinctions on which rational thought relies, showing that "in man, the interior is also the exterior, that the extremity of subjectivity blended into the immediate fascination of the object, that any ending was the promise of an obstinate return" (HM, 518). Here Foucault is gesturing towards the "doubles" that we will see become thematic in OT. The voices of madness evoked by Hölderlin, Nerval, and Artaud reveal the tensions within the concept of the human being, as both subject and object of knowledge. In psychology, these same truths are revealed but here, Foucault tells us, reflective thought "protect[ed] itself, affirming with growing insistence that the mad were nothing but objects, medical things" (HM, 519). Viewed through the lens of scientific objectivity, these truths split into irreconcilable antimonies, resulting in conflicting interpretations of madness within psychology (HM, 520–1).

HM can also be seen as an account of exclusion, starting with the Great Confinement and leading to more subtle exclusions in the asylum.[8] This is a picture of modernity in which power and rationality exclude, banish, or subjugate freedom and madness. This reading assimilates Foucault's view to

---

[7] Whitebrook in Gutting (2005, 329).
[8] For example, Gutting describes the classical experience of madness in terms of "rigorous exclusion" (Gutting 1989, 83).

Weber's theory of modernity as an age of relentless rationalisation, bureaucratisation, and specialisation which imprisons us in an iron cage. Koopman plausibly argues against such a reading, suggesting that Foucault wants to negotiate the tensions between conceptual couples like power and freedom or rationality and madness (Koopman 2013). If modernity is based on a logic of exclusion, the remedy would be liberation, whereby madness is liberated from reason and freedom liberated from power. Yet, as we will see from the "repressive hypothesis" in chapter 6, Foucault thought such liberation was simply enslavement by another face of power. Madness and rationality, freedom and power, presuppose each other. "Man and madman are bound by an impalpable connection of truth that is both reciprocal and incompatible" (HM, 529). Foucault's account of scientific modernity charts the tendency to purify such problematic concepts (Koopman 2010, 2013). While exclusion means banishment or expulsion, purification is "a process in which two kinds of practices rigorously isolate themselves from one another, such that the purification of madness and reason amounts to the simultaneous production of both madness and reason in such a way that they cannot admit of admixture with one another" (Koopman 2013, 157). It would seem that this process of purification entails the production of criteria for the application of concepts which make increasingly finer and stricter distinctions so as to not admit any ambiguity or overlap between categories. Conceptual pairs like reason and madness are opposites that cannot admit of admixture yet are defined in terms of their opposition. Their purification is a process of "inclusion through separation" (Koopman 2013, 157).[9] Presumably this means that modernity separates in order to know and control better and hence subsume within itself.

However, while it is plausible that purification, both conceptual and practical, has assumed dominance in modernity, this doesn't mean that exclusion has been surpassed. As Giladi points out, scientific naturalism deals with problematic objects—normativity, meaning, modality, intentionality, and so forth—by either stripping them of anything that cannot be captured solely in terms of its own vocabulary (purification) or by dismissing them as non-genuine phenomena (exclusion) (Giladi 2020). Notwithstanding the persistence of exclusionary practices, an account of modernity as the increasing dominance of purification is consistent with Foucault's account of the increasing intensity of power that we will see in his genealogies. The response to purification is what Foucault will later refer to as experimentation or transgression.

---

[9] This account of modernity as purification has also been elaborated by Latour who, like Foucault, attacks the ideology of progressive emancipation. For Latour, this "emancipation" is achieved by the purification of nature via its separation from society. Latour's modernity compulsively purifies fact from superstition and liberation from oppression, yet always produces admixtures or "hybrids" (Latour 1993).

In HM, this response is framed in terms of re-establishing the lost dialogue between madness and non-madness.[10] He laments that "compared to the incessant dialogue of reason and madness during the Renaissance, classical internment had been a silencing" (HM, 496). He wants to restore that interaction. He commends psychoanalysis for recognising the primacy of dialogue, but castigates it for turning it into a monologue.

ARCHAEOLOGY AND THE HISTORY OF SCIENCE

Foucault's archaeologies are histories in a sense that demands elaboration. Archaeology's basic premise is that during any given period, a shared unconscious framework both constrains and enables what can be said and thought. This framework—variously referred to as the historical a priori, the *episteme*, the archive—structures observations, concepts, and discourses by ordering and determining what can appear as objects of knowledge.[11] It determines the possibilities of what can be seen, thought, and said in any particular historical period. By an historical analysis of these structures, we come to recognise the contingent construction of our own thinking and the possibility of thinking differently. We are both less free than we think because we are unaware of what constrains us, and more free because these constraints are open to historical change. Foucault envisages a history of things commonly thought to be outside history, such as reason or the subject. His archaeologies challenge the present by historicising these supposedly timeless and universal structures which are assumed to underlie history's multiplicity and contingency (BB, 3).

Foucault's archaeology extends the French tradition of history and philosophy of science to specifically address the complexities of the human sciences. Like Bachelard, Foucault's histories comprise various threads within different regions of scientific work which resist a unified development of rationality (AK, 4). Like Canguilhem, Foucault softens Bachelard's epistemological ruptures, allowing a degree of continuity where continuous concepts are displaced and transformed within discontinuous theories (AK, 5).[12] By excavating basic organising concepts,

---

[10] Talcott sees this re-establishment of dialogue as Foucault's response to Bachelard's notion of the split between an epistemology of the scientific spirit and a poetics of the material imagination resulting from an all-too-imperious reason excluding the irrational at all costs (Talcott, 2019, Ch 6).

[11] Foucault introduces the term "historical a priori" to express the role of discursive formations in conditioning the thought that goes on within them whilst at the same time itself being subject to historical shifts (AK, 143–4). The term "episteme" refers to the unconscious structures which underlie the social production of scientific knowledge (OT, xxiii; AK, 211).

[12] Bachelard characterised epistemological ruptures as decisive counter-intuitive breaks of scientific knowledge from common sense and previous scientific theories which required new concepts, for

archaeology cuts across apparently discrete disciplines, thus extending Canguilhem's "history of the concept" to the determination of the nature and extent of possible knowledge within the broader *episteme*. Unlike Bachelard and Canguilhem who dealt with established scientific disciplines, Foucault cannot adopt the normative stance of current science. This is because archaeology examines discourses which have not reached, and may never reach, the threshold of scientificity where a discipline achieves the status of eligibility for veridical expression. Within human sciences like psychiatry or criminology, with objects such as madness and criminality, the norms of the best current science cannot be used to discern the scientific from the non-scientific. Archaeology uncovers discursive practices that underlie a corpus of knowledge that aspires to the status and role of science (AK, 210). It reveals an underside of reason which is obscured by orthodox historical accounts of science that simply assume the inevitable growth of reason towards the present.

Foucault's accounts of change are layered and complex. Attempting "to free historical chronologies and successive orderings from all forms of progressivist perspective", he excises any suggestion of progress to reveal the movement of history in its own right, uncluttered by teleology (PK, 49). He eschews straightforward causal accounts of how, for example, politics or economics determines consciousness. The archaeological method is structuralist in the sense that it describes the theoretical coherence of discourses among themselves in a given period (EW2, 285). However, by being embedded in historical narratives, it is more than structuralist. Resisting determinism, archaeology reveals the development of science as discontinuous and unpredictable. This emphasis on discontinuity would be familiar to readers of Thomas Kuhn. There are, however, instructive differences.[13] Kuhn's discontinuity is embodied in his notion of scientific "revolutions" in which crises disrupt "normal" science (Kuhn 1996). After reaching a climax, crises subside and the scientific community re-establishes a consensus which ushers "normal" science back in to solve problems within the bounds of the new paradigm. In contrast to Kuhn's emphasis on consensus, Foucault looks for unarticulated preconceptual structures that regulate normal science. Rather than socially achieved paradigm shifts explaining the continuity or discontinuity of theories, Foucault remains at the level of the underlying structures of which we remain unaware. Instead of conscious achievements, Foucault's transformations are spontaneous widespread events lacking any models.

---

example, the development of quantum theory (Gutting 1989, 14–15).
[13] Hacking provides a preliminary comparison between the role of discontinuity in the works of Kuhn and Foucault (Hacking 2002, 87–89).

While Kuhn draws on his model to explain why science is at once progressive and yet not as rational as we are tempted to think, Foucault employs his notion of discontinuity strategically to grasp hidden relations and regularities (AK, 31–2). He doesn't offer psychological notions like crises as explanations of change but seeks to avoid "synthesising operations of a purely psychological kind (the intentions of an author, the form of his mind, the rigour of his thought, the themes that obsess him, the project that traverses his existence and gives it meaning)" (AK, 31–2). By undermining the privileged role of the human subject, he decentres the subject from its traditional explanatory role in history. This enables him to get at the structures which determine our thinking and speech, yet are beyond the grasp of thinking and speech. He wants to show how the science which arises from these structures comes to dominate and shape us. For Foucault, it is the transformation of concepts which are not subject to conscious awareness and control that makes possible the development of individual scientific disciplines.

Since this history of transformations is not visible in individual scientific disciplines, the archaeologist must go to a more fundamental level of conceptual history to reveal the concepts which define the possibility of its first-order concepts. Foucault can't simply align divisions within his analyses with the intentional products of human subjects—essays, books, or the *oeuvres* of particular authors within particular disciplines. Nor can he assume that the works of different authors are related by means such as transmission, influence, tradition, or the "spirit of the age". By seeking a more fundamental layer of history in which subjectivity is displaced, Foucault recalls Canguilhem's "history of the concept" in which the same concept can continue in a series of discontinuous theories.[14] He also recalls Canguilhem's disdain for the notion of the "predecessor" and the tendency among historians of science to explain scientific innovations in terms of their anticipation by earlier thinkers. The problem is that the use of such "precursors" frequently ignored the conceptual differences between similar formulations. However, Foucault's archaeology also goes beyond Canguilhem's thought because his groupings reflect the deeper movement of knowledge in which subjectivity drops out of the account altogether.

This deeper movement is located within broad "discursive formations" in which boundaries between, and new divisions within, conventional groupings are blurred (AK, 32).[15] Discursive formations are groups of

---

[14] Foucault acknowledges the role of Canguilhem in demonstrating this (AK, 5).

[15] This idea of movement within a broader discursive formation allows the possibility of different disciplines developing common terms to form hybrid boundary languages. For example, biochemistry could be considered to have developed out of a hybrid language drawn from both biology and chemistry, two very different sciences. While this approach avoids radical Kuhnian

"statements" unified by rules that govern their formation in relation to elements such as objects, cognitive status, or concepts (AK, 89–98). Although a sentence or a proposition can make sense or be true by merely conforming to the rules of grammar or logic, a series of signs is only a statement if it is related to an associated field of other statements. Inserted into a particular rule-governed system, a statement performs a function. Foucault tells us, for example, that an affirmation like "species evolve" is not the same statement in Darwin and in Simpson (AK, 117). The same proposition performs a different role in different rule-governed systems. Rules of formation of objects include norms which characterise objects (say, as scientific objects), rules about the authority to determine which objects are within a discourse, and "grids of specification" which impersonally classify objects, such as mental patients or criminals. Other rules address contexts from which statements must originate in order to be taken seriously, the formation of concepts, and the formation of "strategies" (themes such as "evolution of species") that develop within the discursive formation. These enunciative rules have no ideal status which can be distinguished from their actual instantiations.

Discursive formations not only delineate particular disciplines, scientific or otherwise, but extend beyond these disciplines to include other knowledges (AK, 203).[16] The boundary separating scientific discourses from what belongs purely to the lifeworld remains blurred. Scientific discourses construct everyday experience by constituting subjects and objects as part of the ordinary lifeworld. Rather than the lifeworld serving as the fundamental level underpinning particular scientific theories, Foucault's archaeological level structures not only scientific discourses but also subjective experiences. Since there are links between science and political, economic, and religious ideologies within a discursive formation, ideology is a more or less natural accompaniment to science (AK, 204). Foucault doesn't see ideology necessarily excluding scientificity (AK, 205). Ideological bias and scientific objectivity are two entwined threads arising from the location of several discourses in

---

incommensurability between disciplines, it also grants an incommensurability between one *episteme* and another although, as Foucault's histories show, this incommensurability is never stable. These ideas have also been explored by figures such as John Dupre, Ian Hacking, Peter Galison, and Nancy Cartwright. See for example, Galison and Stump (1996).

[16] Foucault's understanding of the relationship between discursive formations and sciences requires understanding of the special sense he gives to the distinction between the two French terms for knowledge, *connaissance* and *savoir*. *Connaissance* denotes a particular corpus of knowledge, formal disciplines such as biology or economics, which might be found in "scientific books, philosophical theories and religious justifications" (FL, 13). *Savoir* "refers to the conditions that are necessary in a particular period for this or that type of object to be given to *connaissance* and for this or that enunciation to be formulated" (Translator's note, AK, 16). The particular scientific discipline is the locus of *connaissance*, whilst the discursive formation is the locus of *savoir* (AK p. 201).

a common discursive formation. Because the archaeological level of knowledge is not open to conscious scrutiny, but requires analysis, this entwinement cannot be easily untangled. Science serves as ideology "in so far as science, without being identified with knowledge, but without either effacing or excluding it, is localised in it, structures certain of its objects, systematises certain of its enunciations, formalises certain of its concepts and strategies" (AK, 204). Similar to Habermas's colonisation thesis, Foucault sees scientific discourses restructuring other discourses in discursive formations beyond science's strictly limited domain.

Foucault's account of the constitution of madness, illness, or crime in knowledge of different periods, includes complex practices embedded in external processes, not the mere exercise of the naked eye aided by a theoretical vocabulary.[17] His rejection of a clear distinction between ideology and science is seen in BC, where he is at pains to argue that the development of clinical medicine was not merely a shift from fantastic imaginings to careful observation. Attacking positivist understandings of scientific method as merely the observation of the immediately apparent, he portrays the emergence of modern medicine and its institutions as a series of adjustments in which the perceptual and discursive are interwoven. BC tracks the different techniques by which disease was transformed from an essence that plays out on the surface of the body to an experience of it as a dynamic, changing process located deep within individual bodies. The development of the language of normality and abnormality accompanies this shift.

Of Foucault's work, BC comes the closest to Canguilhem's "history of concepts", especially in the second half of the book where Foucault is increasingly concerned with concepts. While in HM he lacked a set of accepted concepts to analyse the history of madness, in BC he can draw upon generally accepted understandings of disease. He shows that the development of clinical medicine is not simply a matter of precisely matching things seen with speech, but of pushing the "foamy line of language, to make it encroach upon that sandy region" of clear perception for which words can't easily be found (BC, 169). Archaeology reveals that what appears to be nothing more than fidelity to what is given to "pure" observation is actually a mode of perception based on a complex interpretive structure. This interpretive structure was able to emerge with the pathology clinic and Bichat's dissection of corpses which focused on tissues rather

---

[17] Although Foucault's account complicates the observation-theory relationship, it does not straightforwardly lead to Hanson's "theory-ladenness of observation". Concepts rather than theories serve as interpretive elements. See Hanson (1958) and also Rajchman (1988).

than organs.[18] In this way, it becomes possible to describe the likely paths of a disease as it develops. A disease is no longer something that comes from outside the body, but is now a modification of life itself, the "silent work" of death as the organs wear out (BC, 158). By emphasising this interpretive grid rather than simply the application of more careful observation, Foucault does not undercut modern medicine's scientific status. Rather his critique suggests specific ways of calling into question its self-understanding by undermining positivist understandings of scientific method as merely observation of the immediately apparent.

By bracketing subjective intentions, truth-values, and even whether statements make sense outside their context, archaeology is able to display movements within discursive formations as events (AK, 7; Rabinow and Dreyfus, 49). Consider the opening pages of BC, where Foucault cites without comment a bizarre account by the eighteenth-century doctor Pomme, who "treated and cured" a hysteric by giving "baths, ten to twelve hours a day, for ten whole months", observing the membranes of the internal organs expelling themselves until the nervous system, and the heat which sustained it, dessicated (BC, ix). This account is part of a before-and-after image that Foucault typically employs to highlight differences between either side of an epistemological rupture. By their radical incommensurability, such differences prompt the reader to see the depicted events in the light of their underlying concepts. There is no common measure between Pomme's account and our experience. Whole bodies of discourse can be incommensurable because their systems of possibility underlying what they count as candidates for truth are completely at odds. These differences disrupt the smooth narrative of accumulated knowledge and reveal history, which includes our present, as being open to alternative possibilities.

Archaeology doesn't naively attempt to reconstruct the past "as it really was", but rather consciously and self-reflectively assembles, orders, and organises its traces. Foucault wants to write "a history of the present" by explicitly and self-reflectively organising the past in a way which highlights the historicity of the present, the contingency of construction of what appears timelessly given (DP, 31). While conventional history explains difference away by smoothing it over, Foucault dramatises difference in order to throw the present into relief as a historically specific form of rationality. However, historical reality, although ordered by the archaeologist, is not

---

[18] Talcott plausibly argues that, just as Canguilhem asserted the importance of a positive treatment of error in accounting for the normative orientation of thought, Foucault asserts that Bichat's pathology clinic had advanced medical knowledge not by resolving an epistemological question, but by treating death as a technique by which to know the living (Talcott 2019, chapter 7).

reducible to a function of that ordering. Concepts like madness or criminality are not simply socially constructed but have a real material basis in things such as physical bodies and behaviours. Foucault's "present" refers to concepts and practices that are current, yet constituted by past events in ways we don't realise. A "history of the present" lays bare that constitution and its consequences in a way that opens up freedom for changing the present.

ORDER AND THE SCIENCES

Avoiding non-discursive practices almost entirely, OT treats discourse as independent and autonomous. Foucault's archaeological investigation into the human sciences is framed as an analysis of the experience of "order" which constitutes the world.[19] This order eludes the consciousness of the scientist and yet is formative of scientific discourse. Against such constructivism, it might be objected that differences between things don't rely on our conceptual ordering but are discovered to already be there. Foucault would argue, however, that although differences are real enough, we pick out particular differences, rather than others, on the basis of a particular ordering. In structuring a domain of kinds of individuals, we determine what counts as the same individual, what counts as different individuals of the same kind, and what counts as a different kind of individual. While the resultant categories are not arbitrary or unreal, neither are they entirely independent of us. Relative to an order, the reality or non-reality of entities and kinds is determinate. However, independence from any order whatsoever renders reality indeterminate. By having a certain order, we can make true statements about objects so ordered. However, that order could have been different, if not for us being as we are. We can see this by looking at previous *epistemes* which order the world incommensurably differently. Foucault wants to bring this experience of the existence of order to his readers. It is an experience that implies an otherness which escapes the order imposed by scientific discourses. It "always plays a critical role" by opening up space to reveal the contingent basis on which knowledge becomes possible, sciences are established, and rationalities are formed (OT, xxiii). By evoking an idea of otherness, a realm outside the discursive order of things, OT grasps the Foucauldian themes of limits, transgression, and freedom.

---

[19] With "order", Foucault is referring to the underlying set of possibilities which both enable and constrain concepts to divide the world into objects about which we can make true and false statements. "There is no similitude and no distinction, even for the wholly untrained perception, that is not the result of a precise operation and of the application of a preliminary criterion" (OT, xxi).

In OT, Foucault charts how the Renaissance *episteme* gave way to the Classical *episteme* in the mid-seventeenth century, which, in turn, formed the background from which the modern *episteme* emerged. Each *episteme* is governed by an experience of order based on a different principle or historical a priori.[20] Modern thought was not shaped by a series of progressive modifications and improvements. Instead, it was completely transformed by the "mode of being of things, and of the order that divided them up before presenting them to the understanding" (OT, xxiv). The Classical *episteme* had embraced a principle of ordering based on representations of relationships of identity and difference, where resemblances were analysed and differentiated by strict criteria. In principle, complete certainty was enabled by this classification of representations in terms of differences and identities mirroring the order of things in the world (OT, 61). Relations between things weren't seen as just projections of a hidden structure, but as resulting from their very nature (OT, 72). Foucault argues that, prior to Kant, there was no account of how consciousness came to have a capacity to form representations. Consciousness or thought was *necessarily* representative. However, towards the end of the eighteenth century, representation's role as the unquestioned self-justifying starting point for knowledge became problematic. Then, in a total epistemological transformation, the modern scientific disciplines emerged (OT, 227).

According to Foucault, the modern *episteme* orders things by *analogies*, such as *functions*, between organic structures (OT, 236–7). Order no longer depends on the identity of elements, but on the identity of invisible relations between elements within the object. What makes things stand out as objects is no longer necessarily represented.[21] An object is no longer what it is because of its place within a pre-existing ideal system of classification, but because of its existence as a discrete structure given by its place in history. Rather than increased objectivity, more precise observation, or more rigorous reasoning, it is History, the historical a priori of the modern *episteme*, that radically restructures thought to enable the emergence of new empirical sciences. Economics, biology, and philology are not grounded on representations but on something beyond representation that emerges historically—labour, organic structure, and the system of inflections.

As historical a priori, History plays a hidden role, both constraining and enabling what can be said and thought (OT, 237). It has both epistemological and ontological implications. It is both how we know things, "the space

---

[20] It is likely that Heidegger's notion of the epochal nature of truth which "holds complete domination over the phenomena that distinguish the age" influenced Foucault's thought that a shared unconscious framework conditions what can be said and thought within an epoch (see Heidegger 1977, 115–154).

[21] Lungs and gills, for example, are closely related by virtue of performing the same function, despite their markedly different properties (OT, 288).

from which things come to knowledge", and "the mode of being of all that is given to us in experience" (OT, 237). Han suggests that this dual role can be understood by seeing Foucault's position as analogous to, though not identical to, Kant's understanding of causality (Han 2005, 592).[22] Foucault sees History as the principle by which things are constituted precisely as the things that they are. We are bound to construe things as historical by the fact that we are governed by History. However, Foucault's historical a priori differs from Kant's categories. History, as historical a priori, comes to have, but may eventually lose, this status. The stability of Kant's transcendental structures is purchased by the commitment to notions of universally shared ahistorical cognitive faculties and the world-in-itself. Foucault is committed to neither notion. He insists that all we ever have is things that appear to particular historically located human beings, according to the current historical a priori.

Foucault sees Kant's critique as ushering in modernity by its pivotal role in the eclipse of the Classical *episteme* (OT, 263). Kant asks about the conditions for representation in general, showing us that thought is not necessarily grounded in representations, but in something more fundamental—the synthetic activity of the thinker. By losing transparency and becoming problematic, representation now becomes an object of inquiry, thought of first as something *in itself* and then as *about something else*. According to Foucault, this new order resulted in the fragmentation of the field of knowledge. A new distinction emerges between analytic and synthetic knowledge whereby maths and logic are sharply divided from the empirical sciences. The third form of inquiry is philosophical reflection, which seeks a unified account of the nature of reality and the grounds of knowledge. Representation, no longer the unquestionable form of thought and knowledge, becomes a principal object of philosophical inquiry. Together, these three domains of inquiry—the analytic, the empirical, and the philosophical—are the three dimensions comprising the modern *episteme* (OT, 378). Although they represent three irreducibly different forms of knowledge, Foucault notes that one domain can borrow methods from another or take another as its object of inquiry (OT, 375). The human sciences have no place within modernity's epistemological trihedron but are insecurely located "in the interstices of these branches of knowledge . . . in the volume defined by their three dimensions" (OT, 379). They must thus be understood in terms of their special relations with all the dimensions of modern knowledge. Foucault understands modern sciences as decisively breaking with their classical counterparts, their emergence reflecting the autonomy and mutability of knowledge at the archaeological level. I will now briefly sketch Foucault's understanding of economics and biology and their related human sciences. I have not included

---

[22] For Kant, causality was neither in the mind alone nor in nature alone, but was constituted objectively in nature by the application of the subject's pure categories of understanding (Kant 1998).

Foucault's discussion of philology and its related human science, since this brief outline is only intended to illustrate the general pattern of the emergence and place of the sciences within the modern *episteme*.

The Empirical Sciences

According to Foucault, the Classical *episteme* analysed value solely in terms of exchange within a system of representation where money represented commodities. The decisive break was made by Ricardo, who presented labour as not merely something exchangeable, but as the sole *source* of value, although outside the system of exchange. Commodities had value because people worked to produce them (OT, 277). What creates value is the accumulation of labour through the entire production process (OT, 258). Economic history could now be understood as a linear causal series. Productivity of labour is related to forms of production—tools and machinery used, division of labour, capital invested, and so forth. Since forms of production are themselves products of previous labour, current value is the result of a series of overlapping causes stretching back indefinitely (OT, 278). With human beings as economic agents, economics can be understood historically. Human beings are not merely bearers of representations but also are subject to the factors which caused those representations, such as external threats or natural scarcity. Ricardo views economic history as a history of increasing want as populations increase and resources diminish. Marx more optimistically heralds a new consciousness in which the arrangements previously attributed to nature are recognised as historically produced. Ricardo and Marx share a view of economic life as the historical struggle of human beings to survive through their labour. Since both theories are founded on the epistemic break initiated by Ricardo, Foucault provocatively claims they share the same fundamental mode of thought (OT, 280–6).

Modern biology emerges from a similar transformation. Its specific object is life, no longer merely one category of natural things. For modern taxonomy, what is important is the functional similarities of organs, not the identities and differences of phenomenal properties. Life in its imperceptible purely functional form provides the basis for a classification which depends, not on surface phenomena, but on elements hidden from view. Cuvier's understanding of organic structures in terms of their functional roles, before and independently of taxonomy, enabled a discontinuous classification of species to replace the continuous classification, based on the tabulated identities and differences of the Classical order.[23] Species differentiation in modern

---

[23] For example, in modern biology, "vertebrates and invertebrates form absolutely isolated subareas, between which it is impossible to find intermediate forms providing a transition in either direction" (OT, 296).

biology is explained by the particular manner in which each species is linked to its environment, which is no longer merely a setting for the predetermined essential natures of species. Biology views the essential nature of a species as *causally* dependent on its environment (OT, 298). As a consequence, life becomes essentially historical and tied to *time*. Living things are now forced into discontinuous groups formed by their relation to the environment at the time of their formation. Cuvier's innovations thus set the stage for Darwinian evolution.

Labour, living beings, and languages are now understood as essentially historical realities which develop according to laws essential to their being what they are (OT, 319). The emergence of the new empirical sciences involves a suite of new concepts, methods, and objects (OT, 275). Life, labour, and language were not objects waiting to be discovered, nor were they merely concepts. Rather, they were what Foucault calls "quasi-transcendentals" for contemporary thought (OT, 272). They function in the empirical sciences to provide the conditions of possibility of the subject's representational experience. Since the modern empirical sciences each have their own quasi-transcendentals, they have lost the cohesion and the possibilities of completeness and certainty which they had in the Classical *episteme*. Knowledge thus becomes partial, tentative, and disjointed.

The Human Sciences

Foucault sees the human sciences as aspiring to, or borrowing from, all three dimensions of knowledge established when the homogeneity of classical knowledge fragmented. However, the fundamental difficulties in defining the human sciences are their relations to the empirical sciences and philosophy (OT, 382). The human sciences borrow models and concepts which divide them into three interlocking "epistemological regions", each corresponding to an empirical science. The psychological region is linked to biology and is concerned with humans as living beings, in a capacity which "opens itself to the possibility of representation" (OT, 387–8). Psychology is not simply concerned with stimulus-response mechanisms like biology, but how such mechanisms are linked to representation. The sociological region is linked to economics and is concerned with the way in which "the labouring, producing individual offers himself a representation of the society in which this activity occurs, of the groups and individuals among which it is divided, of the imperatives, sanctions, rites, festivities, and beliefs by which it is upheld or regulated" (OT, 388). Sociology doesn't aspire to identify law-like regularities in commodity transactions, but asks how things are valued and represented as commodities in the first place. Each empirical science contributes a pair of concepts in terms of which the objects of the particular linked human science can be grasped.²⁴ Psychology takes from biology the conceptual pair of *functions* regulated by *norms*. Sociology borrows from economics the conceptual pair of *conflict* governed by *rules*. While the conceptual pair borrowed from each empirical science tends to dominate the human science linked to that empirical science, all conceptual pairs can operate in all human sciences.

Like philosophy, the human sciences are concerned with human beings as subjects and objects, knowers whose representations constitute their world and enable them to know themselves as objects within that world. However, the human sciences treat representations as products and processes of *unconscious* structures. Functions and conflicts, organised by rules and systems, can all be represented without appearing in consciousness. Even though a particular society, individual, or culture has no awareness of such representations, the human sciences can speak meaningfully of the *function* of a social practice, a *conflict* within an individual psyche or, to include literary criticism as the third human science, the *meaning* of a myth (OT, 394). By bringing to light functions, conflicts, and meanings, human sciences can account for the ways in which humans represent the fundamental realities of life, labour, and language which appear within the empirical sciences as determinants of humans as objects. These representations appear unconsciously as objects constituted by human beings who structure functions, conflicts, and meanings within the organising concepts of norms, rules, and systems. By the employment of these higher-order organising concepts, the human sciences move towards the unconscious, and are able to show how human beings as subjects represent the very forces that determine them as empirical objects. These concepts structure the entire field of the human sciences (OT, 393-9).²⁵

Foucault regards the human sciences as not strictly sciences but as belonging to a domain of knowledge having a legitimate place in the modern *episteme*. The title "science" comes only from the models they borrow. While the human sciences lack the predictive and explanatory capacity of empirical science, they are not mere opinions. They can yield objective knowledge by employing epistemologically sound methodologies. However, due to the interlocking of their interpretive models, the boundaries of the human sciences become blurred and "intermediary and composite disciplines multiply endlessly, and in the end their proper object may disappear entirely" (OT, 390). Methodological controversies frequently arise from disagreements about which constitutive model is most appropriate. The human sciences are intrinsically unstable,

---

[24] The models borrowed from empirical science are *constitutive*. "They play the role of 'categories' in the area of knowledge particular to the human sciences" (OT, 389).

[25] I am indebted to Gutting's succinct exegesis of this section of OT (Gutting 1989, 208-213).

treating as their object what is, in fact, their conditions of possibility. They are always animated by a sort of transcendental mobility. They never cease to exercise critical examination of themselves. They proceed from that which is given to representation to that which renders representation possible, but which is still representation. (OT, 397)

The human sciences are "constantly demystifying themselves" (OT, 397). However, their instability, blurred boundaries, and proliferation of approaches does not derive from the complexity and difficulty of their object, human beings. It is due to their position in the modern *episteme*, where they are pulled in different directions by their relationships to the empirical sciences, mathematics, and philosophy (OT, 380). Instability does not prevent the human sciences having a critical role in relation to the empirical sciences. They can lay "an invincible claim to be the foundation of [the empirical sciences], which are ceaselessly obliged in turn to seek their own foundation, the justification of their method, and the purification of their history, in the teeth of 'psychologism', 'sociologism' and 'historicism'" (OT, 377).[26]

Beyond the human sciences, Foucault identifies three broadly structuralist "counter-sciences"—psychoanalysis, ethnology, and a type of structural linguistics—as occupying a privileged epistemic position. They exhibit "a perpetual principle of dissatisfaction, calling into question, criticising and contesting what may seem well established" (OT, 407). They disrupt established disciplines and settled beliefs across the range of human sciences. Foucault's sketchy account of the counter-sciences suggests that he is searching for a perspective that will enable a more radical critique of reason. I will not discuss them further because their significance is little more than an undeveloped precursor to the critical role developed by Foucault's genealogy, to be discussed in chapter 6. It is worth noting, however, that Foucault's "counter-sciences" roughly align with Habermas's "critical sciences" in KHI, both in the type of science and their critical role.

At this point, it is worth taking stock of Foucault's analysis by considering how he transforms Canguilhem's approach. In searching beneath the play of theoretical formulations to get at basic conceptual differences and similarities, OT is largely a straightforward application of Canguilhem's history of concepts. We thus see Foucault undermine notions such as precursors based on superficial resemblances and the idea of smooth progressive accumulation of true facts. But he also transforms Canguilhem's history of concepts by linking apparently distinctly different disciplines between which he identifies conceptual affinities. This moves us to the level of the *episteme*, where

---

[26] This critical function can be seen in ongoing debates about the empirical sciences generated by the sociology of scientific knowledge.

individual disciplines can't fully define the terms within which the historian must understand them. By writing a history of scientific disciplines that are defined through their relation to other areas of knowledge, he undermines the privileged role of those disciplines. The possibility of the development of any given discipline is now based on concepts shared with other disciplines which are more fundamental than any single discipline's first-order concepts. Scientific disciplines are now subject to transformation over time in a way not controlled by any individual discipline.

The Analytic of Finitude

I have discussed the modern *episteme* primarily in terms of the historical a priori of History which Foucault introduces in chapter VII of OT. In chapter IX, he turns to another historical a priori—Man, the object of the human sciences. With "Man", Foucault is not referring to humanistic value concepts like a universal moral core, the dignity of the human person, the priority of a substantial "I", or the freedom of a human subject. Nor is he referring to biological, social, or psychological notions. He is specifically referring to human beings understood in a way not possible prior to the modern *episteme*. Man is an "empirico-transcendental doublet", the transcendental condition of all knowledge of the world and, at the same time, a being in that world which can be known as an object which renders those conditions evident. Man is both object and subject. Man is autonomous and rational, yet at the same time this autonomy and rationality is the product of forces beyond his awareness. As such, Man does not designate a particular empirical object, but rather a question pursued by modern philosophy which evades any stable solution.

Man's emergence is directly linked to the emergence of representation as an object of inquiry. The representative condition of representations now appears to be outside representations, in the subject (OT, 259). What makes things thinkable is not part of what they are. Reality is represented only by the contingent faculties of Man. Man is now a subject among objects, a subject not only seeking to understand the world of objects but also understanding himself in his very capacity as subject-object. Man emerges with Kant as playing two roles, both the foundation of knowledge and the object of that knowledge. In order to be grasped empirically, science needs the transcendental arrangement of the human faculties (OT, 259). Although life, labour, and language *exist* independently of Man as transcendental subject, they can only be *known* by the activities of that transcendental subject. This is Man in his foundational role. However, Man is also an empirical object determined causally by life, labour, and language within the field he himself has opened up. Man is not the object of any particular science but the surface effect of the superimposition of three empiricities—life, labour, and language.

Biology, economics, and philology all show that Man is limited by the various processes which enmesh him in the world. Science uncovers the laws that constitute Man, charting the causal products of life, labour, and language that go before him and shape his representations (OT, 342). But rather than being trapped by this finitude, Kant turned it to his advantage, arguing that the very factors which limit knowledge and restrict it to the forms of space and time and the conceptual framework of the categories are, at the same time, the conditions for the possibility of knowledge (OT, 343). It is precisely these constraints that enable objects to appear in the first place. What is specific to modernity is the doubling of Man to become both the transcendental subject and the empirical object of knowledge. This doubling is inherently ambiguous because the distinction between the subject and the object of knowledge is grounded in the same being, Man. Man both separates and unites the empirical and the transcendental. The transcendental subject is the *condition* for knowledge of the empirical object, yet Man is one being.

From Kant onwards, philosophy has been employed to define on what grounds representations are possible and to what extent they are legitimate. Kant thought that it was precisely on the basis of universal and necessary conditions that we constitute objects, which could otherwise not appear to us. While the empirical sciences of biology, economics, and philology elucidate the laws by which Man is constrained and enabled by life, labour, and language, so life, labour, and language constrain and enable Man in coming to know those laws. The human sciences are specifically directed to, and take as their object, the constitution of Man's subjectivity by these laws. Philosophy is also directed to the question of how Man constitutes the world of objects of which he is a part. It undertakes an "analytic of finitude", an account of the relationship between Man as transcendental knower and Man as empirical object of that knowledge. The problem is that the "positive" and the "fundamental" sides of Man must be viewed as the *same*, since Man is one being, yet *different*, since the fundamental grounds the positive.[27]

One response by philosophers following Kant, such as Comte and Marx, has been the reduction of the transcendental to the empirical, such that the empirical conditions of the subject are put forward as the conditions of the subject's knowledge. This reduction takes two forms. First, knowledge is explained in terms of the processes within the body which are involved in the production of knowledge (OT, 347). Second, knowledge is explained in terms of historical, social, or economic conditions. However, both explanations uncritically assume that empirical knowledge of the body, society, or

---

[27] The term "fundamental" designates the actual existence of life, labour, or language in its role as founding or constituting knowledge, while "positive" designates this knowledge itself, as constituted by the transcendental activity of Man.

economics is just given, as though imprinted on a passive subject, and then used to ground knowledge generally. Foucault refers to this sort of response to the analytic of finitude as "positivist". The alternative to this is "eschatology", which claims the truth of scientific and historical accounts of empirical objects on the basis of the truth (once achieved) of our philosophical discourse about knowledge.[28]

Another response is the *cogito*-unthought double which expresses the reality of Man as both an experiencing subject and the never fully understood object of that experience. The modern *cogito* cannot ensure epistemic immediacy and self-certainty because the prereflexive conditions of knowledge obfuscate what appear as evident truths of reflection. Philosophers and human scientists seek clarity by subjecting obscure factual conditions to philosophical scrutiny, in a never-ending task of making present what is absent in the *cogito*, thinking the unthought. The problem is that gaining knowledge of the unthought in Man is conditioned precisely by what remains unthought (OT, 352). The sciences of life, labour, and language can reveal previously hidden ways in which our beliefs, desires, deeds, and words are conditioned, but this knowledge is itself subject to these very conditions. It is not clear whether "I" is what I consciously survey or is what is unknowable that always conditions, behind my back, such a survey. Man is formed by a complex network of background practices which he can never fully grasp, and yet he is the possibility of their elucidation. He is a product of a history whose beginning he cannot reach and at the same time he is the writer of that history. This tension, between what we think we are and the unknown which conditions that thinking and being, forms a tight circle from which we cannot extricate ourselves.

The *aporia* built into Kantian epistemology and the pressures of its unstable doublets manifest in the voracious dynamism of the modern scientific will to knowledge. Foucault suggests that a craving for explanation generates a never-ending compulsion to keep uncovering what remains hidden in our nature (OT, 353). By incorporating this unthought into his thought, Man gains insights into his actions. Under the banner of truth, thinking the unthought appears as a sort of political action, promising eventual liberation as all that lies below our thoughts, deeds, and words is brought to consciousness. Philosophical thought demands resolution of the question of Man with strategies like positivistic or eschatological reductions or by attempts to make the unthought fully explicit. It seems to me that it is science that provides the content of such strategies, appearing to liberate by enabling an objective stance on thought which suspends its force and makes clear its causes. However, thinking of human thought or action in scientific terms, for

---

[28] "Eschatology" is a thinly veiled reference to Marxism, in which "the true discourse anticipates the truth whose nature and history it defines" (OT, 349).

example, in terms of evolutionary psychology or neurophysiology, neutralises the normative dimension of thoughts and actions and makes them mere events that happen. These are not neutral redescriptions that leave things as they were but, as Foucault says, they "cannot help but liberate and enslave" (OT, 357).[29] The problem arises from the attempt to make explicit something that, by its very nature, eludes reflection. Foucault notes that reflection doesn't merely reproduce an original experience but changes its given nature (OT, 357). Reflection is always already a theoretical attitude that involves an objectification of that which is its object. All expression, any attempt to put something in words, is an objectification which changes its object.

In the playing out of the analytic of finitude, philosophy falls into complacency (OT, 372). Beyond the strategies of reduction, clarification, and interpretation employed by the analytic, Foucault offers no further developed or more fundamental solution. To move beyond this, we need to remove anthropology as a ruling category. Kant is clearly implicated in anthropology. Foucault notes that in his *Logic*, Kant added an ultimate question to the three critical philosophical questions, which are "referred back to this fourth [question], and inscribed, as it were 'to its account': *Was ist der Mensch*" (OT, 371).[30] Here Kant, on Foucault's reading, places the question of an essential human nature on the philosophical agenda of modernity as the most fundamental question. Kant's question remains active, constituting and supporting an anthropology from which we are still not free (OT, 371).

The Death of Man

Man, Foucault tells us, is "an invention of recent date. And one perhaps nearing its end" (OT, 422). However, "these are not affirmations; they are at most questions to which it is not possible to reply; they must be left in suspense, where they pose themselves, only with the knowledge that the possibility of posing them may well open the way to a future thought" (OT, 421). Foucault clearly isn't merely providing a neutral description of what he thinks is possible. He is urging our current thinking to go beyond itself. He wants to overcome our "twisted and warped forms of reflection" by "the unfolding of a

---

[29] Foucault should not be seen as rejecting science as dogmatic or authoritarian. As Talcott points out, while Foucault thought that new scientific knowledge cuts and wounds us, he sees it as also transformative in awakening us from our dogmatic slumbers (pers. comm. Talcott 2020; see also Talcott 2014).

[30] In his introductory lectures on logic, Kant writes: "The field of philosophy . . . may be reduced to the following questions: 1. What can I know? 2. What ought I to do? 3. What may I hope? 4 What is Man? The first question is answered by Metaphysics, the second by Morals, the third by Religion, and the fourth by Anthropology. In reality, however, all these might be reckoned under anthropology, since the first three questions refer to the last" (Kant 1963).

space in which it is possible to think" (OT, 373). Gutting argues that Foucault owes us reasons, which Gutting presumes to be found in non-discursive practices, which were bracketed in OT. While the decentring of Man will change our understanding of knowledge, Gutting questions why Foucault urges change to a merely epistemological concept (Gutting 1989, 224). To respond on Foucault's behalf, I would say that the "death of Man" thesis should be seen as a performative speech act which opens up more freedom by articulating a previously unthinkable possibility. We can also cash out the thesis as a call to freedom by reflecting on the negative consequences of the concept of Man seen in HM, where human subjectivity becomes the object of a scientific gaze saturated with domination. This connection of knowledge and power will become thematic in Foucault's genealogies, where the human sciences are linked to processes of normalisation.

A further concern of some commentators is whether the "death of Man" thesis implies the death of the subject more generally. Clearly there is much at stake. Concepts such as intentionality, agency, autonomy, responsibility, and self-reflexivity are all linked to the idea of the subject and would be endangered. But these are not Foucault's targets. The death of Man is a very specific claim that need not map onto the elimination of the concept of the subject per se. Foucault is specifically problematising Man as the ambiguous simultaneously transcendental subject and empirical object of knowledge. By problematising Man, Foucault certainly doesn't endorse Kant's dualism, but neither does he reject it. He *problematises* it. In so doing, what he rejects is the hypostatisation of Man as a theoretical essence which, in the form of the analytic of finitude, requires one to resolve this riddle and find the truth as either a positivist or an idealist or in terms of some other as yet unimagined truth. Rather than "false", I think Foucault would view the concept of Man as problematic, ambiguous, and unproductive. It mires us in confusion and lends itself to the sorts of abuses we saw in HM. It traps us in endless circles, leaving no room to think critically. Foucault may well be an empiricist, but he doesn't reject the transcendental. "In all of my work I strive instead to avoid any reference to this transcendental as a condition for the possibility of any knowledge. I try to historicise to the utmost to leave as little space as possible to the transcendental" (FL, 98). This is not so much a flat rejection of the transcendental as a methodological choice.[31] It is a pragmatic choice that is appropriate to his object, the present situation. In this situation, the

---

[31] Ten years later, discussing discontinuity, Foucault notes that "Canguilhem stresses the fact that for him identifying discontinuities does not have to do with postulates or results; it is more a 'way of proceeding', a procedure that is integral with the history of the sciences because it is called for by the very object that the latter must deal with" (EW2, 471). It seems to me that Foucault's marginalising of the transcendental is also a "way of proceeding", that is, a methodological choice.

constraints which Foucault wants to draw to our attention arise from the hypostasised doubles of Man aspiring to provide an authoritative basis for the human sciences, which lurch from positivism to idealism. The concept of Man is an obstacle to changing the present, "to imagine it otherwise than it is, and to transform it not by destroying it but by grasping it in what it is" (PT, 108). Foucault is, however, unable to say exactly what can or should be done or thought beyond the present, since this is not a question he thinks we are able to address (OT, 272). So, rather than offering a solution to the riddle, we must start from where we are, bringing to light the conditions of possibility for our thinking and acting and posing questions which "may well open the way to future thought" (OT, 421). Foucault mentions one development which suggests the possibility of freedom, the re-emergence of a certain genre of avant-garde literature typified by the works of Mallarmé (OT, 327). As Rajchman points out, Foucault saw in avant-garde literature the possibility of another way of thinking which constantly tries to think outside the bounds of thought (Rajchman 1988, 116).[32]

Foucault wants to show, not only how the discursive limits of scientific knowledge and experience are constituted, but also what escapes those limits. In later interviews, he describes his own works as "fictions". He doesn't mean that they are untrue. What he means is that his works share the aim of fiction, which is not concerned with bringing things under an explanatory schema or showing how something was historically necessitated. Rather, it is a matter of offering an experience which shows how things might be transformed in ways otherwise unimaginable: "My problem is to construct myself, and to invite others to share an experience of what we are, not only our past but also our present, an experience of our modernity in such a way that we might come out of it transformed" (EW3, 242).

Foucault's archaeology isn't gesturing towards some difficult "truth" of the subject. Nor does it envisage an alternative theory of human nature. Any theory of human nature could potentially be stultifying. The one that stultifies our present is the one played out in the analytic of finitude. Rather than having an essence, human subjects are always constituted by and embedded in contingently evolving historical social conditions. In a 1978 interview, Foucault said, "Men are perpetually engaged in a process that, in constituting subjects, at the same time displaces man, deforms, transforms and transfigures him as subject. In speaking of the death of man in a confused, simplifying way, that is what I meant to say" (EW3, 276). Foucault's death of Man thesis is the recognition of this ongoing displacement, deformation, transformation, and transfiguration, a recognition which is itself transformative.

---

[32] Rajchman notes that, as this genre took hold in universities, Foucault came to regard this as the "sacralisation" it had originally intended to resist (Rajchman 1988, 116).

He doesn't urge a "solution" to the analytic of finitude, but an overcoming by self-transformation. By their external structuralist perspective, Foucault's proposed counter-sciences aim to dissolve, rather than solve, the role of the concept Man. Rather than seeking something "deep" or "hidden" in human nature, Foucault calls Man into question by exposing the conditions that make his reality as a representing subject possible, thus avoiding taking Man as a fundamental category (OT, 413–4). Foucault can only point to the exhaustion of the *episteme* and the future Nietzsche offers as both a task and a promise, "the threshold beyond which contemporary thought can begin thinking again" (OT, 373). We will see this future articulated in Foucault's later work as the ongoing undermining of essentialist thinking and a move towards a vision of relations of power which are sufficiently fluid for subjects to reverse.

CONCLUDING THOUGHTS

The diverse corpus of Foucault's archaeology has the feel of a thinker still finding his way. The emancipatory *telos* which animates his entire project only becomes clear from the perspective of his late work, particularly his idea of philosophy as a self-transformative exercise, which I will discuss in chapter 7. From this perspective, Foucault's problematisation of the concept of Man aims to open a space for illumination by exploring the unthought in both theory and *praxis*. The problem is that, in posing the question of the demise of Man, it is not clear where Foucault sees himself located. Is his archaeological account inevitably and completely determined by the epistemic order underlying his own thought?[33] Or has he found a vantage point from the cusp of a new *episteme*, from where he can see a new dawn lighting the horizon? It appears not, since he claims that *epistemes* form a barrier beyond which it is impossible to think.[34] If we take this claim seriously, Foucault's critique seems as provincial as any other discourse, depriving him of any vantage point from which to critique the human sciences and philosophy. But Foucault later backs away from this claim.[35] If we accept this as a clarification, we can see that he isn't necessarily undermined by the historical contingency of his own position. We needn't identify Foucault's account of *epistemes* with a notion of hermeneutically sealed worldviews which are defined in contrast to some pure state of affairs untainted by any standpoint. The truth of his analyses is

---

[33] Rabinow and Dreyfus raise such concerns (Rabinow and Dreyfus, chapter 4).
[34] "In any culture and at any given moment, there is always only one *episteme* that defines the conditions of possibility of all knowledge, whether expressed in a theory or silently invested in a practice" (OT, 183).
[35] In the Foreword to the English edition of OT, Foucault cautions that the work is "a strictly 'regional study'" and that such terms as "thought" or "Classical science" refer "practically always to the particular discipline under consideration" (OT, x).

founded on the experience of his historically located situation, which he can acknowledge as being bound by its constraints like any other *episteme*.

Foucault's literary style of exposition might appear inconsistent with what some critics think is his ambition to provide a robust theoretical account of human nature backed by philosophical justification. Rabinow and Dreyfus, for example, claim that Foucault's archaeology is really attempting to provide a structuralist scientific theory, an attempt that fails, eventually leading him to abandon this project and take up genealogy (Rabinow and Dreyfus, chapter 4). Foucault's aim is not to develop scientific theory.[36] In fact, the scientific pretension of a theoretical understanding of human nature is precisely the target of his critique. While Foucault was drawn to the "scientific" perspective that structuralism seemed to offer, he had very significant differences with it that he became increasingly frustrated trying to make clear to his critics.[37] He insists that his proposed structuralist "counter-sciences" cannot be regarded as neutral bodies of scientific knowledge because they are tied to specific cultural practices, psychoanalysis to the doctor-patient relationship and ethnology to Western dominance at a certain historical moment. Psychoanalysis and ethnology both reveal "truths" that are not objective truths. Psychoanalysis reveals, for example, singular truths which only become manifest in actual transformations of particular subjects. Seen from the perspective of Foucault's later interest in practices of self-transformation, archaeology can also be regarded as critically oriented by its quasi-structuralist stance, but just as limited in its scientificity.

Archaeology dramatises the instability and contingency of knowledge which shifts, at the level of the discursive formation, in ways of which we remain unaware. Yet despite Foucault's dramatisation of their contingency and instability, we must not forget that the human sciences remain sufficiently stable within their *episteme* to ensure their authoritative power. This stability is conditioned by massive historical inertia which archaeology aims to unsettle. Archaeology develops a reflexive relationship to the contingencies that make us what we are, so that we can begin to transform the seemingly natural structures to which we are subjected. This approach does not aspire to scientific objectivity or philosophical justification but is rather an attempt to transform the present by coming to terms with concrete historical data in a way that prompts us to recognise and think differently about contemporary experiences of oppression and exclusion.

If we regard the fact that archaeology doesn't meet standards of scientificity or philosophical justification as a failing, this is perhaps because we are

---

[36] Foucault claims "my discourse . . . is avoiding the ground on which it could find support" (AK, 226).
[37] See Foreword to the English edition (OT, xv).

applying the wrong standards. Rather than asking what archaeology can justifiably *know*, we should ask what archaeology can *do*. By describing the ways in which scientific discourses produce subjects as their objects, archaeology can loosen the grip of those discourses. Enlisting the knowledge axis of the subject's constitution (the axes of power and the self will have to wait for Foucault's genealogies and ethics), archaeology is critically oriented towards the present, potentially transforming it by transforming us. This orientation does not entail the use of concepts such as *episteme* as totalising descriptions to explain why the past had to be as it was, or why the present must be as it is. Rather, such notions serve to help us escape the illusion that we cannot think otherwise than we actually do. They are critical devices in a process which, by priming our sense of contingency, destabilises tightly held epistemic commitments that we are barely aware of. Notwithstanding this orientation, we can only deal with problems as they appear to us as contingently located beings with no absolute vantage point. Seen in this light, no philosophical justification is required, or even possible.

OT's critique of humanism and its philosophical underpinnings posits Man as something to be surpassed. Here is Foucault's desire for freedom, to break free of whatever constrains us by seeing how it constrains us. In his archaeology, we see the beginnings of an approach which he was not able to clearly articulate at the time. Rather than the scientific approach that some of Foucault's critics suggest, his project involves problematising the human sciences by undermining their sense of universality and necessity. Foucault is not so much putting forward determinate claims as opening possibilities for other conceptions of rationality by a self-transformative process which experimentally destabilises our certainties. He doesn't see social-scientific knowledge as the outcome of a progressive accumulation of beliefs converging on a static reality against which it is tested. Nor are scientific disciplines autonomously bound by their own internal logic. The scientific disciplines which Foucault is concerned with to some extent constitute their own reality and interact with non-scientific discourses as part of a broader formation of knowledge, both propelled and constrained by anonymous unconscious forces located in the rules of the *episteme* and its historical a priori which, to a large extent, determine what can be said, seen, done, and thought in a historical epoch.

In OT, any connection with the non-discursive is assiduously avoided. The axis of knowledge floats free in its own self-contained world. In contrast, Foucault's early archaeologies (HM and the first part of BC) included non-discursive practices and sometimes offered these as causes of change, though in strictly local and limited contexts. In AK, published three years after OT, Foucault circles back to these earlier works, by telling us that archaeology concerns not merely discourses and knowledge but also relations between "institutions, political events, economic practices and processes" (AK, 162). On this understanding, archaeology is seen as laying out the complex interactions between the discursive and non-discursive to reveal the hidden constraints of what we can say, do, or think in any particular historical situation. Simply to be aware of the existence of such constraints as "unknown unknowns" is emancipatory both in its deflation of an excessively hubristic sense of autonomy and in its provocation to think the unthinkable. However, AK not only looks back to the earlier archaeological works but also forward to his genealogy, in which Foucault folds archaeology's knowledge axis into a vast web of non-discursive practices. Turning to Nietzsche, he will bring reason into relation with power, providing a motor for historical change and giving real force to the idea of the constitution of the subject.

Chapter 6

Science and Power

The tumultuous period following the events of May 1968 not only led to Foucault's active engagement in political issues but also turned his thinking more directly to questions of power. At his inaugural lecture at the Collège de France in 1970, Foucault linked knowledge to power by introducing the notion of a "will to truth" (Foucault 1981). Conventional understanding suggests that science is exactly that form of knowledge which abstracts from the particularities of interests, prejudices, and power in order to reach universal truths. Against this notion, Foucault's genealogy, developed in his lectures during the 1970s and in his two books *Discipline and Punish* (1975) (hereafter DP) and *History of Sexuality, Volume 1* (1976) (hereafter HS1), showed power and scientific knowledge as inextricably entwined. Genealogy chronicles how successive configurations of power and knowledge, tangled together within discourses, social institutions, and practices, interact to govern individual bodies and populations by constituting both objects and subjects.

This chapter will examine Foucault's account of scientific knowledge and its entanglement with power. I will first discuss how Foucault's genealogy builds on his archaeology. Next, I will consider the emergence and dissemination of modern "power/knowledge", illustrated by Foucault's account of psychiatry. I then examine the constitution of subjects, objects, and concepts by scientific power/knowledge. This sets up the basis to discuss Foucault's understanding of natural sciences, the normalisation of society, the "carceral archipelago", and the concepts of biopower and governmentality. I conclude by considering what critics have considered as Foucault's "normative confusions". In this chapter and the following chapter, I will draw on both Foucault's published works and his lectures at the Collège de France. These lectures complement his published works and enable a broader understanding of his research and the direction of his thought.

FROM ARCHAEOLOGY TO GENEALOGY

Archaeology and genealogy share an analysis of history emphasising the singularity of events. Rather than a complete innovation, genealogy is a return to and an extension of the approach of the early archaeological works like HM. Now, however, the simultaneous analysis of discursive and non-discursive practices becomes an explicit method. Neither power nor knowledge can be analysed alone, since they condition each other and are mutually generative. However, neither is reducible to, nor identifiable with, the other. We have already seen that Foucault repudiates the clear distinction between science and ideology, a distinction that assumes the possibility of a discourse governed solely by truth, free from power (PK, 102, 118; AK, 203–5). For Foucault, scientific knowledge—and not merely its applications or choice of research questions, but its actual content—cannot exist outside power. By developing the concept of power/knowledge, Foucault can analyse knowledge and power as they transform and destabilise each other, thus challenging the strict divide between them. "Perhaps", Foucault muses, "we should abandon a whole tradition that allows us to imagine that knowledge can exist only where the power relations are suspended and that knowledge can develop only outside its injunctions, its demands and its interests" (DP, 27).

In OT, Foucault restricted himself to describing systems of thought, explicitly bracketing explanation of changes from one *episteme* to another (OT, English foreword, xiv). Genealogy returns to what was bracketed, looking to concrete events and arrangements, to explain thought as not the product of thought alone, but contingently developing from many small accidental causes, empirically accessible though now forgotten. The conditions of possibility under investigation are no longer the underlying rules of discursive formations, but a network consisting of "a heterogeneous ensemble of discourses, institutions, architectural forms, regulatory decisions, laws, administrative measures, scientific statements, philosophical, moral and philanthropic propositions, in short the said as much as the unsaid" (PK, 194). This *dispositif*, or apparatus, provides the conditions of possibility for the emergence of scientific objects. An interactive fine-grained multiplicity of factors, the *dispositif* generates changes beyond intention or awareness. Foucault's nominalism appears in a heightened register in his genealogies where, drawing inspiration from Nietzsche, he connects scientific discourses to mechanisms of power and explains the origin of what we take to be natural in terms of a concatenation of contingent forgotten events, such as the emergence of personages like the doctor, institutions like the asylum, social practices like the examination, objects like the panopticon, and concepts like delinquency. In Foucault's genealogies, we cannot fully separate the existence of an object of knowledge from the various practices through which we encounter and deal with it.

Rather than appealing to an ideal autonomous realm of thought or language, genealogy's "grey, meticulous and patiently documentary" historiographical method holds metaphysics in check with its analyses of the singularity of things, ideas, practices, and experiences in terms of "events" (EW2, 369). Events are unique and always caused by the coincidence of a number of causal series. Events undermine the propositional knowledge for which things fall within categories, so that we are always dealing with the same thing. "Madness" is not always the same thing (BB, 3). What we now call "sexuality" doesn't align with the "pleasures" of the Ancients, or the "flesh" of the Middle Ages (HS1, 3). By seeing phenomena such as the birth of the prison in terms of "events", Foucault is drawing attention to what lies behind the self-evident entities and continuities that form the familiar background to our thought. "It is a matter of shaking this false self-evidence, of demonstrating its precariousness, of making visible, not its arbitrariness but its complex interconnection with a multiplicity of historical processes, many of them of recent date" (EW3, 225). The problem is that we cannot see what to do because we are blinded by the apparent self-evidence of one way of seeing. The self-evidence of our present must be ruptured so that we can see through it to recognise what is intolerable or unacceptable in our present.

By adopting Nietzsche's claim that knowledge is an "invention", Foucault brings it into relation with interests and power (EW3, 6–8). Invention implies motives beyond knowledge for its own sake. For Nietzsche, these were ulterior motives, animal drives to survive and dominate. Such motives taint knowledge, stripping it of its claim to purity. Knowledge imposes order on a chaotic world so as to dominate it (WK, 203–4; EW3, 9).[1] Against this picture, we want to insist that truth and knowledge are precisely things not invented. We don't choose what is true and false, like an inventor chooses what and how to invent. Foucault's talk of "invention" challenges us to abandon the conventional perspective of a subject of knowledge and adopt a more distanced view. From such a perspective, we can see that what appears given and natural is human, all too human. It is not particular truth claims that are invented, but the framework of rules which distinguish truth from falsity. Within that framework, truth is not arbitrarily produced (EW3, 230), and the framework itself is only arbitrary to a certain degree, since the truths which it produces reflect a perspective constrained by the recalcitrant nature of things.

---

[1] The imposition of this order means that knowledge necessarily "simplifies, passes over differences, lumps things together, without any justification in regard to truth. It follows that knowledge is always an overlooking" (EW3, 14). I have adopted Kelly's translation of *"meconnaissance"* as "overlooking" rather than the original translation's "misconstruction", which does not fully capture Foucault's meaning (Kelly 2009, 167 n.20).

There is a reality external to the ordering framework, even if that reality cannot be separated from the framework through which we access it. However, knowledge doesn't mirror the world, nor is it an instinct (EW3, 8). Rather, it is "produced because the instincts meet, fight one another, and at the end of their battles finally reach a compromise" (EW3, 8). Inscribed at the level of instincts, knowledge comes from struggle and is compromised, provisional, and temporary.

This theme of struggle is well illustrated in Foucault's 1973–1974 lecture series Psychiatric Power (hereafter PP), which describes the emergence of scientific knowledge from fields of heterogeneous elements—various actors, institutions, objects—and its temporary stabilisations and ongoing transformations. Foucault charts the struggles that constituted psychiatry, which arose from the difficulty of proving madness through demonstration. These struggles reached a climax in the second half of the nineteenth century with a play of power between patients and doctors (PP, 12, 133). An important breakthrough had come around 1850 with the appearance of neurology, which it was thought could locate the mechanisms of madness in the pathological anatomy of general medicine. Around the same time, the figure of the hysteric appears, offering clearly specified symptoms typical of an organic disease, which however defy being fixed to an organic substratum (PP, 135).

A neurologist needs stable, coded, legible symptoms. Hysterical attacks must be regular and develop according to a typical scenario sufficiently similar to a neurological illness, but also sufficiently distinct (PP, 340). If these conditions can be fulfilled, the psychiatrist can become a real doctor, diagnosing a real illness. And the hysteric can claim the right to be properly ill, and thus entitled to be free from asylum discipline (PP, 311, 316). But because this looks like collusion, a broader pathological framework is needed. Since there were no anatomical lesions to be found, Charcot looked for an event that could be assigned as a cause and developed the concept of "trauma"—a violent event, a blow, a fall, a fear, a spectacle—an event transcribed into the individual's cortex, like an invisible pathological lesion which must be discovered in order to be certain that the hysteric was suffering from a real illness.[2] This required that hysterics recount their lives in order to bring to light this underlying event that persisted and continued to activate their symptoms, demonstrating that they suffered from an illness best treated by neurologists (PP, 308). Here we see medical discourse elaborating a theoretical object in a process made possible by the hospital structure and its techniques of subjection.

---

[2] Jean-Martin Charcot (1825–1893) was a French neurologist and professor of anatomical pathology best known today for his work on hypnosis and hysteria.

In a counter-move against this power of objectification, the hysterics poured out all the details of their sexual lives in words and gestures. They resisted objectification by giving the doctor more than he bargained for, because to demonstrate that hysteria was a genuine illness, doctors needed an account free from excessive sexual elements. The hysterics "finally [got] the better of the neurologists and silenced them" (PP, 322). Foucault's account brings to light the interconnectedness of knowledge, power, and resistance and how a shift in one precipitates a shift in the others, provoking an endless struggle in which victory is only a momentary respite and there are constantly changing alignments of forces and strategies. The struggle between doctors and patients "permeated the whole history of 19th century psychiatry" until it was finally conceded that Charcot had produced the hysteria he described and that the neurological body had eluded psychiatry (PP, 191–2). However, the hysterics' victory was countered by further tactical moves. The task was now not so much to eliminate the psychiatric power which produced hysteria as to apply it more precisely and effectively. The dominant approach, represented by psychoanalysis, sought to give neurotic symptoms full expression, but in a manner disconnected from the effects of power. It was thought that the power relation between patient and doctor could be sufficient to bring forth symptoms while not producing them. Thus, we have the psychoanalytic conventions of the analyst's silence and patient's discursive freedom resulting not in an absence of power but "a power that cannot be caught in any counter-effect, since it has withdrawn entirely into silence and invisibility" (PP, 343).

Foucault's account of psychiatry as struggling with a multiplicity of forces to forge an object of knowledge and power can be extended into the twenty-first century, when we see psychiatry still struggling to establish itself as an authoritative science with agreed objects, concepts, and methods. Anne Harrington recounts how psychiatry's turn to biology in the 1980s usurped the authority of the Freudians who had emerged as its leaders in the years following the Second World War (Harrington 2019). Mental illness became increasingly seen as a combination of genetic, developmental, and environmental factors which come together in a way best understood by advances in brain biochemistry and neuroanatomy. However, Harrington argues, this development was not marked by any new understanding of the biology of mental illnesses, but was rather a vision of psychiatry's identity and destiny (Harrington 2019, xiv). Rather than being bound to any particular biological finding, the role of this call to biology was to support the claim that mental disorders were medical disorders. This is the same sort of struggle that Foucault had described in his account of psychiatry's emergence. As Harrington recounts, in the 1980s, "psychiatrists leveraged their new biological identity to push back against psychologists and social workers seeking professional privileges like prescribing rights" (Harrington 2019, xvii).

By 1988, psychiatry's transformation into a biological discipline seemed complete. It has however, become increasingly clear that it couldn't deliver on its promises and that "it overreached, overpromised, overdiagnosed, overmedicated and compromised its principles" (Harrington 2019, xiv). Harrington cites Thomas Insel, director of the U.S. National Institute of Mental Health, who commented in 2013 that "*all* of psychiatry's diagnostic categories were still based, not on any biological markers of disease, but merely 'on a consensus about clusters of clinical symptoms'" (Harrington, xiv-xv). However, over this same period there was a spectacular expansion of mental disorders recorded in "psychiatry's bible", the *Diagnostic and Statistical Manual of Mental Disorders* (DSM). The 1952 DSM I recorded 106 forms of mental illness. DSM II (1967) recorded 182, DSM III (1980) 285, and DSM IV (1994) 307 types of mental illness. In 1997, an influential polemic against the DSM argued that it was a guide created through struggle and compromise between many disparate insider interests which enabled mental health professionals to be reimbursed by insurance companies and drug companies to sell drugs (Harrington 2019, 268). Harrington's analysis is very Foucauldian. We see a discipline aspiring to the authority of a real science. Without the express or conscious intention of any individual actor, it expands its reach and power by struggles and compromises with other actors such as the public, patients, drug companies, insurance companies, and the law. Like Foucault's account of hysteria, we find that, by their framing within scientific discourses, mental illnesses are not independent of those discourses.

As with his archaeologies, Foucault's genealogical histories are not moved by subjects, either individuals like Charcot or collective subjects like the hysterics. To this non-subjectivism, genealogy adds struggle and combat to destabilise foundations of knowledge by an account of historical beginnings "capable of undoing every infatuation" (EW2, 372). Foucault's ironic derisive critique opposes the confidence of positivist histories of scientific progress, where truth is assumed to be an existing, though unknown, state of affairs towards which knowledge inexorably moves (WK, 208).

THE EMERGENCE AND DISSEMINATION OF MODERN POWER/KNOWLEDGE

Conventionally, "power" is understood as the exercise of intentional acts upon others. Foucault's analysis reveals what this conventional notion misses. Rejecting the notion of power as a homogeneous and unitary thing, an essence capable of vast explanatory reach, he assumes that power does not always take the same form and that different forms of power can both coexist and conflict with each other. He doesn't provide a theory of what power is, but an "analytic" which offers perspicuous examples of power's operation by inquiring into "the how" of "this grand, all-embracing and reifying term" (EW3, 336). He asks how power flows, how it structures social relations and constitutes scientific objects and discourses.

Foucault's historical account traces the emergence of the dominant forms of modern power from the earlier sovereign power (EW3, 336). Sovereign power is held by an individual or collective body and operates negatively by prohibiting or seizing things, bodies, and life according to a binary system of permitted/prohibited (HS1, 136). However, from the early eighteenth century, this sovereign power declined in the face of complex social, political, and economic relations which had developed. According to Foucault, it was complemented and surpassed by a more efficient "disciplinary power" which operates automatically and invisibly (HS1, 88T). This modern disciplinary power is not, like sovereign power, "acquired, seized, or shared, something one holds or slips away" (HS1, 94). Nor is it concentrated in individuals, a class, or a state. Rather than static, oppositional, and repressive, it is dynamic, productive, and omnipresent. Modern power is inherently relational and expansive, always using opposition to increase its own circulation and saturation.

In DP, Foucault's elaboration of disciplinary power takes the panopticon prison as its exemplary manifestation. Whilst Foucault employs the panopticon to illustrate the operation of disciplinary power, he makes it clear that discipline permeates all modern institutions. "Is it surprising that prisons resemble factories, schools, barracks, hospitals, which all resemble prisons?" (DP, 228). The significance of the prison for Foucault is not that it forcefully imprisons criminals. This is simply a continuation of the old sovereign power in ameliorated form. What is crucial is the introduction of constant surveillance aimed at producing docile bodies by compelling prisoners to internalise this surveillance and see themselves as subjects of correction (DP, 70–95). Discipline reverses the spectacle of sovereign power by rendering the subject constantly visible. It thus enables prompt responses to any deviation from the normal by the application of continuous corrective pressures. It operates behind the scene, mediated by alignments of a heterogeneous field of elements. In the case of the panopticon, a range of elements—the architectural layout, various actors, documents, tools, instruments, skills, practices, and rituals—interact to maintain continuous disciplinary surveillance. None of these elements or relations between them are completely stable (HS1, 94). However, by operating against a background assumption that other social actors act in more or less predictable ways, power unwittingly forms a circuit which relays anonymous power that has originated elsewhere.[3] Power is reproduced by the re-enactment of these alignments over time. Genealogy reveals the construction, maintenance, and reproduction of these heterogeneous relations upon which power depends. It charts the development of knowledges which have not loosened themselves from the grip of power, but instead have become increasingly locked into its large-scale organisation, in what Foucault calls "power/knowledge" by which potentially all domains are opened to scientific investigation and control (see EW2, 388).

The disciplined body, Foucault tells us, has become an essential component in the operation of a general mechanism of power/knowledge in modern society (DP, 26). Although it is the focus of the most minute scale of local practices, it also remains tied to large-scale organisations of power (DP, 25). Larger-scale practices such as state or class domination still operate, but they are only the effects of a more pervasive power comprising multiple fine-grained "capillary interventions" (HS1 85, 95; PK, 99). Power therefore needs to be analysed in an ascending manner, starting with the social institutions where disciplinary mechanisms arise which later congeal into larger structures. Unlike Marxism, Foucault's genealogy does not dismiss racial, gender, or cultural politics as inconsequential in relation to larger-scale manifestations of power like class struggle. Foucault thus analyses scientific knowledge as inextricably entangled with power by examining interactions at the local level, where a web of multiple force relations constitute a "micro-physics" of power (DP, 26). Since power emerges from multiple interactions—supporting, cancelling, amplifying, undermining, relaying, extending etc.—it cannot be analysed in terms of a binary opposition between rulers and ruled (HS1, 92).

Disciplinary power not only constrains but also produces and enhances knowledge, health, wealth, and efficiency. The individual herself is a product of power, not just raw material which pre-exists the power that moulds her (DP, 170). Discipline operates by a "subtle coercion" of the individual body—its movements, attitudes, gestures—as an object to be manipulated. It is meticulous, exhaustive, all-pervasive, and continuously exercised, constituting bodies as docile and useful by resolving resistances and forming and refining skills and aptitudes (DP, 138). With bodies separated or aggregated, conceptually or concretely, subjects are constituted through practices of both discursive objectification and bodily manipulation (DP, 141–62). As an example, we could consider the physical separation and partitioning within schools, prisons, and workplaces and how these enable an increase in the knowledge and control of groups and individuals, which, in turn, provides a

---

[3] For example, an employer writing a reference is not simply the holder of power exercised over the employee. Her action is an exercise of power only because of its connection to a vast range of disparate elements, for example, the ongoing maintenance of the institution of referencing and recognition of the authority of her position. These unstable social alignments can be reversed.

basis for further knowledge and a still further refinement of spatial controls. Material subjection enables theoretical objectification and the emergence of human sciences like criminology, pedagogy, and psychiatry. These sciences further facilitate the refinement of disciplinary technologies which mark bodies with qualities that make them distinct individuals who appear in a grid of relations, functions, and ranks in a networked segmented space. These disciplines render individuals' characteristics visible by slotting them into categories such that individuals appear endowed with certain types of self-evident essential properties. Through this "assignment of subjective positions", individuals are allotted roles in the social world which provide a range of possibilities in specific arenas of subjection to, and exercise of, power (AK, 107).

Foucault attributes the success of disciplinary power to three interacting instruments—hierarchical observation, normalising judgement, and the examination. These are all exercises of both power and knowledge which have no separate concrete existence. Power relations are not outside, but immanent to, not only knowledge but also relationships such as economic or sexual relationships (HS1, 94; STP, 2). However, these relations do not simply reduce to power and power cannot fully explain them. The panopticon prison, Foucault's well-known model of hierarchical observation, illustrates the economy of disciplinary power (DP, 202–5). Observational power is directed towards the inmates, those unavoidably visible objects, for whom this objectifying power is invisible. This invisibility arises not from ideology or false beliefs, but from the spatial configuration of the panopticon and disciplinary power's imperative of efficiency (DP, 187). The observational gaze operates as a network in which every individual is at the same time both object and subject of power (DP, 176–7). The prisoner is watched but because he is never sure when he is being watched, he never ceases to watch himself.[4] The warder is the watcher and yet he too is watched. Although the prison governor is at the apex of the prison's hierarchical structure, "it is the apparatus as a whole that produces power and distributes individuals in this permanent and continuous field". On one hand, power "is everywhere and always alert" and on the other hand, it is "absolutely discrete, for it functions . . . in silence" (DP, 177). In its omnipresence, power becomes invisible.

Judgement directs observation to what is relevant, just as what is observed directs judgement. Judgement "measures in quantitative terms and hierarchises in terms of value the abilities, the level, the 'nature' of individuals". This judgement measures incremental deviations from a norm, automatically introducing

---

[4] The prisoner is a subject in two senses: "There are two meanings of the word 'subject': subject to someone else by control and dependence, and subject tied to his own identity by a conscience or self-knowledge" (EW3, 331).

"the constraint of a conformity that must be achieved" (DP, 183). The slightest deviation from the norm is corrected (DP, 177–84). For Foucault, "normality" emerged as a fundamental category of behaviour and identity in the nineteenth century. At the same time, "abnormality" emerged and became increasingly associated with degeneracy, as we will see. While normalisation imposes homogeneity, it individualises by introducing all the shadings of individual differences (DP, 184). The techniques of observation and judgement combine to form the examination, an exercise of power over individuals and an experiment to find their truth. It came to take the forms of "tests, interviews, interrogations and consultations" that reproduced and extended power/knowledge relations throughout all disciplinary institutions by tying the power of subjection to the knowledge of individuals as objects (DP, 184–5, 226–7). The examination enabled people to be classified, assigned, and used more efficiently. It produced new properties of people, signs of these properties, and knowledge of who possessed them—for example, low IQ or a history of unemployment. By recording and filing information, individuals could be tracked, their results feeding into subsequent examinations as well as being aggregated at population level to establish or further refine norms. The examination gave rise to the human sciences (EW3, 59).

Although disciplinary power arose in isolated institutional islands—prisons, factories, schools, barracks, hospitals—it was disseminated by "a sort of general parasitic interference with society . . . a disciplinary society replacing a society of sovereignty" (PP, 66). Although many practices of sovereign power remained in place, they now functioned through the social cohesion provided by disciplines operating at a different location and scale. Foucault's 1974–1975 lecture series Abnormal (hereafter AB) illustrates this dissemination in its discussion of the diffusion of psychiatry beyond the asylum. Without resting on any single institution, psychiatric normalisation extended itself by interactions between institutions—church, state, schools, prisons, law, and medicine. Psychiatry underwent mutations and spawned other sciences. These normalising disciplines came to permeate the entire social body (PP, 86). We can see how the conceptual and normative structures embodied in these disciplines have permeated modern cultures. Today, psychologistic categories constitute subjects and objects to form the barely recognised background of interpretations of selves and others.

Foucault explores this diffusion through the genealogies of the figures of the "human monster" and the "little masturbator", produced by particular configurations of power/knowledge. The principle of the monster's emergence is found in shifts in the power to punish, which from the nineteenth century on was not applied merely according to the criminal act, but increasingly according to the criminal's nature. This new disciplinary logic required the scientific study of human beings and the stipulation of the modern legal subject, a universal subject endowed with the capacity for autonomy and freedom, but also the capacity to be trained and corrected. Human beings both discover their own nature and manipulate it to achieve their potential. By recourse to this universal subject, the development of scientific knowledge about humans became possible. According to Foucault, legal procedures—witness accounts, investigations, examinations—produced the truth regime of the sciences represented in a courtroom by sanctioning types of truthful discourse to justify penal practices. The interaction of police, judiciary, and psychiatrists determined the rules for the formulation of valid propositions which generated the discipline of psychiatry.

To be constituted as a proper science concerned with the protection of society, psychiatry needed to show that it could detect danger not apparent to others. Thus drawn to the problem of criminal insanity, psychiatrists argued that motiveless crime was impelled by an irresistible inner force—instinct—that could break through all barriers of morality and reason, even the desire to live (AB, 119–21). The concept of instinct becomes the vector of abnormality, on the basis of which psychiatry not only brings serious disorders, but also minor irregularities, within its ambit as mental illness (AB, 132). This expansion is assisted by a field of new symptoms, consisting of deviations from a range of norms situated on the axis of voluntary/involuntary. With all conduct located on this axis, ultimately "there is nothing in human conduct that cannot be questioned by psychiatry" (AB, 160). Psychiatry could be linked to organic and functional medicine through an understanding of neurological disturbances of voluntary conduct. This meant that psychiatry was able to exploit the norm in the sense of a principle of "functional regularity . . . as opposed to the pathological . . ." However, the explosion of symptoms included a growing array of conduct which deviated from the norm, now understood as "a rule of conduct, informal law, principle of conformity as opposed to disorder, strangeness, eccentricity". These two uses of "norm" are conflated, mutually adapted, and partially superimposed (AB, 162). Thoroughly underpinned by this interplay, psychiatry treated the most common everyday conduct as both deviation from a rule of conduct and pathological dysfunction. Supported by this play of meanings, psychiatry became medico-judicial in its nature and everyday practice.

The lineage of Foucault's other figure, the "little masturbator", can be traced back to medicine's taking over control of sexuality from the Church. From the mid-eighteenth century, a great anxiety had arisen about child masturbation which, like sexuality generally, was increasingly seen to be within the province of medicine. By the late nineteenth century, sexuality is seen as the root cause of all forms of abnormality and, in its intrusion into the family, psychiatry takes up the figure of the little masturbator. The two figures—the human monster and the little masturbator—now merge to become the modern character of the abnormal individual. Psychiatry is able to expand, to intervene in both the family and the judicial system by unifying instinct and sexuality. It develops discourses, methods, analyses, concepts, and theories which combine sexual instinct not only with illness but with an entire spectrum of abnormality, from minor irregularities within the family to monstrous crimes (AB, 277). Here we see modern scientific power/knowledge, operating autonomously and independently of the intentions of actors, embedding, maintaining, and expanding itself in an ongoing feedback loop. "Modern power fashions, observes, knows and multiplies itself on the basis of its own effects" (AB, 48). Psychology, psychiatry, and criminology were originally located within disciplinary institutions such as hospitals, prisons, or asylums which needed more effective discourses, knowledge, and practices. Such disciplines developed as specialisations with their own methods and standards of evidence, not simply from the progress of reasoned discovery but in close association with disciplinary technologies. Foucault argues that, as they swarmed out of these institutional islands to invade the entire social body, they retained the link to their origins as disciplinary technologies directed towards producing docile, useful bodies (DP, 305).

In the eighteenth century, a threshold was crossed after which "the formation of knowledge and the increase of power regularly reinforce one another in a circular process" (DP, 224). By their tightened relations, power and scientific knowledge were now mutually constituted. Along with new forms of social control, new sciences emerge with new objects like the homosexual or the delinquent (DP, 138). More extensive finer-grained knowledge enabled a more continuous and pervasive exercise of control in a sort of political-epistemic homeostasis around a norm. The continuous expansion of the human sciences was enabled by the clinical sciences of the individual, which provided a vast compilation of documentation to further refine disciplinary techniques. By capturing individuality in documentation, the examination homogenises individual features into codes. Anticipating his concept of "biopower", Foucault discusses the constitution of populations by comparative systems that measure "overall phenomena, the description of groups, the characterisation of collective facts, the calculation of gaps between individuals, the distribution in a given 'population'" (DP, 190). Documentation becomes a resource for ongoing and more precise examinations, enabling the emergence of new discourses. Elements can now be correlated and organised "into comparative fields making it possible to classify, form categories, determine averages and fix norms" (DP, 190). The examination constitutes the individual as describable and analysable, a constant object of a permanent body of knowledge, "a reality fabricated by this specific technology of power" constituting individuals as "cases" which are objects of both knowledge and power (DP, 194). Human sciences produce not only theoretical objects, but also desires, forms of experience, and kinds of bodies. Talk of inner motivations, personalities, predispositions, and desires led to an expanded and more intense subjective life. Individuals began to consider themselves as objects of attentive observation.

THE CONSTITUTION OF THE SUBJECT

So far we have considered Foucault's account of modern power/knowledge, along with genealogies charting the emergence and dissemination of specific sciences from an agonistic field of shifting relations between actors, objects, concepts, and institutions. I will now examine how power/knowledge constitutes subjects, before turning to the constitution of concepts such as "sex" and "sexuality". Foucault describes DP as a "history of the modern soul", this soul deriving "its bases, justifications, and rules" from the "scientifico-legal complex" (DP, 23). By "soul", Foucault is not suggesting a metaphysical illusion, but something very real, a historical reality corresponding to "a certain technology of power over the body" (DP, 29). While the term "soul" usually connotes that which is inalienable and authentic, freed from the distorting influences of power, Foucault insists that this soul is already constituted by power, outside of which there is no subject. The modern soul is produced by the functioning of power/knowledge that observes, supervises, arranges, judges, and corrects. It is an effect of a certain type of power and the reference of scientific knowledge. However, although this knowledge is not a mere illusion, neither is it the discovery of a "real" essence.

Foucault doesn't think that human progress consists in the gradual emancipation from oppressive power of "natural" qualities of rationality or autonomy. While Kant thought that autonomy required acceptance of transcendentally revealed limits on reason, Foucault sees such autonomy as contingently produced by discipline, in other words, given by "second nature". As an object of disciplinary sciences, human beings appear as beings able to both discover nature in themselves and manipulate it in accordance with what they see as their potential. This reflects the anthropological dualism discussed in OT—positive knowledge (discovered by the human sciences) and eschatological knowledge (which urges manipulation)—which is intrinsic to discipline. It simultaneously involves positive knowledge of natural finitude and normative knowledge of a promise to be fulfilled. On one hand, a body is disciplined to make it obedient and useful. On the other hand, this same subject appears to be free. Thus "the 'Enlightenment' which discovered the liberties, also invented the disciplines" (DP, 222). The modern subject is a historical achievement, produced through power/knowledge that examines and judges it against norms which introduce "all the shading of individual differences" (DP, 184). The modern individual is the effect and object of an overlapping of power and knowledge, by strategies of discipline backed by the human sciences.

The delinquent is one such effect of power and object of knowledge. Foucault argues that around the time the prison emerged, illegalities had to do with conflicts between social classes. By constituting the concept of a criminal underclass comprising easily identified and controlled "delinquents", these conflicts could be managed. According to Foucault, it is not science which first discovers delinquency, which the prison then addresses. Prison, supported by human sciences, creates delinquents, picking out and isolating them physically and conceptually as the "abnormal" elements of society. In a field of concepts, institutions, and objects, delinquents are constituted as something to be managed. The prison persists because "in fabricating delinquency, it gave to criminal justice a unitary field of objects, authenticated by the 'sciences', thus enabling it to function on a general horizon of truth" (DP, 256).

One way to grasp this notion of "fabricating" subjects is by considering what Hacking calls "interactive kinds" (Hacking 1999, 103–109). Unlike natural kinds, the concepts of which can be applied retrospectively to situations occurring prior to the emergence of those concepts, social objects seem to emerge along with their concepts. Hacking, as an example, argues that although women have long been fleeing their homelands, it may not be right to always call them "women refugees" because this kind of person has only recently emerged from a social matrix comprising institutions such as nations, concepts such as rights, objects such as borders, and the actions of advocates, judges, and politicians. A woman fleeing her homeland is not a refugee until she falls within the appropriate categories, which involves appropriate interactions with objects, actors, and institutions. Categorisation derived from scientific knowledge is then applied to bodies, moulding them to norms and changing the way people are seen and see themselves. Subjects of such categorisation may, perhaps unconsciously, modify behaviour to conform to the expected stereotype, or disguise or exaggerate actions like Charcot's hysterics did. Changes in behaviour may then require a refinement of the original categories. This is what Hacking calls the "looping effects" of interactive kinds, the kinds that the human sciences take as their objects.

Elgin criticises Hacking's historical nominalism, claiming that we are correct to say that the kinds and objects of human sciences, just like those of the natural sciences, existed prior to the schemata of categorisation by which we now recognise them (Elgin, chapter 10). Women refugees existed long before their description, just as dinosaurs existed before theirs. Against Elgin, we might say that how we are recognised by others and recognise ourselves is in part constitutive of what we are. We are always to some extent what we, and others, recognise us to be. In relation not just to human kinds, but to social reality more generally, this sort of looping relation is most clearly apparent in money, where a structure of recognition built into practices and institutions is constitutive of certain bits of paper, metal, and electromagnetic patterns being money.[5] In relation to kinds of people, what we accept as "real" or "natural" seems to be a matter of degree which ultimately rests on how firmly certain kinds are embedded in a network of social practices. We may say that Charcot's hysterics did not exist until he produced them, but do we want to say that those attracted to members of the same sex were not homosexuals before the category was invented?

It is in the spirit of this problematic that Foucault urges us to discover how "subjects are gradually, progressively, really, and materially constituted through a multiplicity of organisms, forces, energies, materials, desires, thoughts etc." (PK 97). Provocatively, he claims it was only in the nineteenth century that "the homosexual [became] a species" (HS1, 43). Before this, power/knowledge had not aligned the contingent configuration of "organisms, forces, energies, materials, desires, thoughts etc." required for the homosexual to exist as today. Arguably, it is not only the homosexual or delinquent that has emerged. Hacking thinks that who we are is not just what we do or have done, but also depends on the possibilities of what we *could* have done (Hacking 2002, 107). If it wasn't possible to be a delinquent or homosexual before the nineteenth century, it also wasn't possible to be a non-delinquent or heterosexual. Such unregistered shifts in what we are have been generated by the increase and spread of discipline which reinforces itself by structuring situations as occasions for more discipline. A well-disciplined individual absorbs not only the specific capacities of training but discipline as such. The population submits to normalisation by various disciplinary techniques within schools, factories, hospitals, or offices. However, we are not entirely trapped within a power regime. A "reverse" discourse can revalue things by insisting that what has been ruled abnormal is normal, or vice versa. Discourses are "tactically polyvalent" and able to be used for different purposes, including resistance (HS1, 100). We will see

---

[5] As Stretton points out, beliefs about economic and social systems are working parts of these systems (Stretton 1999). The "looping" characteristics of social kinds is similar to Gunnar Myrdal's theory of circular and cumulative causation which argues that changes in one form of an institution will lead to successive changes in other institutions. These changes are both circular in that they continue in a cycle, and cumulative (Myrdal 1944; see also Argyrous 2011).

Sexuality and Sex

Scientific discourses constitute not only subjects and objects but also concepts like "sexuality" and "sex". In HS1, Foucault maintains that these concepts are not what the human sciences assume them to be. His point of departure is the so-called repressive hypothesis, the idea that beginning in the eighteenth century, sexuality has been opposed, silenced, restricted, and regulated, a situation we must liberate ourselves from. Contrary to this view, Foucault argues that there has been an explosion of discourse on sex since the seventeenth century, beginning with the Counter-Reformation's prescriptions about confession. Practices of self-examination applied increasingly finer distinctions, leading to a self not discovered but constructed by the categories applied. Instead of repression, there is an increasing incitement to discourse (HS1, 34). This discourse posited sex as a powerful and irrational force that required the techniques of confession to control. Confession became generalised throughout modern society, particularly in psychiatric and psychological disciplines. What started as the ritual of confession under church power eventually was reconstituted in scientific terms.

Sexuality, based on private sensations of pleasure, secret fantasies, and dangerous excesses, developed from the proliferation of scientific discourses produced and harnessed by power from the beginning of the eighteenth century. As an interface between concepts, sexuality constitutes a network interconnecting various styles and perversions into a single grid of scientific power/knowledge (HS1, 44, 103). On Foucault's account, any attempt to liberate our "true" sexuality is to simply fall under the sway of another strategy of power. Mechanisms of power produce, through scientific discourses, the ways in which we experience and conceive sexuality. Any form of sexuality, whether "repressed" or "liberated", involves internalising categories and norms provided by power/knowledge. The belief that one is resisting repression only supports domination by concealing the workings of another face of power.

Foucault considers an objection that focuses on the ontological status of sex, that his history of sexuality or what can be said about sex is irrelevant to the biological reality of sex itself (HS1, 150). By ignoring this reality, it is objected, Foucault has not come to grips with the real referent of the discourses he analyses, a precultural embodied given that exists independently of any discourse about it (HS1, 151). Foucault responds by suggesting that rather than being something in reality that anchors the discourses of sexuality, sex is "a complex idea formed inside the deployment of sexuality" (HS1, 152). As the technology of sexuality developed, "sex" came to be posited as a thing in its own right which is "other than bodies, organs, somatic localisations, functions, anatomo-physiological systems, sensations and pleasures" (HS1, 152–3). Though referring to different entities and playing different roles in different domains, the concept of "sex" established an "artificial unity" by lumping heterogeneous elements together. This "fictitious unity" enabled "sex" to serve as "a causal principle, an omnipresent meaning, a secret to be discovered everywhere" (HS1, 154). Foucault is suggesting that sex, an apparently empirical concept, lacks clear empirical criteria yet operates as an explanatory principle for diverse phenomena. He is not denying the existence of bodies, organs, functions, pleasures, sensations, and so on, but rather is arguing that the concept "sex" doesn't pick out a determinate object but rather a number of objects which are conflated and distorted to manufacture a pseudo-object like the hysteric's "trauma". Rather than being some real biological thing, sex is "the most speculative, most ideal, and most internal element in a deployment of sexuality organised by power" (HS1, 155). Sex serves to link the knowledge of sexuality—in sexology, psychiatry, psychoanalysis, and psychology—to the biological sciences and, without making any real use of these sciences, grants sexuality "a guarantee of quasi-scientificity" through this proximity (HS1, 155). Our self-understanding is now bound up with "what was [once] perceived as an obscure nameless urge, requiring us to expend enormous effort and employ great care in coming to know it" (HS1, 156).

Foucault's nominalism and anti-essentialism is in play here. He is particularly resistant to any move to naturalise human nature, to dictate some essential biological determinant of what we are. If a concept such as "sex" provides an authoritative scientific foundation of our identities and desires, then it may serve to legitimate the control of behaviours on the axis of normal/abnormal by their containment, control, and channelling. While power operates through concrete biological bodies unaware of its operation, Foucault doesn't simply accept biological explanations of human behaviours (EW2, 380). His view of the malleability of what might be regarded as biological—sexuality, insanity, or criminality—reflects Canguilhem's idea that biological normality is a socially posited value (Canguilhem 1991, 158–163).[6] Foucault viewed many biological norms as representing not an objective standard, but a desired state of affairs which is irreducible to objective scientific concepts. Social and

---

[6] Thus, Canguilhem argues that, in the human species, height is "a phenomenon inseparably biological and social". Similarly, "the average life span is not the biologically normal, but in a sense the socially normative, life span . . . . This would be clearer still if, instead of considering the average life span in a national society taken as a whole, we broke this society down into classes, occupations, etc." (Canguilhem 1991, 159, 161).

biological life are inextricably bound together. When certain traits come to be identified as "normal", in the sense of what is common or average, this fact can then be adopted as a norm with prescriptive force, which reinforces the trait. One only needs to consider the range of interactions between facts and norms surrounding notions of "fitness" to realise the social malleability of "biological" norms. In a move anticipating Latour's talk of hybrids, Foucault blurs any hard distinction between culture and nature (Latour 1993).

While normative judgements about health are "subjective", they are not mere whims. A doctor may correctly tell a patient that she is sick based on physiological data, even when she feels healthy. But this is only because science has uncovered causal connections between the current "healthy" physiological state and one that any patient would call "unhealthy". This apparently objective diagnosis results from uniformity in judgements about organic illnesses. Everyone feels the pain of appendicitis. No one wants to die. However, the objective nature of such judgements can slide into judgements on which there need be no uniformity, for example, judgements about gender orientation. Here we see the norm, derived from what is frequent, become valued as a standard which must, under certain historical and social conditions, be enforced and maintained. Yet gender orientation is more deeply subjective, more part of the essential identity of an individual than an organic condition like appendicitis, which can be regarded as an accident. Foucault accounts for our most intimate and personal experiences as conditioned by historical events of which we are only dimly aware. He is concerned with the conditioning of these subjective identities which directly shapes what an individual appears to be "by nature", an appearance which obscures the "second nature" of normalisation that has produced her.

It is not, however, that Foucault imagines that we are unconstrained by biological determinations. Rather, he is cautious of the authority of science which hypostatises its objects as already given. This is particularly problematic for the human sciences, in which humans are the objects of knowledge of themselves as knowing subject-objects. Given what I have discussed as the looping effects of interactive kinds in relation to power/knowledge, Foucault is right to be wary of any determinate concept suggested by an epistemology that naively assumes that nature, including our own, simply imprints itself on our receptive souls. What Foucault's genealogies bring into view is that our epistemic activity is a social activity shot through with power relations and that, if these relations were to shift, knowledge would realign towards differently constituted objects.[7] To see this is to realise that things could be different in ways we never expected.

---

[7] Allan Young provides an example of shifting power relations and the social nature of biological norms in his discusssion of Post Traumatic Stress Disorder (PTSD) (Young, 1997). PTSD emerged

THE NATURAL SCIENCES

Foucault's analyses explicitly focused upon sciences such as psychology, sociology, and other disciplines whose status as science was contested. In a 1977 interview, he was cautious about the possibility of a similar approach to the natural sciences. "If concerning a science like theoretical physical or organic chemistry, one poses the problem of its relations with the political and economic structures of society, isn't one posing an excessively complicated question?" (PK, 109). Notwithstanding this reticence, Foucault tells us that, while the human sciences arose from the disciplinary technologies of the nineteenth century, the natural sciences arose from an earlier technology of power, the inquiry (DP 226). The inquiry had been a particular practice of judicial power which emerged in the shift to juridical law. It required the establishment of the objective facts of "exactly who did what, under what conditions, and at what moment" (EW3, 5). While the examination was organised around the norm, the inquiry was directed at what happened through the evidence of those who should know. It migrated into areas not directly connected to the exercise of power, the sciences of observation—geography, botany, zoology, astronomy, and medical sciences (EW3, 4; DP, 225).

While Foucault sees the human sciences as still tied to the power inherent in their institutional origins, he thinks that the natural sciences have largely freed themselves from the power effects of the inquiry (PK, 109). "In becoming a technique for the empirical sciences, the [inquiry] has detached itself from the inquisitorial procedure, in which it was historically rooted". In contrast, "The examination has remained extremely close to the disciplinary power that shaped it. It has always been an intrinsic part of the disciplines" (DP, 226). However, the inquiry is not as benign as Foucault's brief comments here suggest. In PP he tells us that the inquiry's increasingly fine grid of scientific techniques represents a "tightening grip . . . in our kind of society . . . a police kind of investigation into people's behaviour . . . " (PP, 246). This colonisation extends both in depth, as it penetrates the bodies, actions, and thoughts of individuals, and in breadth as "a planetary extension to the entire surface of the globe" (PP, 246). Although this limitless expansion of

---

in the 1980s out of the experience of Vietnam veterans. It was first listed as a mental disorder in DSM III, published in 1980. To use Hacking's term it is "looping", since the diagnosis feeds back into the experience. Its social and scientific construction serves to provide a theory of memory, unify disparate symptoms, confer benefits and status on sufferers, and power on doctors or psychiatrists. Young argues that the definition of PTSD has continually been broadened. Initially, the diagnosis derived from the concrete experiences of men who fought in Vietnam; however, now it has become a universal category applicable without reference to an actual person's biography, but based on symptoms separated from individuals and then aggregated to create a statistical composite reality (Young 1997).

science is based on universal access to the techniques of inquiry, Foucault tells us that the truths of these sciences are "always deep down, buried, and difficult to reach" (PP, 247). In mentioning this "buried and difficult to reach" nature, Foucault is raising concerns that we saw in Habermas relating to the increasing specialisation of science and its remoteness from the lifeworld. Universities, learned societies, schools, and laboratories are "all ways of organising the rarity of those who can have access to a truth that science posits as universal". However, individuals only have an abstract right to be a universal subject, since concretely, such organisations "will necessarily entail rare individuals to perform the function of universal subject" (PP, 247). This "organisation of rarity" is also what Shapin and Schaffer were describing in their account of Boyle's establishment of a society of gentlemen scientists whose gentlemanly codes of civility could underwrite the veracity of experimental findings. In showing how seventeenth-century natural science relied on a conception of universal assent that was in fact limited to a small class of gentlemen, Shapin and Schaffer, like Foucault, link the question of knowledge to the question of the social order (Shapin and Schaffer 1985).

In sketching a separate genealogy of the natural sciences, Foucault is not positing a rigid division between the human and natural sciences. Natural sciences are not discrete, but linked to other knowledges (PP, 246–47; SD, 179). As we saw in chapter 5, the human sciences are located precariously between the three branches of the "epistemological trihedron" constituted by philosophy, mathematics, and the natural sciences (OT, 378). They rely on the natural sciences for their models and guiding concepts. As we will see, at the broadest level of analysis, an interactive web of power/knowledge consisting of all sciences locks together to constitute modern society as "the carceral archipelago" (DP, 297).

It is plausible that Foucault's constructivism remains operative for the natural sciences, although not at the level of the social power which shapes the human sciences. Rather, Foucault would regard natural science as "constructed" in a sense analogous to Habermas's early "deep-seated human interests". He would see the objects and kinds of natural science as not part of a world "in itself" viewed from a perspective outside that world, but relative to a classificatory order that picks out certain similarities and differences as significant for the species and overlooks others. By claiming that the natural sciences have detached themselves from the particular social powers and interests of their origins, he grants the natural sciences their objectivity, while still being able to deny the possibility of a God's eye view and thus allowing room for a critique of scientific naturalism.

The absence of sharp distinctions between the various sciences raises the question of whether an analysis in terms of Foucauldian power/knowledge can be extended to the natural sciences. Joseph Rouse argues it can, drawing an analogy between laboratories and Foucault's disciplinary establishments—hospitals, asylums, schools, prisons, and so forth (Rouse 1987, 1993). Like these disciplinary establishments, laboratories are "microworlds" which isolate what is being investigated from unaccounted causal interactions. Their partitioned spaces allow the objects of scientific research to stand out clearly from the background of surrounding objects and events. They enable the separation and mixing of materials to produce new forms by employing surveillance and tracking of processes, documentation, and standardisation of basic terms, all analogous to various tactics of power that figure in Foucault's accounts of disciplinary power. By introducing order into the phenomena they study, natural scientific practices induce the production of signs. In a manner analogous to Foucault's discussion of the confessional, the objects of investigation are subjected to manipulations aimed, not at changing them, but at revealing their hidden natures.

Analogous to the "swarming" of the disciplines, new technologies emerge from their laboratory microworlds to proliferate throughout society, which must be made amenable to scientific innovations. Laboratory-like practices require adaptations to enable the new scientific technologies to function in environments outside the laboratory. Consider, for example, the plethora of adaptations induced by the creation of sustained electrical currents reticulated throughout urban areas, or the transmission, reception, and processing of electromagnetic radiation. Rouse suggests two characteristic features resulting from the emergence of scientific technologies from their laboratory microworlds into the social and natural environment. First, interactive complexity increases due to the variety of nonlinear connections required between technical systems and their controlling organisation. Second, "tight coupling" or lack of flexibility in connections between technical and organisational subsystems is induced. Thus, as new technologies adapt to the more complex environment outside the laboratory, the natural environment is simplified, controlled, and stripped of some of its own capacity for self-regulation and buffering. As an example, Rouse cites the agricultural "green revolution" which increased yields but only at the cost of removing natural barriers to failure like genetic diversity and requiring energy-intensive and capital-intensive inputs to maintain the necessary environmental conditions.

The analogy between disciplinary establishments and laboratories suggests the existence of deep-seated interests (to use Habermas's formulation) accompanied by exercises of productive power (in Foucault's sense). Science exercises power over nature in order to construct knowledge to address our interest in prediction and control over nature's vicissitudes. However, this disciplinary power not only produces knowledge which, in ongoing feedback loops, further refines discipline. By its naturalistic vocabulary and physicalist ontology, scientific naturalism disciplines how we speak and think of norms, values, reasons, minds and so forth. Foucault generally restricts himself to what he sees as the main danger, the discourses of the human sciences. However, his account of power/knowledge leaves room for a critique of scientific naturalism's hegemonic disciplinary power.

THE NORMALISATION OF SOCIETY

Indispensable to Foucault's reconceptualisation of modern power is the fusion between disciplinary practices, scientific knowledges, and new social structures. The power that emerged in the early nineteenth century not only produced subjects, objects, and concepts but also entire bodies of scientific know-how that permeated society to restructure it under the banner of technological progress. In SD, Foucault sketches a genealogical account of the institutionalisation of technological knowledge, which he describes as a significant normative project, "one of the great instruments of power at the end of the classical age" (DP, 184). The emergence of technological knowledge involved a battle "not between knowledge and ignorance", but between "knowledges in the plural . . . defined by geographical regions, by the size of the workshops or factories and so on" (SD, 179). They struggled against each other in a society where knowing the secrets of technology was a source of wealth.

In this context of struggle, there developed processes that allowed "bigger, more general, or more industrialised knowledges, or knowledges that circulated more easily, to annex, confiscate, and take over smaller, more particular, more local, and more artisanal knowledges" (SD, 179). The state weighed in, introducing norms enabling knowledges to be communicated and compared. A hierarchical classification allowed knowledges to be fitted together, from the most general to the most particular, and centralisation allowed surviving knowledges to be selected, then controlled, transmitted, or eliminated. Accompanying this normalising hierarchisation and centralisation, were a series of practices, projects, and institutions—schools, curricula, standards, inspectors, journals, institutes, and so forth. Disciplinary power was applied to knowledges that became modern technological disciplines, each with its own internal organisation and criteria of demarcation (SD, 181). Knowledges formed disciplines with clear domains of objects and methods, organised into hierarchies within an overarching discipline called "science". Philosophy lost its foundational role and science, understood as a general domain, took over the task of policing knowledges.

From the eighteenth century on, anything born or shaped outside the university had to prove itself on this institution's terms. The amateur scholar ceased to exist. The tighter relationship between power and knowledge gave rise to a new constraint, "no longer the constraint of truth, but the constraint of science" (SD, 185). While statements were previously excluded or included on the basis of their content, "their conformity or nonconformity to a certain truth", after the establishment of the modern disciplines, control is exercised over "the regularity of enunciations . . . who is speaking, are they qualified to speak, at what level is the statement situated, and to what extent does it conform to other forms and typologies of knowledge?" (SD, 184). The format of scientific reports became standardised and specialised vocabularies were adopted. Claims were judged in terms of the speaker's or journal's scientific authority in a particular area of science. The establishment of these regulatory structures enabled science to build knowledge without having to continually return to basic questions.

By portraying a battle between multiple knowledges, Foucault's account brings the construction of technological norms, saturated with power/knowledge, into relation with other norms. This recalls Canguilhem, who notes that, considered within a broader normative project, such technological norms appear relative to each other within an entire social organisation (Canguilhem 1991, 249).[8] We could say that the formation, operation, and maintenance of scientific and technological norms is part of a systematic whole which does not appear in the conscious deliberations of subjects. Rather than the outcome of either choice or coercion, this technological "normalisation" of society involves the internalisation of what becomes a naturalistic background developing according to its own autonomous logic. Although not mentioned by Foucault specifically, this process of scientific-technological normalisation can be seen as nested within his analysis of modernity as the "carceral archipelago" (DP, 293). Industrial technological normalisation is not distinct from the social normalisation which Foucault calls discipline or, which we will shortly discuss, biopower. Techniques of industrial normalisation, such as the Fordist production line, presuppose reciprocal techniques of normalisation of workers, who must internalise the norms required to render them docile and useful. Within interlocking patterns of circular and cumulative causation, the cars that roll off production lines make the suburbs possible, which in turn make the cars necessary. Analysing this opacity of inextricably entwined technological and social norms, Canguilhem drew the conclusion that it is questionable whether "any society whatsoever is capable of both clear sightedness in determining its purposes and efficiency in utilising its ends". The conclusion he came to, in a

---

[8] For example, the standard 1.44 metres rail gauge established a norm reflecting "the best compromise sought among several initially conflicting requirements related to mechanics, fuel, trade, the military and politics", ultimately reflecting "an idea of society and its hierarchy of values" (Canguilhem 1991, 238, 247).

passage reminiscent of Habermas's developmental communicative account, is that the fact that "one of the tasks of the entire social organisation consists in its informing itself as to its possible purposes—with the exception of archaic and so-called primitive societies where purpose is furnished by tradition . . .—seems to show clearly that, strictly speaking, it has no intrinsic finality" (Canguilhem 1991, 252). We will see this theme of ongoing transformation and open-endedness come to the fore in Foucault's final period of work, discussed in the following chapter.

Foucault's argument that human bodily capacities are *produced* by disciplinary power challenges the conception of modernity which sees labour-power as simply given by nature and then taken up by capitalism's various mechanisms. If labour-power or any human capacity is not simply given, we must account for its production. The "carceral archipelago" specifically refers to the proliferation of norms which mould subjects as they internalise those norms. This process is backed by the authority of the human sciences, which provide a veneer of "scientificity" (DP, 296). These sciences are further refined by feedback from techniques of examination, observation, documentation, judgement, and confession, to produce finer distinctions within knowledge and more precise applications of power. This carceral mechanism establishes a continuous gradation passing seamlessly from a transgression of the law to a departure from a rule or an average (DP, 298). "We are in the society of the teacher-judge, the doctor-judge, the educator-judge, the social worker-judge; it is on them that the universal reign of the normative is based; and each individual, wherever he may find himself, subjects to it his body, his gestures, his behaviour, his aptitudes, his achievements" (DP, 304). Modern technologies of the self nurture a self which recognises itself in the normative categories of psychological and psychiatric sciences and their associated disciplines. Students, patients, and factory and office workers are all made subjects of modern power, of which there is no centre, no teleology, just many small uncoordinated causes. In this account, striking in its parallels with Habermas's colonisation thesis, Foucault is sketching a view of the modern subject subjected to forces below the level of articulation by which it is shaped, defined, and reproduced. In response to this situation, Foucault's account argues against global liberation. All politics must be local, even personal. We will see in the next chapter that the struggle against normalisation involves a mode of fashioning an ethical way of being a self. Revolution begins with the self.

BIOPOWER AND GOVERNMENTALITY

In the final lecture of SD and later the same year in HS1, Foucault broadens his analysis to take up the notion of "biopower". Discipline operates on individual bodies by constituting physical routines and structuring perceptual and social grids. Biopower bears on populations. It involves the administration of norms over the population as a whole in order to regulate mass phenomena such as "the problems of birth-rate, longevity, public health, housing and migration" (HS1, 140). Therefore, rather than prohibiting certain forms of sexuality, nineteenth-century biopower regulated them to optimise the values of reproduction, family, and health. We thus find, for example, psychiatrists and doctors, supported by scientific discourses, turning homosexuality into a perversion because of its failure to focus sexuality on the reproduction of healthy families. While discipline emerged in institutions and diffused into society, biopolitical techniques are part of the state's administrative and managerial procedures. These procedures are supported and legitimated by scientific knowledge of the population's characteristics, structures, and trends in order to manage or compensate for what cannot be controlled. At times Foucault describes discipline and biopower as two distinct, though overlapping and complementary, forms of power (e.g. in STP, Lectures 1 and 2). Elsewhere, he includes discipline within biopower as one of the two poles at which biopower operates (HS1, 139). However we characterise the distinction, biopower and disciplinary power are necessarily entwined. Discipline is not applied to already-existing individuals, but rather presupposes a multiplicity. Biopower's monitoring and interventions rely on disciplinary institutions.

The appearance of biopower must be seen in the broader context of modernity, with its increasing medical and scientific knowledge and organisation of production in large urban centres, accompanied by poverty, housing shortages, poor sanitation, crime, and disease. Biopower developed in tandem with sciences of demography, statistics, epidemiology, analysis of the circulation of wealth, and evaluation of the relationship between resources and populations (HS1, 140–41). It is increasingly incorporated into a continuum of apparatuses (medical, administrative, etc.) to regulate deviations from norms. The law operates increasingly as a norm and judicial institutions are increasingly incorporated into the continuum of apparatuses, such as medical and administrative apparatuses, which function in a largely regulatory manner. Thus, Foucault tells us, "a normalising society is the historical outcome of a technology of power centred on life" (HS1, 144).

From the eighteenth century, a "relative control over life" opened up "methods of power and knowledge [which] assumed responsibility for life processes and undertook to control and modify them" (HS1, 142). Foucault's understanding of biopower here is foreshadowed in OT and BC, which both chart a series of discursive shifts within the *episteme*. In OT, Foucault outlines how the concept of life provided the basis for an understanding which depended not on surface phenomena, but on functional elements hidden from view. Living species became understood in terms of the particular manner in which each is linked to its environment (OT, 298). This understanding of individual bodies in terms of an environmental milieu is reflected in biopower's ambitions to foster life at the level of the population, in which each individual is part of an environment for every other individual.

In BC, Foucault described how modern medicine constituted human beings as objects of science, subjected to external norms of health. Rather than just curing ills, medicine "was given the splendid task of establishing in men's lives the positive role of health, virtue, and happiness" by assuming "a normative posture, which authorizes it to dictate the standards for physical and moral relations of the individual and of the society in which he lives" (BC, 34). Medicine adopts the normative stance of an arbiter of standards that defines positive norms of health and conceives of life as the "bipolarity of the normal and the pathological" (BC, 35). This distinction is fundamental to the human sciences of the nineteenth century, which were not only modelled on empirical sciences like biology, but on the positive norms of medicine (BC, 36). These norms "linked medicine to the destiny of states" in the form of biopower aimed at managing populations. Medically informed biopower is aimed at the "benefit of the state . . . the order of a nation, the vigour of its armies, the fertility of its people, and the patient advance of its labours" (BC, 35). Biopower targets the population, characterised by its own processes, laws, and phenomena that include not only disease but also the production of wealth and its circulation. It regulates by "establishing a sort of homeostasis . . . achieving an overall equilibrium that protects the security of the whole from internal dangers" (SD, 249).

Biopower doesn't merely extend existing power, it represents an epochal transformation, "undoubtedly one of the most important in the history of human societies" (Foucault, cited by Senellart in STP, 369). In the modern era, knowledge produced by the human sciences is harnessed by the pastoral state as biopower to reconfigure political action and determine its goals in terms of the life of the population. Foucault tells us that it was biopower that "brought life and its mechanisms into the realm of explicit calculations and made knowledge-power an agent of transformation in human life" (HS1, 143). Prior to modernity, human beings were not fully objects to themselves, things that could be conceptually grasped and subjected to "explicit calculations". However, with the emergence of biopower, there arises an abstract, risk-averse imperative to optimise life. This underpins its right to kill, which is "simply the reverse of the right of the social body to ensure, maintain, or develop its life" (HS1, 136). Thus, according to the Nazis' biological racism, the death of the inferior group would remove a threat to the nation's existence (SD, 256–60). By exclusionary racism, war, eugenics, or euthanasia, biopower ensures by its "explicit calculations" that the life of the population, in a quantifiable yet abstract sense, comes out ahead. If humanity is a thing that can be made, Foucault shows us that it is not made with rational oversight or consciousness, but is the product of many small, forgotten acts of conditioning and resistance. By applying scientific knowledge to populations, humans might succeed in mitigating the vicissitudes of precarious existence but, by the nature of its authority, scientific knowledge also renders pressing problems invisible.

Foucault tells us that resistance to biopower is everywhere, although this resistance has not sought a return to sovereign power. Instead it relies for support on what biopower has invested in, humans as living beings. Biopower's notion of life has been "taken at face value and turned back against the system that was bent on controlling it". In the tactical polyvalence of discourses, resistance takes the form of demands for "the 'right' to life, one's body, health, happiness, to the satisfaction of needs, and beyond all oppressions and 'alienations', the 'right' to discover what one is and all that one can be" (HS1, 145). According to Foucault, these are not rights in any classical juridical sense, but rather struggles framed around the question of life. Similar to the "counter-movements" which Habermas discusses (see chapter 3), Foucault here is referring to the phenomenon of a "new politics" of alternative social movements turning on quality of life, self-realisation, participation, and human rights. It appears however that those who resist biopower are bound to frame their resistance in the terms it dictates, much as sexual liberationists remain trapped within the power technology of sexuality. In the following chapter, we will see Foucault articulate a richer concept of resistance. The shift to the concept of governmentality is a key turning point.

Governmentality, Liberalism, and Political Economy

The sciences underpinning biopower's regulation of populations must be consolidated by being made commensurable, and thus able to serve as a basis for the modern state's decisions and actions. Economics or, to use its earlier name, "political economy" has come to take on the role of a filter through which other sciences—epidemiology, demography, natural resource sciences and so forth—are passed. Exercising political power has come to mean governing a population on the basis of forms of scientific knowledge, the most important one today being economics.⁹ Thus Foucault treats liberalism, not in terms of economic or political theory, but specifically as the art of governing that submits political decision-making to the judgement of the market, constituted as a natural site of veridiction by the scientific discipline of political economy. Before turning to liberalism and political economy, I must mention Foucault's concept of governmentality, a concept that widens the scope of discipline and biopower by including the domain of politics as traditionally understood.

In SD and HS1, Foucault raised the question of biopower in a cursory and generalised form. Although the 1977–1978 and 1978–1979 lecture courses seem to set out to consolidate this account, Foucault reorients the lectures to place the hypothesis of biopower in a broader context. In both courses, biopower is discussed not in terms of the biological and physical life of the population, but within a broader framework of "governmentality". These lectures form a conceptual hinge on which Foucault turns away from the Nietzschean discourse of battle to the problematic of governmentality understood broadly as "techniques and procedures for directing human behaviour. Government of children, government of souls and consciences, government of a household, of a state, or of oneself" (EW1, 81). Governmentality thus includes the ethical question of how one conducts oneself, a theme foregrounded in Foucault's work of the 1980s. The concept of governmentality mediates between power and subjectivity, making possible the consideration of how forms of self-government are articulated in practices of political government. This broad meaning of government does not conceive of the constitution of subjects and state formation as two independent processes but analyses them from a single analytic perspective (EW3, 341). In his genealogical work, Foucault had insisted that knowledge and power could not be analysed separately. Now armed with the concept of governmentality, he will bring a positive concept of subjectivity into this constellation.

Liberalism emerged in the eighteenth century as a specific form of governmentality, a distinct regime of power/knowledge, distinguished from disciplinary and sovereign power by its rationality, aims, and methods. It has not superseded these forms of power, but rather harnessed them. Its major form of knowledge is political economy (STP, 108–09). With liberalism, there also emerged the idea of the economy as a conceptually and practically distinguished space. Liberalism draws on Adam Smith's idea of the individual,

---

⁹ The prestige and authority of economics as a science is attested to by Thornton, who claims that economics attains its scientific status by building and statistically testing closed-system deductivist mathematical models (Thornton 2017, Ch. 13). Based on the assumption of optimising behaviour by rational agents, these tests must employ econometric techniques rather than, or at least in addition to, descriptive statistics. "For the defenders of mainstream economics these simple rules are what make it a science, which is envied and increasingly imitated by the practitioners of less favoured disciplines in the areas of management and social studies" (King 2011, 64).

intending only his personal gain, being "led by an invisible hand to promote an end which was no part of his intention" (Smith 1776, Bk 4, Ch. 3, para. 9). This end serves the public good. The benign opacity of the economy's invisible hand applies not only to individual citizens but also requires government to discern and follow the natural, spontaneous mechanisms of the market, since attempts to modify its mechanisms will risk impairing and distorting them. Rather than impose injunctions or imperatives, the state's role will be to manage these "natural" processes (BB, 31).

Thus, liberalism acts only within the bounds prescribed by that which is taken as nature. This nature constitutes not only the basis but also the boundary constraining government from interfering. By advancing the idea of the spontaneous self-regulation of the market on the basis of natural prices, the market becomes a site of truth, something that must be obeyed because it reflects natural spontaneous mechanisms which must be allowed to appear for its proper functioning. However, liberalism is as much a way of acting as not acting. Liberalism's objective is "not so much to prevent things as to ensure that the necessary and natural regulations work, or even create regulations that enable natural regulations to work" (STP, 353). By acting only in order to allow the market to operate "naturally", it appears that we are not really acting at all but stepping back to allow nature to follow its own course. Liberalism educates state reason by making clear the bounds of its knowledge and determining what it can and cannot do. It is preoccupied with the vulnerability of the opaque processes of population which it seeks to make secure. Liberalism's freedom is regulated by a principle of calculation, such that individuals are expected to measure and calculate risks and take precautions by weighing disadvantages against costs. "Everywhere you will see this stimulation of the fear of danger which is, as it were, the condition, the internal psychological and cultural correlative of liberalism. There is no liberalism without a culture of danger" (BB, 66–7).[10]

In his 1979 lecture series, concluding just weeks before Margaret Thatcher was elected as prime minister of Britain, Foucault introduced an analysis of neoliberal thought. These lectures trace liberalism's transmogrification into neoliberalism and its dissemination into political discourse from as early as the 1950s. Neoliberal thought took many and varied forms, with some forms not opposed to social safety net provisions, such as those favouring negative taxation (BB, 203–04). Despite their prescience and rich insights, Foucault's lectures on neoliberalism are sketchy and speculative intellectual histories.

---

[10] The danger inherent in liberalism recalls Kalecki's argument that sustained full employment is unlikely under capitalism. Liberal governmentality operates best, with a compliant labour force, when it is risky to lose one's job due to the large "stock" of unemployed (Kalecki 1971).

His largely descriptive account need not be read as either an endorsement or a critique of neoliberal theory. Although we cannot expect Foucault's account to address the many directions which neoliberalism took after 1979, his thinking remains a valuable springboard for thinking about our present.

Foucault is adamant that neoliberalism is not "the resurgence or recurrence of old forms of liberal economics" (BB, 117–18). It is not "Adam Smith revived" (BB, 130). Rather than the market being a principle of government's self-limitation, "it is a principle turned against it: It is a sort of permanent economic tribunal confronting government" (BB, 247). The state must now model itself on the economy. "What is at issue is whether a market economy can in fact serve as the principle, form and model for a state" (BB, 117–18). By the extension of market rationality to all domains of human activity, neoliberalism intensifies the market as a site of veridiction which organises, limits, measures, and legitimates the state. "The economy produces legitimacy for the state that is its guarantor" (BB, 84). Even without comprehending the market economy, we must submit to it as we must submit to laws of nature. According to Friedrich Hayek, economic processes are a "spontaneous order" which, like Smith's "invisible hand", we cannot fully comprehend. Since he rules out political intervention in the processes of this spontaneous order as dangerous interference, we have no alternative but to simply submit to the imperatives of the market. "A refusal to submit to anything we cannot understand", Hayek tells us, "must lead to the destruction of our civilization" (Hayek 2008, 211). Hayek's work, which many regard as a founding theoretical underpinning of neoliberalism, purports to eschew the coercion and conflict of politics in favour of the harmonious and mutually beneficial social relations of the market.[11] The influential Chicago School economist, Milton Friedman, developed Hayek's free market ideas into what he regarded as an objective and value-free science.

By criticising state intervention, neoliberalism entails the transfer of government to non-state actors. However, this apparent "retreat of the state" is actually an extension of government that displaces government from formal to informal techniques. While avoiding state governance, neoliberalism uses it to extend competition throughout social life. Neoliberalism is not about "small government" which leaves the economy alone. The state must support the economy, organising its conditions and facilitating its growth. "Government must accompany the market from start to finish" (BB, 121). As a form of generalised reason, the logic of the market economy permeates "the social body including the whole of the social system not usually conducted through or sanctioned by monetary exchanges" (BB, 243, 268–89). As

---

[11] Jessica Whyte offers a valuable account of the intellectual origins of neoliberalism (Whyte 2008).

with Habermas's account of the colonisation of the lifeworld, we see areas previously regulated by non-market forms of social governance opened up to market imperatives. In this vein, Gary Becker, who succeeded Friedman at Chicago, argued that it was possible to generalise the economic form of the market throughout society to include relationships not usually subject to monetary exchange. Economic rationality thus becomes the rationality of all human action.[12]

This radicalisation of *homo economicus* involves positing a fundamental human faculty of free choice which enables economic calculation to overcome established categories and frameworks. Work is seen to require the resources of skill, aptitude, and competence, which comprise the worker's "human capital". Competition is advocated as the dominant principle for guiding human behaviour (BB, 241). As the central principle of market rationality, competition means that political subjects lose guarantees of protection by the state. Labour is transformed into human capital and workers into entrepreneurs competing with other entrepreneurs. What is most striking about Foucault's analysis is that he analyses neoliberalism not as a primarily economic programme, but as a form of governance. Neoliberal governance entails the production of a particular type of subject by fostering self-interest, expressed as the atomised individual's tendency to take risks and compete. Perhaps we can see this production of a particular form of subjectivity in the regime of concepts, models and metaphors promulgated by the academic discipline of economics.[13] Economic theory is also applied directly to bodies through the increasingly insecure nature of work. In an endless looping between description and prescription, neoliberal economics, aspiring to the status of a science, constructs the truth of the subject that it at the same time claims to discover. This subject supposedly negotiates the social realm by constantly making rational choices based on economic knowledge and strict calculation. Society becomes a game in which self-interested, atomised individuals compete for maximal economic returns.

On Foucault's account, neoliberalism is a historical development that seeks to redraw the boundary between the economy and politics by generalising the economic form throughout the social body. It seems plausible then that,

---

[12] Foucault provides a discussion of Becker's theory of human capital, which sought to fill the gap in economic analysis by providing a unified explanation of a wide range of social phenomena. The most striking example that Foucault discusses is the mother/child relationship (BB 229–30, 243–44). Becker treats the time the mother spends with the child, as well as the quality of care, as an investment that constitutes human capital. Investment in the child's human capital produces a return when the child grows to maturity and receives wages.

[13] Although interpretive caution is required, a number of studies in which selection bias was controlled have suggested that students trained in economics tend to be more aggressive, less cooperative, more pessimistic about the prospects of cooperation, and more prone to cheating than students who have not undertaken study in economics (Frank et al. 1993, 1996, cited in Thornton 2017).

after thirty years of its ascendency, the shift to neoliberalism would show up in responses to climate change. In this regard, Oels has suggested that, while climate change started as an environmental issue framed in moral terms, it is now mostly discussed in terms of cost-benefit analysis (Oels 2005, 197). Slocum has also observed a significant shift in how climate change is framed, with both campaigners and administrators shifting emphasis from "saving the climate" to "saving dollars". This shift was accompanied by a neoliberal rationality which saw local administrators address the public not as active citizens concerned with climate change, but as passive consumers of electricity and gas who respond to incentives (Slocum 2004, cited in Oels, 202). Foucault's analysis brings into focus the construction of a particular form of governmentality that restricts thinking and discussion to the market's efficiency, and hence rules out other responses, particularly those that Foucault referred to as "subjugated knowledges" (which we will come to shortly) along with questions of fairness and justice which such approaches may help address (SD, 6–12).

Foucault's analyses of liberalism and neoliberalism identify political economy as the principle form of scientific knowledge that, linked to power, constitutes modern state governmentality. However, the scientific objectivity of political economy requires the maintenance of a certain distance from the perspective and concerns of the state. Political economy is, Foucault tells us, a science which is "external to the art of government and that one may perfectly well found, establish, develop, and prove throughout, even though one is not governing or taking part in this art of government" (STP, 351). In other words, political economy appears as a value-free, descriptive science open to all. But it is also a form of scientific knowledge which government must concern itself with. However, political economy can't generate a detailed programme for governmental action. Although unable to constitute an art of governing, political economy assumes the role of a knowledge which is *"tête à tête* with the art of government", a phrase suggesting entwinement and complicity. Foucault's further comments support this. Political economy is a particular amalgam of science and government, a "kind of more or less confused magma ... an art of government that would be both knowledge and power, science and decision ... two poles of a scientificity that on one hand, increasingly appeals to its theoretical purity and becomes economics, and, on the other, at the same time claims the right to be taken into consideration by a government that must model its decisions upon it" (STP, 351).[14] We could

---

[14] The Cambridge economist Joan Robinson similarly argued for the unavoidable entwinement of the descriptive and the prescriptive in economics. "In the general mass of notions and sentiments that make up an ideology, those concerned with economic life play a large part, and economics itself ... has always been partly a vehicle for the ruling ideology of each period as well as partly a method of scientific investigation" (Robinson 1962).

say, somewhat extending Foucault's meaning, that by seeking to remove politics in favour of market governance which it promotes, neoliberalism is in fact profoundly political. This theme is seen throughout Foucault's genealogy, that of the scientific discourse which appeals to its purity while still entwined with power, a "confused magma" which relies on an interplay between social facts and prescriptive norms. Through such entwinement, political economy assumes an authoritative position as a science endorsing the rationality of markets as rationality per se. While the authority of this science trades on its appearance of purity, its entwinement with power produces the freedom-inhibiting normalising consequences highlighted in Foucault's genealogies.

Foucault analyses the objects of governmentality, such as the state, the population, and the market economy, in a manner similar to other aspects of social reality. "The question here is the same as the question I addressed with regard to madness, disease, delinquency and sexuality" (BB, 19). These objects aren't simply natural objects that are discovered by proper scientific methods, but elements in a power/knowledge apparatus. The point of Foucault's investigations is "to show how the coupling of a set of practices and a regime of truth form an apparatus of knowledge-power that effectively marks out in reality that which does not exist and legitimately submits it to the division between true and false" (BB, 19). To make sense of this, we need to consider how such concepts are constructed within a context of power. The liberal concept of "market economy" is, for example, linked to concepts of naturalness which lend it legitimacy.[15] It has properties exceeding a mere aggregation of exchanges and, like the concept "sex", serves a range of functions. Once recognised, it is objective. However, its existence relies on this recognition. It exhibits the looping effect of Hacking's interactive kinds, where description influences phenomena which loop back to influence description. These looping effects, or effects of circular and cumulative causation, by which beliefs about objects constitute those objects which then influence beliefs about them, are clearly apparent in economics. The carefully calibrated statements made by central bank officials about the state of the economy are not so much intended to report or predict a state of affairs as to influence a state of affairs by influencing what is believed about it. This challenges the current dogma that methodological individualism provides macroeconomics with rigorous foundations.[16] It can be surmised that the persistence

---

[15] By positing the market as a spontaneous natural mechanism, political economy is linked to biology's notion of an inner organisation, characterised by life as a dynamic and abstract principle common to all organisms (OT, 248). The ensuing biological notions of self-regulation and self-preservation are also found in the liberal conceptions of market freedom and autonomy assumed by political economy.

[16] See Thornton (2017, chapter 2).

of such a dogma is explained by the inseparability of a science like political economy from power, combined with the ideological commitment to a belief in their separateness, in other words, in the purity of science.[17]

Foucault destabilises such commitments, insisting that objects such as market economies or delinquents "are not things that exist, or errors, or illusions, or ideologies. They are things that do not exist and yet which are inscribed in reality and fall under a regime of truth dividing the true from the false" (BB, 20). Foucault is saying that these objects don't exist like the mind-independent objects of the natural sciences. However, their appearance can't be explained as illusory or ideological effects of hegemonic practices. Once picked out by concepts and given a role in scientific theories linked to power, these objects are constituted as reality, as candidates for true statements which shape social practices. Foucault is less concerned whether or not something is *really* real or true, than with the power effects of regarding something as "real" or a statement as "true".

NORMATIVE CONFUSIONS

I began this chapter by noting that science is conventionally understood as a form of knowledge that develops methods to abstract from the particularities of interests, prejudices, and power in order to reach universal truths. I will return to this theme in the light of Foucault's notion of power/knowledge and Habermas's criticism that Foucault's work implies a justificatory framework which relies on tacit appeals, presumably to values like Kant's normative notions of dignity and autonomy.

According to Habermas, Foucault's resolutely descriptive stance requires him to bracket any evaluative judgement of practices, leaving him no position from which to evaluate regimes of power/knowledge. However Foucault, according to Habermas, consistently assumes a critical position in relation to modern society, a position of "arbitrary partisanship of a criticism that cannot account for its normative foundations" (PDM, 276). His notion of power/knowledge thus has an implied justificatory framework. However, Foucault cannot acknowledge this because it is ruled out by his totalising critique of modernity. According to Habermas, he commits a "performative contradiction" and stands accused of *crypto-normativism*, the smuggling in of values, contradicting his supposedly value-free stance (PDM, 284). By its refusal to normatively ground his judgements, Foucault's "critique" of the human sciences loses any bite. Nancy Fraser agrees, asking "why is struggle preferable

---

[17] Thornton explains this commitment in terms of satisfying the needs of practitioners and adherents for order, control, and predictability (Thornton 2017).

to submission? Why ought domination be resisted? Only with the introduction of normative notions could Foucault begin to tell us what is wrong with the modern power/knowledge regime and why we ought to oppose it" (Fraser 1989, 29).

Mark Kelly offers a strong defence of Foucault by arguing that he makes no significant normative claims (Kelly 2018). Michael Kelly suggests that a normative premise is there all along. Freedom is presupposed and not undermined by power (Kelly 1992, 365–400). Tulley refuses the terms of engagement of Habermas's argument by suggesting that Foucault's critique is directed towards the very idea of normativity (Tulley 1999). Rather than tease out the various insights and shortcomings of these arguments, I argue that Habermas simply misunderstands what Foucault is doing. I will develop this argument further in chapter 8 when I bring Foucault and Habermas together to examine their fundamentally different orientations towards critique and argue that these differences reflect distinctly different moments within modernity's self-reflexive reason.

Critics frequently point to Foucault's occlusion of progressive aspects of modern democracy, science, technology, and liberal individualism in DP. They refer to his most explicitly evaluative language, for example, in his condemnatory accounts of the "carceral archipelago" with its "insidious leniences, unavowable petty cruelties, small acts of cunning" (DP, 308). Such passages, it is argued, present modernity as a seamless system, articulated by institutions of domination tied to a logic that shapes, controls, and directs every aspect of life (Best 1995, 137; Grumley 1989, Ch. 7). Against this view, I argue that although Foucault undoubtedly underplays the achievements of modernity, he doesn't deny them. He simply walks away from any account framed in terms of progress or regress. "I don't say that humanity doesn't progress. I say that it is a bad method to pose the problem as: 'how is it that we have progressed?' The problem is: how do things happen" (PK, 50). In showing how things happen, Foucault's histories of forgotten contingent events uncover the bleaker aspects of social life. However, he does not intend to present those aspects as a "balanced view" from an ahistorical perspective of history "as it really happened". These accounts are intended to shake the hold that reassuring narratives of history have on us. As bleak and totalising as Foucault's account of the carceral archipelago is, that is not the point of his accounts. It is only in his later work, which we consider in the following chapter, that Foucault is able to finally clarify his project and acknowledge his earlier over-emphasis of the inexorable domination of power.

Even without considering his later works, we can see that Foucault is not simply highlighting what he sees as injustice or cruelty. If in places he does, he more generally *suspends* normative commitments so as to reveal the workings of power/knowledge which he believes we cannot see clearly.

What Foucault wants us to see is not the obvious cruelty or injustice but the play of power/knowledge which constructs categories of thought that compel reasonable people, not so different from ourselves, to accept these cruelties and injustices as reasonable. Likewise, in his genealogies of sexuality, Foucault doesn't emphasise the suffering, or champion the rights, of marginalised sexualities, but is concerned primarily with the larger theme of the fabrication of what we take in human nature to be most natural and the power effects that flow from this. In short, Foucault's genealogy doesn't prescribe what we must think but aims to open space in which we can think. One way in which we can see how Foucault does this is in terms of Koopman's plausible notion of modernity's urge to eliminate ambiguity by purification that I referred to in chapter 5. Foucault is not primarily showing us the truth, falsity, rightness, or wrongness of our ideas about madness, freedom, reason, power, or delinquency. Rather he is trying to open dialogue across boundaries of concepts which define themselves in opposition to each other, to show us that madness is not always madness, freedom not always freedom, that reason has a history, power doesn't negate freedom, and delinquency was invented in the nineteenth century. Foucault's work need not be read as a totalising condemnation, leaving him nowhere to stand. He wants to maintain the tension of ambiguity by showing that we are *both* inexorably conditioned by power *and* radically free and reasonable. Seen in this light, Foucault is not pulling the rug from under his own feet by claiming that all claims are tainted with power and therefore illegitimate. Rather he is claiming that legitimating reasons, his own included, will always be caught up with the power relations of their particular situations and that to lose sight of that is dangerous.

Given this position, it is not clear if, or why, it is problematic that Foucault describes modern society in evaluative terms while withdrawing from any explicit normative foundational framework or criteria. Rather than simply judging certain practices, he is *problematising* their authoritative scientific underpinnings, drawing them out from an apparently unquestionable background to open the possibility of thinking differently. Foucault explicitly locates his own reflection as an attempt to break free of the epistemic sovereignty ascribed to science, which involves the adoption of a position outside its historical and social situation from which truth claims can be validated. Instead, Foucault acknowledges his social and historical embeddedness, offering a conception of power and knowledge without any sovereign standpoint. Any evaluative aspect of Foucault's work does not flow from "normative foundations" but arises from an imaginative response which enlists resistances to power, drawn from Foucault's and the reader's unarticulated backgrounds of experience, which can't be captured within a general framework.

Foucault's genealogies allow us to recognise how subject-objects—the delinquent, the hysteric, the homosexual—are constructed by particular alignments of power/knowledge, not simply discovered as something that was always there. As the most authoritative form of knowledge, it is science, particularly human sciences, from which these power effects flow. Foucault wants to make us aware of this and how things could be otherwise. His critique is "not a matter of saying that things are not right as they are. It is a matter of pointing out on what kinds of assumptions, what kinds of familiar, unchallenged, unconsidered modes of thought the practices we accept rest" (EW3, 456).

Foucault's critique questions the relations between knowledge and power that issue in epistemological certainties that buttress ways of organising social relations while foreclosing alternative ways and limiting the possibility of thinking otherwise. One might reasonably ask what good is "thinking otherwise" if we aren't guided by principles to gauge success or failure. As Butler points out, there are no reassuring answers to give to such a line of criticism, since reassurance is not the aim of critique (Butler 2001). Foucault's critique emerges from an existing crisis within the epistemological field in which we already can't see the way forward. His analyses emphasise that the categories which order social life produce certain realms of unspeakability. It is from the awareness that dominant discourses have produced these impasses that critique emerges as an exploratory experimental practice.

To conclude, I want to frame Foucault's normativity in terms of a question briefly raised in the last chapter—what is Foucault up to? To understand Foucault, we need to uncover a broader pragmatic dimension in his work by asking after his point or purpose, not simply what genealogy *describes* but what it *does*. One thing it does is to render problematic or questionable beliefs that have been unreflectively adopted. However, Foucault's critique need not be considered as merely directed to making particular beliefs questionable—say those about the self, sexuality, punishment, or economics. I argue that it operates broadly across the conceptual terrain by inducing what Karl Mannheim referred to as an "unmasking turn of mind", which does not seek to refute concepts as fictitious, but rather displays their "extra-theoretical function" to strip them of their false authority.[18]

Genealogy doesn't necessarily show the human sciences to be untrue or even dubious, but *problematises* them by stripping them of their automatically granted authority and opening them to broader evaluation. It shows us how power/knowledge constructs objects which, endorsed by scientific

---

[18] For a discussion of Mannheim in this context, see Hacking (1999, 53–56).

authority, become naturalised as objects, such as sexuality, delinquency, madness, or the market economy. Such objects don't exist in an objective scientific sense. They are not independent of our recognition and naming of them. However, once recognised and named, they are real enough. When knowledge of these objects is granted the honorific title of "science", it becomes a foundation on which we build further knowledge and practices. Foucault's genealogy problematises by showing how these foundations were constructed. Genealogies operate as a counterpoint to familiar stories about, for example, liberation from sexual repression, the enlightened reform of punishment, or the progress of reason.

Describing practices which entrap us in repressive regimes of discipline, Foucault's accounts seem to call for resistance. However, he rejects the idea that there is any position outside power from which to resist. Power is always accompanied by resistance, which is never exterior to power (HS1, 95). Resisting power is simply subjection to another power. "Power is 'always already there' [and] one is never 'outside' it". However, this "does not mean that one is trapped or condemned to defeat no matter what" (PK, 141–42). This is clear from the later work in which Foucault distinguishes between, on one hand, the domination "which people ordinarily call power" where asymmetrical power relations remain frozen and, on the other hand, power relations "understood as strategic games between liberties—in which some try to control the conduct of others, who in turn try to avoid allowing their conduct to be controlled or try to control the conduct of others" (FL, 447). Rather than impersonal conflict between opposing forces, the exercise of power in the latter sense is "a mode of action on the action of others". Since "action" implies freedom, power never exhaustively determines a subject's possibilities (EW3, 342). So, although power is everywhere, even in the most harmonious and natural relations, it is not something we are trapped in like Weber's iron cage.

Not only is the status of knowledge an outcome of struggle. It is also a weapon in this struggle. In SD, Foucault opposes his "historicism" to the anti-historicism of "all the great philosophies" and the human sciences, which adopt the pose of the universal human beyond history. Against the tyranny of these "overall discourses", Foucault urges an "insurrection" of "subjugated knowledges . . . of the "psychiatrised, the patient, the nurse, the doctor, . . . the delinquent" (SD, 7).[19] His genealogies are "weapons in the struggle . . . we wage war through history" (SD, 172). They impel action by problematising the power monopoly of established sciences by playing off disqualified knowledges against unitary scientific theories which "filter them, organise them into a

---

[19] By "subjugated knowledges", Foucault means historical contents that have been disqualified as below the level of scientificity, or masked by formal disciplines which subsume them within universal schemas (SD, 7).

hierarchy, organise them in the name of a true body of knowledge, in the name of the rights of a science that is in the hands of the few" (SD, 9). "Global theories" that seek hegemony by curtailing, dividing, caricaturing, and overthrowing non-global discourses are a "hindrance to research" (PK, 81). Distortions and omissions are required to contain the messy diversity of particular concrete things and events within the terms of a unitary science. While the human sciences are Foucault's primary concern, there is no reason why the insurrection could not be directed against the regulatory hegemony of scientific naturalism. By resuscitating historical knowledges submerged within totalising schemas, genealogy does not undermine the content of science nor deny its truth claims. It directs its insurrection against "the centralising power-effects that are bound up with the institutionalisation and workings of any scientific discourse organised in a society such as ours". Genealogies question "the aspiration to power that is inherent in the claim to being a science" (SD, 10).

Despite talk of insurrections and weapons, Foucault's genealogies step back from political engagement. This is clearly seen in his preamble to the first lecture of the 1977–1998 lecture series Security, Territory and Population in which he rejects "an imperative discourse that consists in saying 'love this, hate this, that is bad'". He insists that "the dimension of what is to be done can only appear within a field of real forces, that is to say within a field of forces that cannot be directed by a speaking subject alone and on the basis of his words, because it is a field of forces that cannot in anyway be controlled or asserted within this kind of imperative discourse" (STP, 3). Faced with this disconcerting distance from explicit political engagement, commentators may be tempted to read more into Foucault's work. Behrent, for example, suggests that the 1979 lectures "should be read as a strategic endorsement of economic liberalism" (Behrent 2009a, in Patton 2013). I think Behrent misconstrues Foucault's stance. Rather than endorsement, Foucault's discourse shows a studied lack of political commitment. I agree with Patton that Foucault assumes this stance because he recognises how discourses are implicated in the general regime of power in society and he wants to distance himself from this (Patton 1984). This view is supported by Foucault's criticism of the "universal intellectual" who speaks for all and is regarded as standing outside power relations and opposed to their repressive effects.[20] While Habermas's reconstructions yield a progressive developmental logic marked by normative milestones, Foucault steps back from such normativity to show how power/knowledge operates, for example, how neoliberalism functions as a "politics

---

[20] This universal intellectual perpetuates a system of power which prevents others from speaking for themselves. According to Patton, Foucault rejects "a mechanism of social regulation of discourse which traces utterances back to the founding act of an author, the initiator of a discursive practice". Foucault's refusal to write in these terms should be seen as a "the result of a commitment to a different intellectual ethic" (Patton 1985, 72).

of truth" by producing forms of knowledge, inventing concepts, and shaping styles of thought. What Foucault's reader does with this is up to them.

Foucault's problematisations challenge not so much the universalist aspirations of science, as the accompanying sense of ahistoricity and necessity which serves to legitimate the application of unitary sciences as blunt instruments of social engineering. Instead, he advocates responding to local issues with a specificity that escapes global systems of thought. The "specific intellectual" works "at the precise points where their own conditions of life or work situate them (housing, the hospital, the asylum, the laboratory, the university, family and sexual relations)" (PK, 126). He is not so much concerned with what is true or false, but rather the social and political functions of taking something as true or false. "It is not a matter of a battle 'on behalf' of the truth, but a battle about the status of truth and the economic and political role it plays" (PK, 132). Foucault wants to draw our attention to the preconditions which social, economic, and cultural power establish for what can count as true. He doesn't want to free knowledge from power altogether, since this is impossible, but to loosen the grip of the current hegemony and reveal the possibility of thinking beyond the present.

In this chapter, we have seen the inseparability of power and scientific knowledge, and how particular sciences have arisen within particular configurations of power in which they remain trapped. Rather than power distorting knowledge, we see power/knowledge producing and linking bodies, concepts, discourses, practices, objects, and subjects, forming chains which further extend its reach. In the following chapter I will turn to Foucault's ethics, in which the axis of the subject is drawn into the constellation of power/knowledge. We also see his mature understanding of his project, an understanding from which he claims to "see what [he] had done from a new vantage point and in a clearer light" (HS2, 11).

Chapter 7

Science and the Genealogy of the Subject

FOUCAULT'S BROADER FRAMEWORK

Thus far we have seen Foucault concerned with the limits prescribed by the archaeological *episteme* and genealogical power/knowledge. By bringing these limits into view, he wants to open up the possibility of a perspective beyond them. His critics have pointed to his difficulties in locating himself in relation to these limits (he seems trapped within them) and articulating his own normative standpoint (he offers no reasons for surpassing them). Having foregrounded the inextricable entwinement of science and power, Foucault cannot simply adopt an "external" perspective, as we saw in Habermas's strategy of harnessing science to articulate and ground a normative framework beyond that of the parochial everyday lifeworld. His genealogies of the subject instead open a view of the self's constitution of itself, thus problematising the dominance of scientific rationality as a historically contingent mode of self-relation. By bringing science into view as one possible regime of truth alongside other regimes, Foucault problematises the authority which modernity grants to the sciences as ultimate arbiters of knowledge and conduits of power.

While Foucault's shift to this broader perspective was gradual, the 1979–1980 lectures suggest a decisive turn towards a more active and responsive subject which then remained the central focus of the final period of his work. The four lecture series, from 1980–1981 to 1983–1984, and the further volumes of the *History of Sexuality*, consolidate this as a perspective from which he then does not deviate. In this final period, Foucault's genealogy of the subject traverses ancient philosophy, the Christian pastoral, and the "Cartesian moment", the decisive break that opened up the modern scientific worldview, a break consolidated by Kant's transcendental philosophy and the rise of Man as the subject-object of the modern human sciences.

Foucault's three constitutive axes of the subject—knowledge, power, and self-constitution—are now inextricably entwined. Rather than abandoning power/knowledge to consider only the self-constituting subject, Foucault *complicates* these dimensions by including the subject, now become political, as a locus of both subjugation to, and resistance against, scientific power/knowledge. In this chapter, I will firstly consider Foucault's turn from power to the subject in the 1979–1980 lectures and how his new focus on agency, autonomy, and freedom is consistent with his earlier work. I will then analyse Foucault's notion of "regimes of truth", a concept that both sets up his genealogy of the modern subject in terms of its constitution by truth, power, and knowledge and illuminates how Foucault views science. I will consider Foucault's concepts of ethics, aesthetics, and spirituality as a basis for discussing this genealogy. By tracing the subject's constitution from antiquity to the post-Enlightenment period, we will locate Foucault's view of science and modernity in relation to critique and philosophy.

From Power to the Subject

The title of the 1979–1980 lecture series, On the Government of the Living (hereafter GOL), suggests an intention to return to biopolitics, bringing to bear reflection enriched by Foucault's work on governmentality. However, Foucault had been visiting the United States and was aware of the criticism of his treatment of the subject as a passive object of power.[1] Despite its title, the 1979–1980 course reflected Foucault's awareness of these concerns by leaving behind the earlier themes of biopower and the government of living populations. He now turns to the notion of "government of men by truth", to effect the shift from the "worn and hackneyed theme of knowledge-power" to the subject (GOL, 11). This concept of government now explicitly includes the subject, governing, governed, and self-governed. "Government by truth" involves the subject in its relations to the manifestation of truth, or the ways in which subjects bind themselves to truth. For as long as Foucault was studying the modern West, he had treated the subject as the product of systems of power/knowledge. His archaeologies displaced the subject with discourse or knowledge and his genealogies brought power to the fore. However, prompted by his reading of ancient philosophy, Foucault became increasingly aware that the subject could not simply be analysed as a passive object subjected to power/knowledge but must also play an active

---

[1] Szakolczai (1998) notes the importance of Foucault's discussions with Hubert Dreyfus and Paul Rabinow during his 1979 visit to California.

role in its own shaping. This active role emerges with Foucault's concept of government, which entails an ethical stance of governing oneself and others. Now the subject can turn back upon itself to critically examine the processes by which it was constituted within the three dimensions of knowledge, power, and self-constitution.

Foucault's shift in perspective is reflected in his discussions of the current political situation, which he analyses in terms of governmentality. He now frames the new social struggles he initially identified in HS1 (see chapter 6) as responses to the modern state taking over the pastoral role of the Church and governing citizens by scientific norms. In a text first appearing in 1982, he tells us that these struggles are "not exactly for or against the 'individual', but rather . . . struggles against the 'government of individualisation'" (EW3, 330). Against the "secrecy, deformation, and mystifying representations imposed on people", these struggles resist the identities thrust upon individuals by administrative and scientific categories (EW3, 330–1). Foucault is referring to movements such as gay liberation, indigenous land rights, feminism, and so forth. Somewhat similarly to Habermas's analysis of juridification and the countermovements which resist it, Foucault sees these movements as a form of resistance to scientific power/knowledge which "categorises the individual, marks him by his own individuality, attaches him to his own identity, imposes a law of truth on him which he must recognise and which others have to recognise in him" (EW3, 331). These struggles are not about the truth or falsity of scientific knowledge, but "the way it circulates and functions, its relations to power. In short, the *régime du savoir*". Resistance is directed against subjection—in the sense of making individuals into subjects—by administrative-scientific power/knowledge.

Some critics think Foucault's earlier genealogical accounts of pervasive power are incompatible with this focus on resistance and self-reflexive practices, which they think implies autonomy and independence from power. McCarthy claims that genealogy *treats* "the subject merely as an effect of power" and then goes on to assume that Foucault thinks that the subject *is merely* the effect of power. He concludes that this means that if Foucault is right about power, he has no basis for the subject's autonomy. Alternatively, if subjects are autonomous, then his account of power is wrong (McCarthy 1992, 258). As far as I am aware, Foucault has been careful to never say that the subject *is merely* an effect of power or discourse. By saying that the subject is an effect of power or discourse, he should be understood as emphasising *the extent* to which subjects are conditioned by contingent arrangements of power/knowledge. What we see in his ethics is an option that has always been available to him, to correct what he now recognises as an over-emphasis by highlighting the moment of autonomy within action, even though that autonomy is never without constraints (EW1, 281–2). Although Foucault's genealogy certainly treats humans as passive objects, we shouldn't regard this as his view of the truth of the human condition without remainder. We should recall Foucault's lectures at the *Collège de France* in the years immediately preceding the publication of DP, which laid out the struggles between established institutional powers and the resistances of the marginalised. As we saw with the hysterics, resistance sometimes succeeds, at least temporarily. Immediately after finishing DP, Foucault set to work on HS1, in which we see again that "power is never external to resistance" (HS1, 95).

Foucault never intended to deny the role of the subject, but rather to account for its constitution within a historical framework (PK, 117). In his 1980 lectures at Dartmouth College, he discusses his current work as "a genealogy of this subject [which examines] the constitution of the subject across history which has led us up to the modern concept of the self" (Foucault and Blasius 1993, 202). By means of this genealogy, he can avoid treating "the foundation of all knowledge and the principle of all signification as stemming from the meaningful subject" (Foucault and Blasius 1993, 201). Rather than foundational and fixed, Foucault's subject is relational, dynamic, actively becoming (EW1, 290–1). Two "technologies" constitute the subject, a technology of domination and a technology of the self (Foucault and Blasius 1993, 203). Later that year in Berkeley, he admitted that "when I was studying asylums, prisons, and so on, I insisted too much on the techniques of domination". Now his task was to correct this over-emphasis by bringing technologies of the self into play and treating power as "due to the subtle integration of coercion technologies and self technologies" (cited in McCarthy 1992, 279, n.64). The term "subtle integration" suggests that the two technologies are entwined in such a way that both are present as moments of action.[2] Just as the emphasis on discourse and power in Foucault's earlier work bracketed, but did not preclude, the subject, now the self-reflexive subject does not preclude power.

Foucault's understanding of autonomy is perhaps best seen not as an absence of power but as a movement of power away from normalisation. On this understanding, autonomy would entail a change in the relationship between these two technologies such that the subject was able to step back from normalising power. This process of "desubjectification" involves "new forms of subjectivity through the refusal of this kind of individuality which has been imposed on us for several centuries" (EW3, 336). Neither the subject nor power is done away with. Instead, a new form of subjectivity is constituted that is to some extent distanced from the normalising power supported by the sciences. Foucault sees this process of distantiation as occurring

---

[2] By "moments of action", I mean that although one can be said to "passively" accept power or "actively" resist it, in either case acceptance or resistance is an *action*. Yet, at the same time, the range of possibilities for such actions is always circumscribed by a given context of power.

paradigmatically in thought which he links to problematisation (EW1, 117).³ Thought problematises, disclosing the world in a new light as what is first given as an obstacle is transformed into a question (FL, 421). Foucault offers two ways to respond to power, two different methods of problematisation corresponding to different levels of power. The first, ethics, aims at the micro-level of power located in a subject, while the second, critique, is directed at the level of the anonymous pervasive network of power relations which he analysed in his earlier work.⁴

Foucault's shift in focus to the ethical notion of "care for oneself", whereby the self acts on itself by shaping itself as a moral subject, no doubt prompted him to think of his own philosophical activity in these terms. He characterises his work as "another kind of critical philosophy ... that seeks the conditions and the indefinite possibility of transforming the subject, of transforming ourselves" (Foucault and Blasius 1993, 224, n. 4). Such transformations apply to Foucault himself and are seen in his shifting perspectives. This was no post hoc justification, but rather an insight prompted by Foucault's understanding of his own activity in the light of what philosophy had been and still could be.⁵

As we have seen, Foucault resisted confinement within any one framework, seeking to remain receptive to new kinds of phenomena which need different frameworks (HS2, 6). He sought multiple standpoints to gain a broader view, each having limitations which could only be overcome by further shifts. He doesn't however present the shift to the subject as just a new framework, but rather as the organising principle of his entire previous work. He now claimed that, for twenty years, his objective had been "to create a history of the different modes by which, in our culture, human beings are made subjects" (EW3, 326). This constitution of subjects occurs within three historical dimensions represented by the frameworks of discursive practices, power practices, and practices of the self on itself. From 1980 onwards, in interviews and lectures, Foucault presents his overall project as revealing the contingent historical construction of the subject within these frameworks.⁶ None of these dimensions can be grasped independently because each mediates the others. This understanding of the subject opposes any foundational conception of subjectivity. Rather than possessing a universal and timeless essence discoverable

---

³ "Problematisation" is defined at footnote 2 of chapter 5.
⁴ I am indebted to Mark Kelly for the distinction between these two different analytical perspectives (Kelly 2009, 61).
⁵ McGushin plausibly argues that, after coming across the practices of the self, Foucault better understood his work as not merely diagnostic, but as *etho-poetic*, a term denoting the making of a character, an *ethos* (HS, 237; McGushin 2007, xii–xiii).
⁶ See EW3, 326–7; HS2, 6; FR, 336–8; EW1, 262–3, 318.

by science, Foucault's subject emerges in history as it becomes the object of discursive practices, power, and practices of the self (EW1, 318).

Foucault places his project within the history of philosophy, as part of a subterranean path of critical thought. He identifies this path as a vision of philosophy as way of life which can be traced through aspects of Plato's *Laches*, the Cynics, the Stoics, de Sade, Kant, Hegel, Marx, Nietzsche, and the Frankfurt School. He links the shifts in his theoretical framework to the self-transformation essential to the philosophical life, thus enabling him to situate his own reflection, something he previously had difficulty doing. Now, with the subject not simply subjugated by power/knowledge but actively constituting itself, Foucault can see his own historic-critical reflection as a practice of the self, an attempt "to get free of oneself" (HS2, 8). This involves testing limits to explore what can be changed in oneself and one's present. Whereas Foucault's archaeologies and genealogies emphasise coercive subjection to scientific power/knowledge, ethics articulates a form of resistance. This resistance is not a rejection of science or a denial of scientific truth, but a revaluation of science's authority from which its power effects flow. Foucault's relativisation of science as one form of rationality among others is evoked by his notion of "regimes of truth".

Regimes of Truth

In the 1979–1980 lectures, Foucault talks of "regimes of truth" constraining individuals to certain "truth acts", or roles taken on by subjects in the manifestation of truth. He asks in what ways, by what procedures, in view of what ends, is a subject bound to the manifestation of truth. It is this relationship between the manifestation of the truth and the forms of the subject's involvement that defines "regime". Returning to the analytic of finitude (see chapter 5), but now through the lens of ancient philosophy and its transformations, Foucault poses the question, "How has Western man bound himself to the obligation to manifest in truth what he himself is?" The genealogy that unfolds in these lectures examines "how this double bind, this regime of truth, by which men find themselves bound to manifest themselves as object of truth, is linked to political, juridical etc. regimes" (GOL, 101).

The political connotations of "regime" are intended, and here Foucault anticipates an objection. It doesn't seem right to talk about a "regime" of truth in the same way one speaks of a political, medical, or penal regime (GOL, 94). Truth acts that entail an obligation, such as believing in the Resurrection, are not genuine truth acts, since they involve coercion regarding the non-true or the unverifiable. In the case of genuine truth, there is no need for a "regime" of obligation. The coercive force resides within truth itself (GOL, 95). There is only the need to speak of a regime of truth when something other than truth is involved. Foucault sees this objection as motivated by a view that truth is already there, waiting to be discovered and to speak to us directly without any need for coercion (PP, 235). Against this "philosophical-scientific standpoint", he recalls the archaic standpoint towards truth as a "dispersed, discontinuous, interrupted truth" which only speaks in certain places to certain people (PP, 236). It is a truth that is not waiting for just anyone to discover it but requires favourable moments and propitious places, for example, the oracle at Delphi or the god who cures at Epidaurus. This now lost notion of a truth-event contrasts with the universal ever-present truth of modern science, which presupposes that "there is truth everywhere, in every place and all the time" (PP, 234). Such scientific truth may be difficult to reach but it is still there, waiting to be discovered. Access to truth is never ruled out in principle (PP, 236).

The ancient notion of truth entails that the same truth does not appear for all but depends on how the subject has been constituted. With this ancient notion in mind, Foucault analyses Descartes's meditations, noting that "I think therefore I am" seems "theoretically unanswerable" (GOL, 98). However, behind the explicit "therefore" Foucault locates an implicit "therefore" which issues from the acceptance of a particular framework, a regime of truth which requires a particular kind of subject who is "qualified in a certain way" to be compelled by evidence. Behind every scientific argument, reason, or piece of evidence, there is a hidden prior commitment outside the framework of the regime. While the rules that constitute the regime dictate the division between true and false within the regime, those rules themselves don't follow from a rule but require the commitment of a subject constituted in a particular way. As will be seen, Foucault reads Descartes's *Meditations* as a way of constituting a particular regime of subjectivity that opens up the scientific worldview.

Despite the French language convention of referring to sciences in the plural, Foucault thinks that in terms of regimes of truth, it is legitimate to refer to science in the singular. "Science would be a family of games of truth all of which submit to the same regime . . . and this very specific, very particular regime of truth is a regime in which the power of the truth is organised in a way such that constraint is assured by truth itself" (GOL, 99). In other words, with science, the political nature of the regime recedes, such that it defines itself as both oriented and constrained by truth alone. The objectivating stance of science involves not only perpetual revisability but also methods that cancel the contingency of the particular form of subjectivity required. It seems to me that this feature accounts for what appears as the imperialistic tendencies of scientific naturalism in its insistence on universality and objectivity and the reduction of all values to its own schema, while dismissing that which doesn't fit as illusory. However, from an "external" perspective, the regime of science appears as one among other possible regimes. To overlook this perspective is to adopt the positivist assumption of an existing, though unknown, state of affairs towards which science progresses, and the non-truth of statements and non-reality of objects outside science's regime.

Foucault has long opposed such positivism. The ordering of the world into conceptual categories is an activity that precedes truth, since truth relies on such ordering for its subsequent establishment (WK, 108). In assuming the existence of truth from the outset, a truth that knowledge gradually uncovers, positivism endorses science as an inexorable force to which one should stoically submit. Foucault wants to challenge this assumption and open the possibility of different divisions between the true and the false. In the 1979–1980 lectures, he frames our activity of forming a conceptual order in terms of the constitution of subjects who are not merely subjugated objects, but also constitute themselves. This is ultimately the ethical question of how to care for oneself.

In his analysis of regimes of truth, Foucault eschews the binary opposition of science "in which the triumphant autonomy of truth and its intrinsic powers would reign" to "ideologies in which the false, or the non-true, would have to arm itself or be armed by a supplementary and external power in order to take on, improperly, the force, value, and effect of truth". Rather there is a multiplicity of truth regimes, all of which, whether scientific or not, entail specific ways of linking the manifestation of truth to the subject. While some regimes have a history and domain close to science, others are "quite coherent and complex and very distant from scientific regimes of self-indexation of truth". Science is not defined in opposition to ideology but is "one among many possible and existing regimes of truth" (GOL, 100). Other truth regimes, for example, religious, artistic, psychoanalytic, or moral ones, require the constitution of different subjectivities in order to manifest (GOL, 99).

The notion of a regime of truth carries critical force because it reveals the contingency of the scientific truth regime. There is nothing that necessitates, obliges, or forces us to affirm the constitution of a subjectivity that demands a scientific stance which excludes all other stances. Foucault is not questioning the truth of scientific content. Rather he is challenging the authoritative value granted to scientific truth and the *commitment* to the belief that science imprints the world-in-itself directly onto our passive souls without the process to any extent depending on us. Such commitments block the freedom of thought by freezing subjectivity, character, or *ethos*, amounting to an ethical entrapment which Foucault wants to unmask.

ETHICS, AESTHETICS, AND SPIRITUALITY

To further draw out Foucault's view of science and its relation to other practices like philosophical discourse, we need to examine the role of his ethics regarding self-formation and resistance to scientific power/knowledge. Foucault sees the self as constituted by power relations, yet still capable of autonomous critical reflection and self-transformation. The subject exercises freedom by problematising her behaviour, beliefs, and social context. While there can be no overall liberation from power, local emancipations from particular systems of domination are still possible. Foucault's ethics raises the possibility of resistance by contesting determinations, or "refus[ing] what we are" (EW3, 336). Resistance to biopower involves "a refusal of these abstractions of economic and ideological state violence which ignore who we are individually, and also a refusal of a scientific or administrative inquisition which determines who one is" (EW3, 331). On one hand we resist by "refus[ing] what we are" and on the other hand, we must invent "new forms of individuality" (EW3, 336).

Foucault calls for a new type of ethics, not based on abstract philosophical ideas that obscure the complexity of specific concrete situations, but on the ethics of the self, inspired by Greco-Roman antiquity (FR, 374).[7] This ethics is essentially a mode of self-formation not involving universal ethical codes, like Judeo-Christian or Kantian ethics, that dominate modern Western society, permitting or prohibiting actions. In contrast, antiquity's more individualistic ethics give an important place to self-formation. Care of the self aims at establishing the right relation with oneself, defined in terms of self-mastery, tranquillity, harmony, or joy (HS2, 28). However, the care of the self isn't a form of inward-looking attention. The concept implies care for others, since the relation to self forms part of a network of social relations (EW1, 287). Critics have objected that such a project of care is available to relatively few people and is vulnerable to commodification as a form of new-age individualism which obliterates the connection between personal and social struggle (e.g. Best, 124–9). Though this remains a risk that Foucault would certainly acknowledge, such criticisms are in danger of reproducing the Marxist dichotomy between the political and the personal by which the latter is stigmatised as merely "bourgeois". The attempt to transform oneself is an aesthetic task, but also an ethical task that requires moderation, sublimation, and practical reasoning, It is also a political task that challenges normalising rationalities and institutions.

The care of the self involves "a number of actions exercised on the self by the self, actions by which one takes responsibility for oneself and by which one changes, purifies, transforms, and transfigures oneself" (HS, 11). Such activities include meditation, examination of conscience, checking representations which appear in the mind, and so forth. These activities are not ways

---

[7] Foucault's notion of ethics borrows the idea of practices of the self from the ancient Greeks, but not the actual contents of their ethics (EW1, 256).

of coming to know an object in scientific terms, but rather involve monitoring, protecting, admonishing, curing, rewarding, or cultivating oneself. By means of such exercises, "the subject constitutes itself in an active fashion" (EW1, 291). This active self-constitution, or "subjectivation", contrasts with the self's subjection to the disciplines or biopower.[8] It is the self's own action on itself, striving for principled ethical coherence by undertaking *ascesis* in order to create an *ethos*, "a mode of being for the subject, along with a certain way of acting, a way visible to others" (EW1, 286).[9] While Foucault refers to these practices as "practices of freedom" chosen by an individual, they are "not something invented by the individual" but "are models that he finds in his culture and are proposed, suggested, imposed upon him by his culture, his society, and his social group" (EW1, 291). Although practices might be freely chosen, they are chosen from a pool of possible choices provided by society.

Foucault thinks the culture of care of the self is vitally important for modernity because it provides the resources to challenge the normalisation of the modern subject who is compelled to seek the truth of herself in science (HS, 9; EW1, 255–6). In antiquity, care of the self was bound up with living a beautiful, noble, and memorable life and it was the framework within which one understood the injunction to "know oneself" (HS, 4). Self-knowledge always involved prudence in a particular setting. "Know thyself" (*gnothi seauton*) was subordinated to care for self (*epimeleia heautou*). To know oneself included knowing one's "place" or limitations. It was part of caring for oneself. It was through the self's primary relation of care to itself that the self became constituted as an object of practical, not theoretical, knowledge. In modernity, the situation is reversed. The care of the self can only take place through the knowledge of scientific truths about the self. However, if the modern self were to understand itself as something to be cultivated, it could subscribe to an individualistic and voluntaristic ethics not requiring the support of the normalising power of science. As its own artist, the self would enjoy the autonomy that modernity requires.

Foucault emphasised that ancient ethics was not prescriptivist or universalist but directed at aesthetic qualities such as beauty or nobility (EW1, 261). Like any skill, it was not a matter of learning facts, but a *techne* of artfully constructing, cultivating through repetitive exercises, a certain sensitivity

---

[8] "Subjectivation" is a term coined by Foucault in the early 1980s to capture the idea of the active constitution of oneself as a subject. See McGushin (2007, 304, n. 6); Kelly (2013, 513).

[9] *Ascesis* refers to the practices through which one becomes a subject, producing a certain type of relationship that one has with oneself. *Ascesis* does not necessarily require deprivation or renunciation.

within oneself to oneself.[10] Pierre Hadot thought that Foucault's notion of a culture of the self was *too* aesthetic (Hadot 1995, 211).[11] Hadot emphasises that the ancient care of the self is directed towards "the best portion of oneself", which is "in the last analysis, a transcendental self". Ancient ethics is not concerned with the pleasure one finds within oneself, as Foucault seems to suggest. The happiness of an ethical life consists "in virtue itself, which is its own reward" (Hadot 1995, 207). Hadot thinks that the practices of the Stoics or Platonists not only involve a relationship to oneself but also "the feeling of belonging . . . both to the whole constituted by the human community, and to that constituted by the cosmic whole" (Hadot 1995, 208). Hadot's concern is that Foucault's talk of ethics as aesthetics, stylisation, and pleasure appears to reduce ethics to pleasure. In response, I will firstly clarify one thing that Foucault was *not* talking about. He rejects "the Californian cult of the self" in which "one is supposed to discover one's true self, to separate it from that which might obscure or alienate it, to decipher its truth thanks to psychological or psychoanalytic science, which is supposed to be able to tell you what your true self is" (EW1, 271). Foucault was very critical of the self-absorption and introspection of Western culture. It seems plausible that he employed expressions such as "aesthetics of the self" precisely to distance his concept from the "Californian cult" which assumed a reality knowable by science. An aesthetics of the self suggests the crafting of the self by the self, not according to anything discoverable by science.

Foucault need not be held accountable for an accurate interpretation of ancient ethics, since this was not his intention. Rather he was inspired by, and borrowed key notions from, ancient ethics. But he also included the notion of aesthetics. In Kantian terms, aesthetics is the experience of a common sense, defined by communicability and as inseparable from the desire to communicate. While aesthetic judgements are subjective, aesthetic reasoning aspires to universality (Kant 1987, 159–162). Aesthetics doesn't entail relativism. There is always something at stake in aesthetic judgements that, while not producing universal consensus, still matters. Foucault wants an ethics based on freedom and self-determination achieved by the autonomous exercise of the will. He therefore highlights the relationship between ethics and aesthetic *poesis*. However Foucault's ethics is aesthetic not by being oriented towards the pleasure of beauty, but by regarding our lives as material to be transformed. "This modernity does not 'liberate man in his own being'; it compels him to face the task of producing himself" (PT, 108–9).

---

[10] By *techne*, I mean knowing how to do or make something, something that, like art, can't be specified fully in advance due to its variability and exploratory nature.
[11] Foucault's embrace of ancient ethics was influenced by reading the work of Hadot, a colleague at the *Collège de France*.

Drawing on Baudelaire's aesthetic modernity, he equates modernity with transgression and perpetual transformation, the "feeling of novelty, of vertigo in the face of the passing moment" (PT, 106). In the struggle with modern power-knowledge-subjectivity, ethical self-fashioning cuts across categories of thought as a force that resists a "science of life". Thinking of existence in aesthetic terms releases it from the realm of scientific knowledge, and the endless self-decipherment and subjection to psychological, biological, and economistic norms. In describing an ethics which is individualistic, voluntaristic, non-normalising, and transgressive, it seems apt to include aesthetics, despite the fact that this inclusion fails to square precisely with ancient ethics. Foucault's interpretation of ancient ethics is directed to a present in which, as Hadot himself admits, "transcendental notions of 'universal reason' and 'universal nature' do not have much meaning anymore" (Hadot 1995, 208).

Ancient ethics and philosophy were structured by the relationship between subjectivity and truth, which Foucault called "spirituality".[12] While philosophy is concerned with knowing oneself, spirituality is "the search, practice and experience through which the subject carries out the necessary transformations on himself in order to have access to the truth" (HS, 15). The subject's access to truth is grounded on *ascesis*, on transforming her mode of being as a subject through certain exercises. Spirituality is the price paid for truth, which "is never given to a subject by right" (HS, 15). In order to gain truth, the subject must undergo a sort of conversion, by transforming her way of being a subject, her way of seeing and inhabiting the world. Truth is not experienced as correspondence between beliefs and states of affairs, or between propositions and experiences, but emerges as salvation, happiness, tranquillity, or fullness (HS, 16, 17). This stands in contrast to modern philosophy, where spirituality has been occluded and access to truth is by knowledge gained by method and evidence, akin to science.

Foucault's notion of truth is both diagnostic, disclosing power/knowledge relationships, and *etho-poetic*, transforming the mode of subjectivity. This truth is not acquired by mere acts of cognition. The subject, as that which stands opposed to objects in the world, does not by its nature have an automatic right of access to the truth of those objects or of itself. The subject must transform herself to gain access to the truth. Foucault understands his books as works which transform the subject, opening this changed subject to new truths. They do this by functioning more as experiences than as records of historical truth. The essential thing is not that they operate in a factual mode in the language game of truth or falsehood, but as experiences which "might

---

[12] In Foucault's usage, spirituality doesn't involve any commitments to religious concepts such as supernatural beings or immortality of the soul.

permit an alteration, a transformation, of the relationship we have of ourselves and our cultural universe: in a word with our knowledge" (cited in Best, 138).

As well as his distaste for the grand pronouncements of universal discourses, I think Foucault was equally wary of the sheer power of reasoned argument, a position that anyone who has argued with a good-willed and well-informed friend about philosophy, politics, or morality might agree. That such arguments are often left unresolved (even for those guided by Habermas's ideal speech situation!) suggests that Foucault's wariness is well-founded. I think that rather than being compelled by argument, Foucault urges a more fundamental mode of transformation by experiences from which one emerges to see the world differently. To modern ears, such a notion of truth accessed through spiritual practices sounds like a relic of a superstitious past that we should put behind us. Foucault maintains that the reason why spirituality is so difficult for us to grasp is due to a historical event in which the scientific knowledge of the self came to overshadow the care of the self. Inevitably, our perspective is determined by the framework which science has established and which induces a sense of "retrospective realism" such that we imagine this framework forced upon us by the world itself. In modernity, we gain access to truth by scientific knowledge, as "access to a domain of objects" by objective, methodical thought, logical analysis, evidence, and so forth (HS, 191). The thought of spiritual access to truth, experienced as fulfilment and salvation, no longer has any meaning beyond religion or self-help seminars.

It could be objected that we have access to aesthetic, moral, or religious experiences only by being constituted—formed, conditioned, indoctrinated, educated—as particular sorts of subjects. However, the objection goes, these are not experiences of "truth" in any *literal sense*. These experiences are not "truth apt". However, this move doesn't so much delineate a pure realm of truth as serve to diminish the significance and import of these experiences and the role they play in our lives and social practices. We should consider what Foucault says about psychoanalysis and Marxism which, while "not exactly sciences" possess "at least certain elements, certain requirements of spirituality". Two questions—what the subject must be in order to gain access to truth, and what aspects of the subject will be transformed by access to truth—are "absolutely typical of spirituality [and] are found at . . . the source and outcome of both these knowledges" (HS, 29). Foucault is suggesting that we must *be* a certain way in order to grasp the truths of psychoanalysis or Marxism and that these "truths" will change us. Psychoanalysis and Marxism require a certain receptivity that is more than just attentiveness to facts or logic. The truth involved is not the correspondence of states of affairs to propositions. It involves a sensitivity that will enable us to experience what was totally unexpected and could not be deduced merely from acquaintance with facts arrived at by scientific methods and evidence. This might be the grasping of an autobiographical genealogy in psychoanalysis, or the import of a critique of ideology in Marxism. Foucault refers to an experience that "has the function of wrenching the subject from itself, of seeing to it that the subject is no longer itself" (EW3, 241). It is an experience of apprehending a new network of significances, relationships, and values that change the subject's self-apprehension and the world in ways unforeseen. It is this sort of experience that has been covered over, invalidated by the demand for scientific method and evidence directed at objects independent of our mode of being. As we will see, Foucault argues that a historical event, the Cartesian moment, ushered in this covering-over by constituting a form of subjectivity which enabled the scientific worldview to narrow what can count as true. With its compelling qualities of naturalness and givenness, this new regime of truth went on to occlude the older notion of truth.

THE GENEALOGY OF THE SUBJECT

Having set up the basic concepts underpinning Foucault's final period of work, we can now trace his account of the subject's genealogy which enables him to locate both his own critical practice and science within modernity. I am persuaded by Edward McGushin's reading of this work as an *ascesis*, directed at the transformation of subjectivity, both Foucault's and the receptive reader's (McGushin 2007). This *ascesis* is an urgent and indispensable task.

> When today we see the . . . almost total absence of meaning, given to some nonetheless familiar expressions . . . "getting back to oneself", "freeing oneself", "being oneself", "being authentic" etc. . . . we may have to suspect that we find it impossible today to constitute an ethic of the self, even though it may be an urgent, fundamental and politically indispensable task, if it is true after all that there is no first or final point of resistance to political power other than in the relationship one has to oneself. (HS, 251–2)

This needs unpacking. The "almost meaningless expressions" ("freeing oneself", "being oneself", "being authentic" etc.) obscure the real political question about how we should live, which cannot be decided by a scientific image of our "true" selves. We see, in Foucault's analysis of the repressive hypothesis, the modern obsession with knowing the scientific nature of the self as a sort of prison. The host of proliferating discourses doesn't liberate a "true" self, but only one linked to further machinations of power (EW1,

271). Thus the modern regime of power/knowledge constitutes subjectivities which are necessarily already political. The task is impossible because, by the eclipse of the care of the self, the only truth of the self we have recourse to is the one presented by the authoritative discourse of the sciences. Induced normalisation produces widespread forgetfulness about the question of how to live a good life. Sciences such as economics presuppose "true" goals of human life and society and so preclude from the outset any serious consideration of such goals. Our modern attitude is dominated by an anxiety to find the truth by scientific inquiry into our natures, which will then tell us how to live. The modern subject has transformed itself into a scientific object (EW1, 294). Political resistance requires a new ethics based on a relationship with our subjective self not possible through science.[13]

In outlining Foucault's genealogy of the subject, I will draw on the lectures from 1981 to 1984, which I present chronologically in terms of their genealogy, not Foucault's biography. I will start with the 1984 lectures which show how, as a response to a crisis in democracy in ancient Greece, care of the self came to be linked to *parrhesia* or fearless truth-telling. After examining Foucault's discussion of the Cynics, I will turn to what Foucault calls the "Cartesian moment" when the ancient ascetic practices were left behind and, by a realignment of power/knowledge, the modern sciences emerged. The chapter concludes with Foucault's analyses of Kant, the Enlightenment, and modernity, which frame his views on both philosophy and science.

Socrates, the Cynics, and *Parrhesia*

In his 1983–4 lecture series, Foucault introduces the concept of a "mode of veridiction", or a particular form of truth-telling in which the subject recognises herself and is recognised by others as speaking the truth (CT, 2–3). Foucault notes three modes of veridiction in the ancient world—that of the prophet, the sage, and the teacher-technician. *Parrhesia*, courageous truth-telling, is a fourth mode which Foucault contrasts to the first three (CT, 15). It involves showing others what one is thinking by avoiding any kind of rhetoric, saying what is true because one believes it is true, and having the courage to speak this truth in the face of danger (Foucault 1983). In Plato's dialogue the *Alcibiades, parrhesia* proceeds in the form of an "ongoing ontology of the soul", requiring an account of the nature and condition of the soul. However, in the *Laches*, Plato has Socrates set

---

[13] "Recent liberation movements . . . need an ethics but they cannot find any other ethics than an ethics founded on so-called scientific knowledge of what the self is, what desire is, what the unconscious is, and so on" (EW1, 255–6).

up the *parrhesiastic* scene within which an individual learns to care for himself by focusing on the form of his life, the choices and practices that make up his everyday existence. He learns to step back from himself, making himself into a question, a problem, and an object, not just of knowledge, but of care.

From these two Platonic dialogues, Foucault traces a historical narrative comprising two separate strands within which he inscribes the history of philosophy Both dialogues start on the basis of resisting the prevalent neglect of the soul by taking care of oneself. However, from the *Alcibiades*, "the care of the self leads to the question of the truth and specific being of that which one must be concerned about. What is this 'me', this 'self', we must care about? . . . what do we discover in the mirror of the soul contemplating itself?" (CT, 246). This leads to the Western metaphysical tradition which seeks to *know* the soul as an ontologically distinct reality (CT, 160). Here we see the emerging preconditions for the study of human beings as objects that were to come to fruition in the nineteenth century as the human sciences. However, from the *Laches*, "the care of self does not lead to the question of what this being I must care for is in its reality and truth, but to the question of what this care must be and what a life must be which claims to care about self" (CT, 246). The *Laches* never raises the question of exactly what it is one must care for. Rather "what is designated as the object one must take care of is not the soul, it is life (*bios*), that is to say the way of living" (CT, 127). This discourse, accentuating the living body as subject to constant testing, is taken up by Cynicism.

The Cynics practised a form of *parrhesia* which vehemently rejected the conventional social and political world, aggressively challenging all values and practices. Cynicism obstinately tied the duty to speak the truth to the true life. The conventional world was seen to be in opposition to the truth, which the Cynic modelled on nature, animality, and natural instincts. The Cynic refused to have shame about his body and its desires. He lived in public in near nakedness and poverty, haranguing those who went by. In this uncompromising form of *parrhesia*, involving a life at odds with convention, the bodily presence of the Cynic represents a permanent provocation to the society he finds false. The Cynic makes his life a manifestation of truth. Established philosophical principles of reasoning were displaced not by deploying them in discourse, but directly in the Cynic's way of life, which confronts people with the truth of their own lives in a manner irreducible to discourse. The true life appears as a completely other life, incompatible with the conventional life which it mocks. Incurring the anger of others, the Cynic risks his life by displaying it. This life is political, not leading to inner tranquillity, but militantly calling for another way of life. It is a form of care linked to unrestricted courageous truth-telling "which pushes its courage and boldness to the point that it becomes intolerable insolence" (CT, 165).

Christian asceticism became "the major medium of the Cynic mode of being" from the fourth century (CT, 181).[14] At the same time, authorities entrusted the conduct of souls to pastors. The individual was no longer able to bring about her own salvation. A relationship with God was mediated by the requirement for obedience, leading to a mistrust of the self, which became "the object of an attentive, scrupulous, and suspicious vigilance" (CT, 334). By including the principle of obedience, Christian asceticism forged "a new relation of the self to the truth, a new type of power relation, and a different regime of truth" (CT, 320–1). In this new regime, *parrhesia* was obscured and devalued (CT, 333). By the institution of the confession, the hermeneutic subject is constituted to exercise vigilance over her thoughts by examining and confessing them to a priest, who interprets their true meaning.

Philosophy was gradually purged of radical Cynicism and stripped of spirituality, taming it to become a theoretical discipline, a body of knowledge more than a way of life. Despite the paucity of its textual legacy, Foucault sees Cynicism as a force in modernity, still posing the question of the true life as the radically other life. This legacy is not a doctrine but "much more an attitude and way of being" (CT, 178). Cynicism still asks the "perpetually embarrassing question" about the philosophical life (CT, 234). Foucault argues that philosophy can't be separated from a philosophical existence. It must "always be more or less life exercise". Philosophy is a form of *parrhesia*—courageous, frank, truth-telling—that constitutes its activity as a mode of subjectivity. The philosopher is bound to a way of life, a true life. This distinguishes philosophy from science which, in its objectivating stance, cancels out any subjective commitments except to propositional truth. Anyone can be a scientist if they have the skills and education. Philosophy, however, entails a commitment that has consequences for a way of life.

This idea of a "philosophical life" seems peculiar to modern ears. It is best grasped through the work of Hadot, who regards ancient philosophy as a way of life comprising ascetic exercises aimed at transforming the mode of subjectivity of the practitioner (Hadot 1995). Foucault argues that, although we can still trace its subterranean path through philosophy's history, this understanding is now almost completely lost to us. The philosophical question of a true life was taken over at the end of antiquity by religion, obscuring and covering over the ancient ethical notion of an aesthetics of existence. Later the institutionalisation of truth-telling practices in the form of normed,

---

[14] The idea of bearing witness to the truth was found in spiritual movements such as the mendicant orders throughout the Middle Ages. "The Franciscans with their destitution, wandering, poverty and begging are up to a point, the Cynics of medieval Christianity" (CT, 182).

regulated, and institutionalised science emerged, the result of a historical rupture beginning in the sixteenth century which narrowed the understanding of truth and invalidated what is outside the scientific method (CT, 235). The emphasis on deciphering and renouncing the self is eventually incorporated into the disciplinary structures of the human sciences (EW1, 274).

The Cartesian Moment

Foucault refers to the historical rupture that heralded modernity as the "Cartesian moment". It occurred in the context of tensions between the Church's pastoral care of the self and the political art of governing populations (EW3, 315; STP, 227–48; SD, 25). In the sixteenth century, "the art of government exploded" as a general social problem embracing government of oneself, the poor, families, armies, souls, children, cities, and the state (EW3, 201–2; PT,44). As we saw in chapter 6, biopower arose as a new form of political power which translated the spiritual government of souls into the political government of populations. It was into this context that Descartes inscribed his *Meditations* (STP, 230). The meticulous quality of Descartes's method can only be understood if we bear in mind "that from which he wants to distinguish and separate himself, which is precisely these methods of spiritual exercise that were frequently practiced within Christianity and which derived from the spiritual exercises of Antiquity" (HS, 294).

After the Cartesian moment, philosophy was able to claim back a form of *parrhesia* from Christian spiritual practices, such that "knowledge itself, and knowledge alone gives access to the truth" (HS, 17). The subject was no longer required to transform its being. There were conditions, but these weren't spiritual conditions. The Cartesian moment leads to the government of life and populations becoming the concern of politics, in the form of biopower backed by science. Descartes's method becomes a new form of care of the self that ensures the proper conduct of the mind, thus giving the subject access to certain knowledge of everything that is useful for life.[15] To grasp how this scientific worldview was opened up and how the modern subject was constituted, we need to examine Foucault's analysis of Descartes's *Meditations*. In this 1972 essay, Foucault offered a strikingly original interpretation that bears directly on the history of subjectivity and the fate of the philosophical notion of the care of the self (HM, 550–74).[16] Here we see an earlier Foucault, in the

---

[15] Instead of "the speculative knowledge taught in the Schools", Descartes's new form of knowledge "is most useful in life" and will "make ourselves masters and possessors of nature" (Descartes 1987, 78).
[16] The essay was a response to Jacques Derrida's criticism of Foucault's interpretation of Descartes in HM.

midst of his turn to genealogy, already concerned with the history of subjectivity and the self-constitution of subjects in their relation to truth.

The *Meditations* present themselves as a response to Descartes's awareness of having previously accepted many false opinions as true and his lack of any firm criteria to distinguish between truth and falsity (Descartes 1968, 95). His ultimate motivation is to gain a firm foundation on which to build certain mathematical and scientific knowledge. By systematically employing general doubt to doubt everything, Descartes aims to investigate what access the subject has to certain truth, and how the subject can be certain that it is the truth. Foucault sees Descartes as initiating a particular regime of truth, a particular way of binding the subject to truth. Descartes's *Meditations* are not merely an investigation or an argument, but an *ascesis* that transforms the subject, leading her to a form of self-knowledge which opens up the scientific worldview that then replaces *ascesis* as the access to truth.

Foucault focuses on the passage in the first meditation in which Descartes questions whether he can actually doubt his senses. Certainly one can doubt what is "weak and distant", but can one really doubt what is "vivid and near"? (HM, 554). A "practical syllogism" tells Descartes that he *must* doubt the senses because they have deceived him before, but he simply can't find a way to doubt that "he is sitting here next to the fire" even though reason tells him that he ought (HM, 564). What is required is work on himself, to become a different sort of subject. Foucault argues that Descartes's hypothesis that the subject is dreaming is not merely another proposition in an argument, it modifies the subject (HM, 554–5). The dream hypothesis allows doubt to overcome "the actuality of the meditating subject (the place of his meditation, the action that he is carrying out, the sensations that strike him)" (HM, 565). This extension of doubt to the actuality of the meditating subject, "sitting here, next to the fire", can only take place if the subject's mode of presence to itself can be separated from this "system of actuality". Doubt allows the subject to disclose itself in a way that is distinct from its actual being-in-the-world. By a form of distancing, a subject is posited who no longer needs to distinguish between the near and far or the vivid and weak, but who can view space and time neutrally from a disembodied scientific perspective. The dream appears as though it could be an actual state of the meditating subject, allowing him or her to free himself herself from all representations (HM, 567).

What Descartes is doing in the *Meditations* is undertaking *ascesis*, modifying his subjectivity, just as his readers are being led to do. However, according to Foucault, this *ascesis* serves to establish a relationship between the philosopher and truth that will no longer rely on *ascesis*. After the sixteenth century, "We have a non-ascetic subject of knowledge. This change makes possible the institutionalisation of modern science" (EW1, 279). A particular form of objectivity is instituted, motivated by the concern to let the object show itself as it is, untainted by anthropomorphic, idiosyncratic, or "subjective" constructions and interpretations. But this form of objectivity is also a form of subjectivity which strives to view the object in the right way, by adopting the appropriate methods and posture as a subject. The modern philosopher doesn't have to undertake spiritual exercises and transform herself. She just examines the evidence and follows the method and rules laid down for analysis. Philosophy can reclaim *parrhesia* and govern itself and others according to its own proper rules. Clear and distinct perceptions will enable philosophy and science to analyse things and see how they are composed. This is a new regime of truth, a new way of binding the subject to truth which is linked to a new application of power.

The Cartesian moment philosophically endorses knowledge of self (*gnothi seauton*) and discredits care of self (*epimeleia heautou*) (HS, 14). Self-knowledge, in the form of the impossibility of doubting my existence as a subject, makes "know thyself" into a fundamental means of access to truth (HS, 14). The hermeneutic imperative of the pastoral, to know oneself, continues, now guided by science rather than religion. As we have seen, this imperative is increasingly generalised throughout our "singularly confessing society" in the relationships between an individual and their doctor, psychiatrist, psychologist, or psychoanalyst, the scientific expert who will interpret the truth of ourselves (CT, 5; HS1, 59). Instead of a notion of truth accessed through self-transformation, Cartesian knowledge "is knowledge (*connaissance*) of a domain of objects" (HS, 191). Rather than enlightenment, fulfilment, completion, or salvation, the reward for this knowledge is indefinite scientific progress (HS, 19). Although the new truth regime liberates subjects from the power apparatus associated with the Renaissance *episteme,* it also involves impoverishment by concealment. Discussing psychoanalysis and Marxism, Foucault suggests that the attempt to bolster the scientific status of these disciplines required concealing the relationship between truth and the subject and the spiritual dimension entailed (HS, 29–30).

Accompanying the Cartesian moment, political power is transformed by being no longer concerned with merely abstract subjects of juridical law, but rather with subjects in which it fosters life at the levels of individual bodies and of the population. Political power becomes concerned with people's desires, the uses and conditions of their bodies, how they behave, and their relations with others. The life of individuals and populations is reduced to biology and public hygiene, which is forged into a scientific/technological support for the deployment of this rationality (SD, 182–3, 239–63). Biopower takes over the care of the self, putting forward the biological life of the population as its material to be ethically normalised. Scientific knowledge of government, subjects, and society is central to this project. Knowledge of the state must include wealth, resources, and the population's characteristics.

Political reason tends to be reduced to the administration of life, which is increasingly seen in biological and economic terms. As knowledge becomes regimented, philosophy becomes disqualified. It cannot constitute itself on the model of science, nor can it delineate a distinctive field of study or set of problems.

Kant, Enlightenment, and Critique

In terms of the narrative thread set out in OT, if the Cartesian moment marked the shift from the Renaissance to the Classical *episteme*, it is the figure of Kant who defines the shift to the modern. It is from this point in Foucault's genealogy of the subject that we see his own project in sharper focus. In the wake of the Cartesian moment, the practices of the self migrated into forms of government directed towards discipline and biopower—the biological and economic normalisation of individuals and living populations. Philosophy was shorn of any residual affiliation to spirituality. If it was irrational to think that one could transform one's relation to the truth, it isn't surprising that Kant will discover human finitude in terms of necessary limits, rather than the possibility of breaching limits. Kant gives the Cartesian moment a "supplementary twist . . . which consists in saying that . . . we cannot know the subject. Consequently, the idea of a certain spiritual transformation of the subject . . . is chimerical and paradoxical" (HS, 190). The Kantian subject is the condition of all possible experience and as such lies beyond experience itself. In my discussion of OT (chapter 5) we saw Foucault highlight the oscillation between irreconcilable perspectives of transcendental subject and empirical object in Kant's philosophy. Here Foucault seemed uncompromisingly critical of Kant's transcendentalism. However, Foucault has a much more affirmative attitude towards some of Kant's essays, such as *What Is Enlightenment?* Here he links critique to Kant's understanding of Enlightenment, which involves a striving towards maturity and entails both an obligation and the courage to pursue it as an ongoing task (PT, 41–3, 97–121; GSO, 26–39). Since humans are responsible for their state of immaturity, Enlightenment can only occur by "a change that [they themselves] will bring about in [themselves]" (PT, 100). Here it seems Foucault is reading Kant through the lens of ancient philosophy, specifically the ethical qualities of the self's care of the self and the courage required by *parrhesia*.

Habermas claims that Foucault has two radically different readings of Kant which reveal fundamental contradictions within his work. Referring to Foucault's essay, *What Is Enlightenment?*, Habermas asks "how such an affirmative understanding of modern philosophy . . . fits with Foucault's unyielding critique of modernity" (Habermas 1992, 152). To clarify this apparent discrepancy, we should recall that Foucault's critique is part of a broader tradition of the critique of reason extending from Kant and Hegel through Nietzsche and Weber to the Frankfurt School. We should also note that Foucault discerns a strong contrast between Kant's transcendental critiques, which deal with the necessary and universal conditions of possibility for reason, and this minor essay which addresses the present in which we find ourselves. Foucault arrived at his historical form of critique by a Kantian move—asking after the limits and conditions of possibility of subjective experience. The fact that, in addressing Kant's transcendental critiques, Foucault problematises these limits rather than endorsing them is not so much a negation of Kant's critical project, as a radicalisation constituting a critique of critique. This radicalisation is both an endorsement of critique in the broader sense, exemplified by Kant's essay which inquires into its present situation, and a criticism of critique in the narrow sense, exemplified by Kant's transcendental critiques which inquired into the universal and necessary limits of knowledge.

Foucault's critique is not just a problematisation of regimes of power/ knowledge, but a self-transformative process directed at opening a space beyond present limits. To question truth regarding its power effects and to question power regarding its discourses of truth requires a critical practice based on self-transformation. This "art of voluntary insubordination" ensures "the desubjugation of the subject in the context of what we could call . . . the politics of truth" (PT, 47). By "politics of truth", Foucault means the relations of power that order in advance what can count as true and meaningful. Just as Kant's question about Enlightenment reflected on his present situation, so Foucault reflects on what is particular to his present, two hundred years later. Modernity is not an epoch but, he tells us, an "attitude", an *"ethos"*, a "mode of relating to contemporary reality" (PT, 105). Rather than "a theory, a doctrine, or . . . a permanent body of knowledge that is accumulating", this *ethos* entails "a philosophical life in which the critique of what we are is at one and the same time the historical analysis of the limits that are imposed on us and an experiment with the possibility of going beyond them" (PT, 118).

PHILOSOPHY AND SCIENCE AFTER KANT

While Kant saw philosophy as an academic discipline, Foucault saw philosophy as a way of life. While Kant's project responded to the "scandal of metaphysics", Foucault responded to today's main danger, the growth of modern power (PT, 54). This growing power is inseparable from the growth of scientific knowledge. Foucault sees one legacy of Kant's epistemological critique as a stitching together of scientific positivism and the development of nation states to establish a science which plays "an increasingly determinant part in the development of productive forces and . . . state-type power increasingly exercised through refined techniques" (PT, 50–1). He is referring to the fact that, since the nineteenth century, science has been increasingly integrated with state power, which uses it to both develop productive forces and maintain and legitimise itself. This rationalisation has "effects of constraint and maybe of obscuration, of the never radically contested but still all massive and ever-growing establishment of a vast technical and scientific system" (PT, 55). Blind to their own conditions of emergence within power/ knowledge, Kant's critiques lent science the authority to operate on the basis of the necessary, universal finitude of the transcendental subject, within the secure limits of what could be known. Foucault wants to challenge the power effects that such authority engendered—effects "never radically contested". This challenge employs genealogy to reveal that what can count as true and meaningful is linked to contingent forms of power/knowledge/subjectivity, thus opening the possibility of different relations between the subject and truth beyond those encapsulated by the methods of modern science.

Foucault locates himself in relation to critical theory by outlining the tensions arising between Kantian transcendental philosophy and the Enlightenment understood as an attitude or *ethos*, tensions which lead to increasingly sceptical questioning: "For what excesses of power . . . is reason not itself historically responsible?" (PT, 51). In Germany, the question of the Enlightenment was "shaped into a historical and political reflection on society . . . . In France, it is the history of science in particular that has served as a medium for the philosophical question of historical *Aufklärung*" (EW2, 468). Although the two national styles are very different, Foucault observes that "both groups ultimately raise the same kind of questions . . . . In the history of the sciences in France, as in German critical theory, what is to be examined, basically is a reason whose structural autonomy carries the history of dogmatisms and despotisms along with it—a reason, therefore, that has a liberating effect only provided that it manages to liberate itself" (EW2, 469).

In Germany, the Frankfurt School aimed to "show the connections between science's naïve presumptions, on one hand, and forms of domination characteristic of contemporary society, on the other" (PT, 51). Foucault generally agrees with their approach, crediting Horkheimer as the first to highlight the fundamental problem: "The Enlightenment's promise of attaining freedom through the exercise of reason has been turned upside down, resulting in a domination by reason itself, which increasingly usurps the place of freedom" (EW3, 273–4). However, he finds the Frankfurt School's conception of the subject too permeated with Marxist humanism for him to accept and argues that "what we need to do is not to recover our lost identity, or liberate our imprisoned nature, or discover our fundamental truth; rather it is to move toward something altogether different" (EW3, 275).[17] Foucault doesn't say what this "something altogether different" is. To spell this out would be to restrict the possibilities of freedom by prescribing in advance notions of society or of human nature, which are not things to be discovered, but things made and hence able to be made differently.[18] In France, the notion of the Enlightenment, with all the questions involved, was less influential than in Germany (PT, 52). However, Foucault notes that the problem of the historicity of the sciences addressed by Bachelard and Canguilhem gave rise to the same fundamental question of the Enlightenment, albeit by asking different questions: "How is this rationality born? How is it formed from something which is totally different from it?" (PT, 54). Rather than the Frankfurt School's analyses of the rationalisation of society or culture as a whole, Foucault's background in the history and philosophy of science leads to analyses of rationalisation "in several fields, each with reference to a fundamental experience: madness, illness, death, crime, sexuality and so on". Rather than invoke the progress of rationalisation in general like Weber or Habermas, he analyses specific rationalities (EW3, 329).

Foucault thinks that since Kant, the question of the Enlightenment has been posed primarily in terms of knowledge, most often by asking "what false idea has knowledge gotten of itself and what excessive use has it exposed itself to, to what domination is it therefore linked" (PT, 58–9). He turns this around to ask about gaining access "not to the problem of knowledge, but to that of power" (PT, 59). He wants to locate the connections between power and knowledge. In the 1980s, he added the subject, such that all three elements— knowledge, power, and the subject—must always be analysed together. But in so doing, philosophical critique can't play the role of a science. Philosophy can't "divide the true and the false in the domain of science" (GSO, 354). While science discovers what we cannot see, philosophy allows us to see the same things in a new light which reveals their constitution from our contingent practices tied to structures of power/knowledge/subjectivity. "Where the role of science is to make known that which we do not see, the role of philosophy is to make seen that which we already see" (cited in Kelly 2009, 129). Foucault is also adamant that philosophical critique cannot say what should be done in politics. Nor should it set out to free the subject. Rather, philosophy has to define the forms in which the relationship to self may

---

[17] This is a travesty of the Frankfurt School position.
[18] "In the course of their history, men have never ceased to construct themselves, that is, to continually displace their subjectivity, to constitute themselves in an infinite, multiple series of different subjectivities that will never have an end and never bring us in the presence of something that would be 'man'" (EW3, 276).

possibly be transformed. Philosophical critique analyses the entwinement of power, knowledge, and subjectivity, though from a location distanced from politics, science, and the self.[19] In other words, critique doesn't propose solutions, discover anything new, or legislate on truth or morality. In chapter 8, I will expand on the thought that although Foucault's critique is not prescriptive in any ordinary sense, neither is it merely descriptive. It steps back to render visible new possibilities for the relation of the subject to truth within contexts of power and resistance, thus inviting, rather than legislating or prescribing, political action.

In HS2, Foucault claims that the only worthwhile motivation for his work is a curiosity that doesn't seek "to assimilate what it is proper for one to know, but that which enables one to get free of oneself" (HS2, 8). Philosophy is "the critical work" of "the endeavour to know how and to what extent it might be possible to think differently, instead of legitimating what is already known". It explores what might be changed in its own thought, "through the practice of a knowledge that is foreign to it". This foreign element provides "the assay or test by which, in the game of truth, one undergoes changes" (HS2, 9). This practice is a resistance to the modern philosophical neglect of the self, which has now passed over into the hands of doctors, psychiatrists, counsellors, and scientific knowledge. This is what Foucault would see as the main danger today—a configuration of power-knowledge-subjectivity, a regime of truth bound up with science, especially the human sciences, by which subjects are subjugated. This configuration is itself the residue of past resistances, such as Descartes's resistance against the Scholastic philosophy of the Renaissance *episteme* or Kant's resistance against the scandal of metaphysics. Grown far beyond the context of the problematisations from which they emerged, these regimes of truth have hypostatised into hegemonic forms of domination which call for further resistance.

Foucault would insist there is no set of practices and discourses that can be considered final, true, and universal. In a veiled reference to Habermas, he insists that critique is not "a matter of identifying general principles of reality", timeless universal principles from which we can discern "what is true or false, founded or unfounded, real or illusory, scientific or ideological, legitimate or abusive" (PT, 59, 60). There is no essence, no "natural" form of human life or sociality, that science or philosophy can discover. Philosophical critique shows this, not by rejecting scientific truths, but by revealing the contingent constitution of constellations of power/knowledge/subjectivity as preconditions for the emergence of such truths. In this

---

[19] "Philosophy as *ascesis*, as critique, and as restive exteriority to politics is the mode of being of modern philosophy. It was, at any rate, the mode of being of ancient philosophy" (GSO, 354).

chapter, we have seen Foucault expand his analytic framework to embrace the axis of active subjectivity, inextricably entwined with the archaeological axis of knowledge and the genealogical axis of power, to offer a broader account of our historical constitution. In chapter 8, I bring Foucault and Habermas into dialogue with each other by drawing on the radical reflexivity of modern reason and revealing two distinctly separate but interdependent standpoints within its bounds.

Chapter 8

Science, Philosophy, and Modernity

The claim that science encompasses all we can know is not a scientific claim. It is a metaphysical claim which has risen to prominence as science increasingly prescribes the boundaries of our thought. Foucault and Habermas both reject scientific naturalism's claim that what we can know of nature is limited to its scientific image. They both resist the authority which proclaims science's purity, masks its power-effects, and insulates it from challenge. Both take the pressure off ontology by asking not so much what is real, but rather what is the social function of treating something as real. This capacity to step back to a more distanced view has roots in both critical theory's incorporation of scientific perspectives and genealogy's historical problematisations. But there are important differences. To see Habermas's and Foucault's projects in relation to each other, a broader framework is required. This chapter will elucidate this framework by teasing out their projects in terms of two tendencies within modern reflexive thought.

I shall begin by raising the question of reconciling the projects of Habermas and Foucault. I will then consider self-reflexivity and its radicalisation within modernity in order to characterise what I call the strategies of "discovery" and "self-transformation" used by each thinker. I will then examine the imbricated dialectical relationship between these strategies by considering them in terms of necessity and contingency, the Enlightenment, and the role of normative foundations. I will return to the question of reconciliation before extending this analysis to highlight how Foucault's genealogy and Habermas's critical theory yield differing accounts of the development of science, and how both approaches open new possibilities.

THE RECONCILABILITY OF
HABERMAS AND FOUCAULT

Can, or should, the projects of Habermas and Foucault be reconciled to offer a combined view? Perhaps we can take parts of each and fit them together to form a single project which could inform our politics or morality. A good deal of interpretive effort has gone into this endeavour. Benhabib, Hoy, McCarthy, Allen, Cooke, and Koopman have all provided nuanced interpretations which contribute to the project of crafting a viable critical theory encompassing the best insights of both Habermas and Foucault (Benhabib 1986; Hoy and McCarthy 1994; Allen 2008; Cooke 2006; Koopman 2013). While I fully support this aim, mine is different. In this chapter, I want to understand Foucault and Habermas more in their own terms, and to make use of the tensions that arise to cast light on our modern scientifically informed rationality. My approach is best seen by comparison with Koopman's, which I drew on in chapter 3 to interpret Habermas's universalism in terms of ongoing processes of universalising claims across different contexts. Now I want to consider Koopman's "strategy of delegation" which takes the best of Foucault (his genealogical problematisations) and the best of Habermas (his normatively-oriented critique) to craft an improved critical theory. This strategy relies on reading the work of both thinkers as including a descriptive "explanatory-diagnostic aspect" and a normative "anticipatory-utopian aspect".[1] Koopman delegates Foucault's genealogical problematisations to the explanatory descriptive task of critical theory because he sees Foucault's project as divided between genealogical problematisation, which he does well, and ethics, which he does poorly. I want to problematise this distinction by adopting a broader, more pragmatic understanding of normativity. Initially, I only need to discuss Foucault, since similar sorts of considerations apply to Habermas.

In the final years of his life, Foucault frequently stressed the unity of his project, usually in terms of the contingent historical construction of the subject along three intertwined axes—knowledge, power, and the self-constituting subject (EW1, 262–3, 318; EW3, 326–7; HS2, 6; FR, 336–8). Notwithstanding Foucault's proclaimed unity, Koopman's division between genealogical problematisation and ethics has initial plausibility. First we problematise, then we respond to this problematisation. As diagnostician, Foucault undertakes the intellectual work of assembling problematisations of the ways in which we have been constituted as subjects. Then as ethicist, he provides a basis, not a good one according to Koopman, for his readers to respond to these

---

[1] This definition of the two moments required for any rigorous critical theory draws from Benhabib's account (Benhabib 1986, 226). For discussion of the two moments in relation to Habermas see p. 96 chapter 3.

diagnoses by working on, caring for, and transforming themselves. We should note, however, that the care of self involves a set of practices by which one cultivates oneself. I would argue that such practices include the actual reading of Foucault's genealogies, through which we come to recognise ourselves *in an entirely unforeseen way*, as subjects subjected to modernity's regimes of power/knowledge. This doesn't involve learning about a problem which we then, as a separate step, do something about. Like the problem addressed by Wittgenstein's concept of "perspicuous presentation", the problem here is that we can't see the problem (Wittgenstein 2003, 122). I think that it is along these lines that McGushin sees the late work of Foucault as opening up "an ethic that represents a new way of practicing philosophy . . . the activity of reading and thinking about these as such is already a practice of care, a conversion of regard towards oneself" (McGushin 2007, xxi).

On this basis, I argue that Foucault's genealogy, including his earlier archaeologies, is itself an *ascesis*, an exercise that has the potential to transform the reader's subjectivity. Referring to HM, Foucault tells us that he seeks "to construct" himself, and to "invite others to share an experience of what we are . . . an experience of our modernity in such a way that we might come out transformed" (EW3, 242). It is not that we firstly diagnose modernity and then, on the basis of this diagnosis, go about transforming ourselves. Similarly, in OT, Foucault refers to his analysis as an "experience" of order (OT, xxiii). This experience challenges categorical frameworks in a reformation of subjectivity analogous to a religious or psychoanalytic experience. After grasping the import of the subject's historical transformations by power, knowledge, and self-constitution, we can no longer think of our possibilities in quite the same ways. Such self-transformation involves a certain mode of receptivity that amounts to an ethical orientation held prior to normative deductive reasoning. It emerges not from some new insight into practical reason, but from problematisations of it. Certainly, "genealogy" and "ethics" differ in their meanings. But as exercises in a philosophical life, they are inseparable. Understanding genealogy as self-transformation in this way undermines the clean division of Foucault's work into genealogical problematisations and the ethics which responds to problematisations.

It is not clear to Koopman "how the normative commitments that Foucault dug out of antiquity can be made suitable for modern moral living" (Koopman 2013, 201). But the ancient practices Foucault discusses are not intended as moral recommendations. They are genealogies of the self's formation of itself and play the role of problematisations by virtue of the fact that this self-formation, like our own, has a history. The fact that Foucault doesn't develop a normative account is not an "innocent weaknesses" as Koopman suggests, but a deliberate stance necessary for his genealogy (Koopman 2013, 216). It is necessary because, I will argue, it offers an observer's perspective unavailable to participants in normatively governed social practices. Foucault is clear that his critique cannot tell us what to do the way politics does (GSO, 354). Nor can it provide moral prescriptions. He is not concerned with what we can or should do, but with the possibilities of what we can or should do. Genealogical problematisation is not, as Koopman suggests, a separable part of Foucault's work that can be straightforwardly delegated the role of providing data for Habermas's normative reconstructions that answer the question of what we should do (Koopman 2013, 222).

Foucault criticises the "universal intellectual" who speaks on behalf of others to answer such questions. It is only at the coal-face of actual situations that "specific intellectuals" should respond. However, the problem that Koopman sees is "not that Foucault does not tell us how to live, but rather that Foucault does not tell us enough about how we might set about the task of figuring out how to live for ourselves" (Koopman 2013, 213). I think that Foucault is not concerned with figuring out how to live, but with the possibilities of living. He also can't tell us how or what to think. Self-transformation involves a displacement of our established self-identities, which only we can allow to happen to us. If self-transformation entails clearly seeing the fact that no-one can tell us how to transform ourselves, then there is nothing Foucault can tell us. He can invite, urge, gesture, provoke, and so forth, but he can't tell us in the way that Koopman wants. Such a telling would be self-contradictory. Like the Cynics, his philosophy is a way of life by which the ideal of freedom is shown, not said.[2] This non-discursive critical practice takes the form of a "true" life shown by one's bearing and resolution, in Foucault's case, the embodiment of principled resistance as *parrhesia*. Saying inevitably projects universal categories but showing offers an *experience*. Transformation does not come about through the force of the better argument or reasons justified by foundations, but through experiences which change one. Foucault emphasises that what is important is that his books "function as experience, for the writer and reader alike, much more than as the establishment of a historical truth" (EW3, 243).

Just as Foucault's project can't be straightforwardly divided into a genealogical and an ethical component, neither can Habermas's normativity be separated from the rest of his project. I will not discuss Habermas here, since my discussion of Foucault's ethics will serve my purpose of problematising Koopman's strategy of delegation. This strategy serves Koopman's purpose by erecting a partition between the normative and non-normative. However, this is not my purpose. I want to dissolve this partition by teasing out a broader notion of normativity. To do this, I need to go beyond the Foucault-Habermas

---

[2] I follow Wittgenstein's thought that ethics belongs to that realm where things cannot be said but only shown (*Tractatus Logico-Philosophicus*, 4.1212, 6.421).

debate framed by Foucault's supposedly problematic crypto-normativity and Habermas's supposedly authoritarian universal normativity.

I will therefore draw upon ordinary language, in which "normativity" refers to the social phenomenon of designating some actions or outcomes as good, desirable, or permissible, and others as bad, undesirable, or impermissible in relation to a standard against which claims, behaviour, or outcomes are judged. Norms can be merely conventional, like etiquette, or can hold inexorable sway over us, like norms of rationality. Norms can be explicit like the Ten Commandments or implicit, like the rules of grammar which we follow without being aware of following them. There is no clear distinction. Behavioural regularities to which we conform without thinking of ourselves as following rules may over time emerge from the background to become explicit foundations for conscious deliberation, judgement, and action. Alternatively, norms originally promulgated by explicit rules may retreat into this background. We should therefore see the concept of normativity as referring to something unstable, leading to constant slippage between the prescriptive and non-prescriptive, something Foucault discusses in considering the use of the word "normal" as, on one hand, "a rule of conduct, informal law, principle of conformity as opposed to disorder, strangeness and accentricity . . . " and, on the other hand, a statistical average with absolutely no prescriptive force (AB, 162).

My point is that we cannot simply choose to walk away from normativity understood in this capacious sense. Certainly Foucault's descriptions operate by suspending the normative force of certain norms while describing the historical conditions by which they are socially constructed. But this doesn't mean he is "non-normative" or that Habermas's explicit normativity is "more normative". Rather than pretend we can escape normativity, we should tease apart its threads to see what Foucault and Habermas are doing. By employing a broad conception of normativity, we can see within its layering, not only the problems surrounding reconcilability but also the roles of science and philosophical reflection in modern reason. To that end, I will now turn to the self-reflexivity of modern reason as it alternates between the perspectives of participants in social practices and of observers of those practices.

REFLEXIVITY AND ITS MODERN RADICALISATION

As *participants* in social practices, we are sensitive to, and respond to, sets of norms. For example, as moral agents we praise or blame others for actions for which we hold them responsible. But we can also become third-person *observers* by bracketing the norms governing such practices. This applies not only to morality, but also to the social practice of reasoning. For example, a psychologist might observe the norm-governed activity of a reasoning subject while bracketing that subject's rational norms. We can also observe ourselves as objects in much the same way. Habermas is correct in noting that the perspectives of participants within a shared lifeworld and those of observers who adopt an objectivating attitude towards that lifeworld's objects are inextricably interlocked (Habermas 2007b, 35; BNR, 169). "Imbricated" might be a better term. Accompanying each step backwards to a more distanced observer's view, the participant is always already present yet latent, requiring an abrupt shift in the subject's standpoint to become manifest. The anthropologist studying the moral practices of another culture may be a detached observer, yet still remains a participant in a vast field of unquestioned background norms, including scientific norms. Wherever we find an observer, there is always already, in the background, a participant responding to an unthematised set of norms. What we call an "observer" is one who foregrounds this type of relation to a certain "target" object or practice. Whether we are participants or observers depends on the thematisation of this relationship to an object or practice.

We moderns assume the possibility of adopting an observer's perspective from which we reflexively question our own received beliefs and values, so gaining critical distance from inherited norms and roles. Thus we see Foucault portray modernity's heightened reflexivity as propelled by a never-ending compulsion to keep uncovering what remains hidden (OT, 356). This necessity to pursue and uncover a background of continually receding conditions of possibility is "profoundly bound up with our modernity" (OT, 357). For Habermas, modernity *"has to create its normativity out of itself.* Modernity sees itself cast back upon itself without any possibility of escape. This explains the sensitiveness of its self-understanding, the dynamism of the attempt, carried forward incessantly down to our time, to pin itself down" (PDM, 7). In chapter 3, we saw Habermas, in his social evolutionary account, trace this tendency as yielding an accrual of potentially emancipatory differentiations within the lifeworld. Yet modernity's heightened contrast between the participant and observer perspectives can be problematic. What it is to be a rational agent from a participant's perspective seems to be in tension with the causal conditions of such agency from an observer's perspective. This tension cannot be resolved in favour of either perspective as being more fundamental, but rather requires shifting between the two perspectives. It is not clear that both perspectives bring into view the same objects. Is language really just marks and noises, the mind just the brain, or normative force just behavioural regularities? As Foucault's Nietszchean account shows us, these questions depend on a prior division of the true from the false to supply the criteria for what counts as reality. Both Habermas and Foucault criticise the stance of the detached scientific observer who proclaims the primacy of this stance. However, both would acknowledge that while an observer's "external" perspective doesn't necessarily cancel or trump a participant's perspective, it can bear on it in significant ways, even leading to revisions of lifeworld intuitions.

In terms of this framework of imbricated perspectives, both Foucault and Habermas necessarily occupy the positions of both participant and observer. The difference is in how each engages in the play between alternating perspectives, where one perspective is thematised while the other is suppressed. Adopting a radical reflexivity, Foucault wants to maintain this movement with an ongoing "rational criticism of rationality", problematising whatever appears as ahistorical and universal (FL, 353). This entails not only criticism of a particular form of rationality, but criticism of that criticism, and further criticism of *that* criticism. The rationality that is criticising is, at the same time, always potentially the object of further criticism, though this is never criticism of rationality per se. We find this trope of never-ending movement throughout Foucault's work. The self that does the caring is itself a potential object of care. The "distinctive feature of philosophy as a discourse *of* modernity and *on* modernity" is that it inquires into itself (my emphasis, GSO, 13). The modernity which inquires into itself is also the object of that inquiry. Since we are always in the position of beginning again, critique of the present requires permanent reactivation. Any stopping point seems arbitrary, because at that point one can always ask, "What standpoint lies behind and conditions my current standpoint?" Notwithstanding this perpetually shifting *ekstasis*, Foucault does make a stop to problematise the settled lifeworld of the present, wherein lie today's main dangers. Tomorrow may be different.

Habermas finds Foucault's "radical critique of reason" unnerving (PDM, 336–7). He thinks we need to nail at least some things down. From these two attitudes flow two diagnoses of modernity. Foucault wants to move on by opening a space to think again, though what we are to think he cannot tell us. Habermas nails the problem down as fundamentally one of balance, where the inhuman forces of systems have turned on their human creators to attack what makes them human. My conjecture is that, caught within the play of participant and observer perspectives, both Habermas's and Foucault's responses to modernity unavoidably reflect this structure of imbricated perspectives which constitute modern scientifically informed reason. These responses are strategic. Habermas deploys a strategy of discovery, Foucault a strategy of self-transformation.

DISCOVERY AND SELF-TRANSFORMATION

In order to gain an initial foothold on these respective strategies, I will briefly recapitulate the life experiences that have animated and ultimately shaped the two projects. I hasten to emphasise that I don't intend a speculative psychologistic reading of Foucault and Habermas, nor will I rely on biographical sources to definitively determine interpretations of their projects. Rather, I have drawn on a number of sources which contextualise their projects in terms of their personal, political, social, and cultural milieu in order to provide interpretive clues and a sharper focus on what each is attempting to do.[3] Let us recall that at the age of fifteen, like most of his contemporaries, Habermas was drafted into the Hitler Youth. He later described his father as a "passive sympathiser" with the Nazis and admitted that he had shared that mindset. This complacency was shattered by the Nuremberg Trials and the exposure of Nazi crimes. His horrified reaction constituted what he described as "that first rupture, which still gapes" (Jeffries 2017). Having read Heidegger enthusiastically, he was again shocked to find Heidegger unwilling to account for his involvement with a criminal regime (BNR, 18). It is easy to understand that Habermas would be anxious to defend a form of reason capable of resisting the forces of irrationality which he had seen around him. Unsurprisingly, he invested cautious faith in a project of extending and clarifying reason and science in an attempt to anticipate their limits and possibilities. Scientific technology promised freedom from material want and the human and social sciences offered better understandings of the crises faced by modern societies. Nonetheless, he remained aware of the risk of new forms of enslavement by the narrow instrumental reason which his Frankfurt School predecessors first criticised. He turned to a theory of communicative rationality, based on an "ethics of discourse", to promote and justify a more adequate conception of democracy. Although he takes science seriously, Habermas's prior commitment is to the lifeworld and its communicative structures, which he sees as being eroded by scientism, technocratic consciousness, and systemic imperatives. He wants to safeguard the lifeworld's communicative structures, which are so vital for open and free societies. In this endeavour, he will pursue a *strategy of discovery* of what is stable and universal and, in a qualified sense, necessary. Habermas's strategy offers a degree of certainty as a hedge against history's radical uncertainty.

To discover a firm basis to defend the lifeworld, Habermas initially adopts an observer's stance by incorporating science as a view external to the everyday lifeworld. By drawing on generalised empirical knowledge to reduce the context-dependency of understanding, he can analyse symbolically pre-structured objects and events. By reconstructing the learning processes in which societies evolve through a series of stages, Habermas discovers the progressive development of communicative capacities. Speculative developmental theories of human communicative competences drawn from

---

[3] For Foucault, see Eribon (1991), Macey (1995), Szakolczai (1998); for Habermas, Muller-Dohm (2016), and Specter (2010).

developmental psychology and anthropology provide plausible and universal empirical theories. Future science will either confirm these theories of what we are or show that we are different in some determinate way. After initially adopting a scientific observer's stance to discover these universal conditions which are necessary to us being what we are, Habermas then seamlessly shifts to the participant's perspective, endorsing the evolutionary sequence he has discovered in terms of developmental *progress*. The "linguistification of the sacred", the emergence of three discrete "worlds" with corresponding validity spheres, and the development of systems that coordinate action "behind our backs" are not just events that have happened but also positive advances which he endorses.

Critical capacity is thus restored to critical theory by a participant stance which brings science to bear on contemporary social crises, enabling a normative stance from which to diagnose the ills of modernity, such as the "colonisation of the lifeworld" or the "fragmentation of culture". Habermas will also draw on the authority of science to declare certain "foundations" on which his critique rests and which he urges us to accept. By "foundations", I am not implying an epistemological theory which commits him to "foundationalism", but simply mean a basis on which we should all agree, a point where argument should stop. For example, while we can continue to argue over the ideal speech situation and the value of undistorted communication and so forth, Habermas urges that we don't. These ideals are put forward as foundational. As a participant in the philosophical game of argument, Habermas uses these normative foundations to build a comprehensive social theory.

Foucault's biography is equally emblematic of the twentieth century. As a gay man growing up in 1940s France, we can imagine the internal conflicts generated by coming to understand and articulate his sense of self in a way opposed to the accounts of scientific experts. In response, his work sought to liberate the human capacity for self-formation and choice. His genealogies portray knowledge and power inextricably locked together to produce categories, objects, and subjects. Despite its omnipresence, power/knowledge is contingent and always susceptible to resistance. In this spirit, Foucault saw his challenge as revealing in "what is given as universal, necessary, obligatory, what place is occupied by what is singular, contingent, and the product of arbitrary constraints" (PT, 113). He was, like Habermas, well aware that rebellion may not lead to emancipation, but just new forms of domination, so he doesn't say that "everything is bad, but that everything is dangerous . . . the ethico-political choice we have to make every day is to determine which is the main danger" (EW1, 256). Foucault points to the constant risk of the authoritative truths of science being drawn into constellations of power/knowledge where they may have negative consequences. What is dangerous may or may not turn out to be bad, but we had better stay alert to risks. Rather than seeking to discover the ideal in order to eliminate all danger for all time, we should operate from our provincial perspectives to discern the main danger every day. This discernment occurs in specific contexts and cannot be further legitimated by any appeal to science or general principles abstracted from that context (PT, 115).

Foucault did not, like Habermas, articulate theoretical structures to posit universal regulative ideals for the good life. He didn't have that much faith in reason. Such faith itself would be a form of complacency, against which we must always be vigilant. Wherever knowledge is too secure, it must be challenged by revealing its origins—who constructed it, for what purposes, and with what interests. His *strategy of self-transformation* is directed towards "no longer being, doing or thinking what we are, do or think" (PT, 114). Whatever is given as natural—madness, delinquency, sex, morality, and even truth—is the way it is only because our conceptual and practical activities have constructed it that way. But Foucault is not pursuing a purely negative critique. What we thought was necessary is revealed to be contingent and with this revelation comes freedom for new orders to emerge. Foucault is not affirming change for change's sake. He is urging an ongoing struggle of transformation which problematises our most secure commitments and opens up fresh perspectives and freedoms.

Foucault explicitly adopts an observer's descriptive stance towards knowledge, bracketing validity claims and only addressing the question of truth to demonstrate its power-effects. He eschews commitment to scientific categories, in order to go beyond the modern consciousness structured by these categories. His genealogical accounts trace the course of concepts and discourses under different truth regimes as they interact with each other and with the non-conceptual and non-discursive. His fine-grained nominalism reveals the plurality and contingency of events and things. Foucault's generalised notion of all-pervasive power/knowledge enables us to see an ensemble of forces that relay power. On their surfaces, Foucault's texts steadfastly hold to the stance of a distant *observer*. He shows no commitment to norms or conceptual categories, or at least none that he can admit to. What may appear to be categories of nature are our own creation. Species like the homosexual, the delinquent, or the insane are, in all their diverse historical incarnations, our own inventions. Our norming, conceptualising, naming, and categorising interact with the things thus normed, conceived, named, and categorised, not only things but also people, classes, kinds of people, and ideas. Yet lying beneath the calm surface of detached description lies a participant, passionately committed to his own historically contingent norms of freedom and justice.[4] Foucault is particularly concerned with

---

[4] Habermas captures this nicely in his description of what most impressed him on meeting Foucault: "the tension, one that eludes familiar categories, between the almost serene scientific reserve of

the unassailable power-effects of the human sciences which ripple quietly through the social order, restructuring relations and reconstituting both subjects and objects.

The strategy of each thinker is best seen by considering the "point" or overall aim of their projects, what each wants to *do*, rather than what they want to *describe*. Foucault is not advancing a theory, pursuing an argument, proving a point, or doing science. We can only grasp his point if we see that what he does is oriented by a norm of self-transformation. Foucault's overall aim is not guided by a desire to describe events that have occurred, but by an orientation towards emancipation, a norm that gestures towards the possibility of another world, as indeterminate as that may be. Unlike this ongoing exploratory self-transformative critique, Habermas has a more determinate notion of emancipation. The point of his project is seen in how he identifies himself with the Enlightenment project, "consist[ing] of the relentless development of objectivating sciences, of the universalistic foundations of morality and law, and of autonomous art, all in accord with their own immanent logic", a process which "encourage[s] the rational organisation of social relations" (Habermas 1996, 45). By the use of fallible reconstructive science, Habermas pursues a strategy of discovering a hidden logic of social emancipation that he wants to endorse and nurture.

Habermas' and Foucault's incommensurable strategies have been variously described by commentators in terms of universalism/nominalism, mind-independence/construction, necessity/contingency, foundations/practices, objectivity/consensus, transcendental/empirical, and so forth. These pairs don't all reduce to the same underlying phenomena. However, the projects of Habermas and Foucault can be characterised in terms of two seemingly irreconcilable moments suggested by these terms which line up more or less consistently with one or the other strategy. I say "more or less" because I don't want to overstate Foucault's detachment nor Habermas's "transcendentalism". The participant and observer stances which incessantly play beneath the surfaces of these strategies are unavoidably intertwined, one explicit and thematised, the other implicit and suppressed. Like the two sides of a coin, only one side can be visible at any time.

Both Habermas and Foucault are aware of the tension between their strategies, which they see in some sense as problematic or paradoxical. Habermas wants to show that Foucault is caught in "a self-referential denial of universal validity claims" (PDM, 286). Foucault wants to highlight what Habermas puts forward as universals as being actually singular and contingent, products of

---

the scholar striving for objectivity on the one hand, and the political vitality of the vulnerable, subjectively excitable, morally sensitive intellectual on the other" (Habermas in Kelly 1992, 150).

arbitrary constraints. This tension between strategies arises from their different aims. Habermas wants to discover what lies below the surface of communication and thread it through a story of human development, both phylogenetic and ontogenetic, to yield milestones marking a logic of development that we can recognise as progressive. He wants to employ these milestones as standards to guide our ongoing progress. Foucault wants to transform himself and his readers by showing "how it is that subjects are gradually, progressively, really and materially constituted through a multiplicity of organisms, forces, energies, materials, desires, thoughts etc." (PK, 97). Subjects are "fabricated", put together from historically sedimented systems of knowledge, sets of rules and norms, and ways of self-relating (DP, 194). Just as there is no thing-in-itself of pure madness, there is no pure subject, no "I" or "me", prior to the descriptions given to a person. Nothing, including our self-descriptions, is either one thing or another except that history has made it so. And this is a history of struggle: "The history which bears and determines us has the form of a war rather than that of a language: relations of power, not relations of meaning" (PK, 114). Every way in which I can think of myself as a kind of person has been constituted within a web of historical events.

Necessity and Contingency

By initially adopting the observer's perspective, Habermas discovers what must be necessarily presupposed. Then, as participant, he endorses his discoveries as having determinate practical application to the good life. Foucault sticks doggedly to his radical observer's perspective, always questioning what seems to be necessarily presupposed by relativising such presuppositions as being presupposed only by creatures such as Man who, in any case, may soon disappear (OT, 421–2). These two strategies can be formulated in a way that highlights what is at stake—necessity versus contingency. Foucault wants to draw attention to the constitutive role of concepts, which open up the world in certain ways rather than in other ways. In Nietzschean terms, the world is radically contingent because we "invent" it and we could invent it differently.[5] Habermas, more focused on discovering what exists, would resist this. Certain standards, such as the pragmatic presuppositions of communicative action, are necessary as standards of reason and argument (OPC, 21–105). This is a conditional necessity. *If* we are to be the sort of creatures that we are—reasoning, arguing, progressing towards post-conventionalism—*then* it is necessary to accept certain presuppositions underlying our practices. Habermas is unequivocal in his view that reconstructive sciences

---

[5] Chapter 6 gives an account of Foucault's Nietzschean sense of "invention".

aim to elucidate underlying universal presuppositions as necessary conditions of possibility of any communicative interaction whatsoever. Societies pass through a sequence of sociocultural changes, comprising a "logic of development". No stage can be passed over and each stage presupposes previous stages (McCarthy 1978, 247). According to this logic of development, the communicative competences that emerge in modernity represent the determinate ways in which we have become, or are becoming, what we are. The extent to which we depart from this logic of progress is determined by the "dynamics of development", the empirical, contingent factors. With the expression, "logic of development", suggesting both a form of necessity and of progress, Habermas aims to discover both the truth of what we are and of what we ought to be. Whether this truth ends up involving Habermas's idealisations and milestones, reconstructed from developmental sciences, or something as yet undiscovered, there is still a determinate way that we are, and are becoming, which it is the task of fallible science and philosophy to discover.

While Habermas's notion of developmental dynamics accommodates the fact that the way we are is at least partially conditioned by contingent empirical factors, Foucault highlights this as instability. He would most likely suggest that Habermas's "logic of development" along with his principles and idealisations are themselves further power-plays and cannot have the status Habermas grants them. He is not interested in the determinate way the world is or we are, which science slowly uncovers, but in the way in which our scientific concepts order or "invent" the world. While we cannot simply choose the world we want, in some sense, we contingently constitute the world and ourselves as subject-objects within it. By showing how subjects-objects are put together, so that we could undo them if we chose, Foucault reminds us of our possibilities and responsibilities.

Habermas's emphasis on universalisability frequently occludes the role of contingency and complexity. Foucault's emphasis on contingency and complexity almost always ignores universalisability.[6] My view is that the difference between these emphases—on necessity in Habermas' project and on contingency in Foucault's—is not a difference in kind, but in practical commitment. We should therefore look to the roles those terms play in each strategy. The necessity invoked by Habermas's idea of a "logic of development" functions as an appeal to accept certain fundamental aspects of our condition as indicating the direction in which social, cognitive, and rational progress should proceed. For example, in the course of their development, humans make increasingly finer

---

[6] Foucault doesn't deny universality but holds a notion of universality close to that of Habermas as reinterpreted in chapter 3. "That [thought] should have . . . historicity does not mean it is deprived of all universal form, but instead that the putting into play of these universal forms is itself historical" (FR, 335).

differentiations between types of validity.[7] By implying necessity, Habermas urges us to stop wasting time on useless possibilities and endorse what appears sufficiently certain as a stable foundation on which to build a useful social theory and praxis. If future science shows that this appearance is false, then we will need to accommodate whatever is shown. Foucault's emphasis on contingency functions to disrupt this picture. We can't discern any "logic" of development, but only constant struggle. Progress or regress doesn't even come into view for the distanced genealogical observer. Habermas's trajectory of progress is only a projection of the human, all-too human. And why should we endorse the sort of creatures which we are as desirable? Foucault wants to escape limits imposed by what appears necessary in order to open up new possibilities.

While Habermas's fallibilism admits that the world could always be some determinate way that our descriptions fail to capture, Foucault presses the view that the world is not any particular way at all, apart from our contingent descriptions of it. Habermas's fallibilism implies that the world is a particular way independently of our descriptions of it. We are tempted to say that science converges on describing this way the world is by the falsification of descriptions that fall short of it. But Foucault could point out that this is already a particular description of the world *imagined* as fixed, independent, and determinate. While Habermas's fallibilism readily admits that the world could be different *to* the way we think it is, Foucault might suggest that the world could be different *if* our thinking was. This seems implied when he suspends commitment to the "truth regime", the framework of categories which precedes the division of the true from the false. In PP, he reminds us of a truth that is not already there waiting to be discovered (PP, 235). Against the "philosophical-scientific standpoint", he recalls this archaic notion of truth that does not appear for all but requires the subject to "be changed, transformed, shifted, and become . . . other than himself" (HS, 15).

Aiming to discover universal presuppositions for communicative action, Habermas's fallibilism leaves room for correction by future sciences. However, by suspending commitment to any particular truth regime and specifically to the "philosophical-scientific standpoint", Foucault's strategy of self-transformation suggests not so much discovering what is hidden but, by a transformed subjectivity, having one's eyes opened to new significances and connections. "Where the role of science is to make known that which we do not see, the role of philosophy is to make seen that which we already see" (cited in Kelly 2009, 129). Foucault wants to reveal our constitutive activity as requiring a particular form of subjectivity tied to a particular constellation of power/knowledge. What is taken as truth changes not only in response

---

[7] Chapter 3 discusses Habermas's account of the differentiated rationalisation of society.

to scientific discovery but, more fundamentally, in response to shifts in subjectivity. It is because concepts are realigned to form new constellations and refer to differently constituted objects that Foucault can claim that "the homosexual [became] a species" only after the nineteenth century (HS1, 43).

If I am right that the difference between what we call "necessary" and what we call "contingent" involves a pragmatic choice that depends on our purposes, it appears that the difference between the strategies of discovery and self-transformation is not a matter of truth or rightness. As a pragmatic question, there is nothing that can settle it given the different purposes to which Habermas and Foucault put their inquiries. Whilst different, the strategies of discovery and self-transformation are not in conflict. Clearly, we must reach determinate conclusions to settle the irritation of doubt and engender the confidence to act. And clearly, we must continue transforming ourselves by problematising.

Discovery, Self-Transformation, and the Enlightenment

As discussed in chapter 5, Habermas and Foucault are broadly within the Kantian Enlightenment tradition but engage in two contrasting forms of philosophical reflection on the limits of thought and action. For Habermas, the possibility of freedom is provided by certain unavoidable aspects of human existence, such as the norms of communicative action which are required to make sense of our shared world and coordinate our actions within it. He therefore articulates these norms to consciously guide the Enlightenment project. Embracing a sliver of certainty, he seeks what is stable and universal. Foucault radicalises Habermas's standpoint by challenging what we think is stable and universal. By revealing what we thought was necessary, universal, and ahistorical to be contingent, local, and historical, he opens new perspectives which can loosen the hold of domination (EW1, 283). Foucault not only suspends commitment to truth, but even the conceptual framework of ahistorical universal categories which precede truth.

Habermas's approach is derived from Kant's concept of *critique* in its recognition of the limits of reason and its critical-transcendental power to ground claims to truth and normative rightness. Substituting a fallibilistically conceived philosophy of intersubjectivity for Kant's philosophy of the subject, Habermas wants to rehabilitate the essential kernel of Enlightenment humanist thought, in which he sees a liberating communicative openness. Modernity is the incomplete project which must not be abandoned (Habermas 1996, 38–55). While he recognises the potential of reason to guide modern societies towards emancipation, this potential is being obstructed by the one-sided nature of a form of rationalisation which impoverishes reason's norm-governed communicative dimension. Committed to fundamental Enlightenment values, Habermas wants to surpass the negative critique of the previous Frankfurt generation. Challenging the lifeworld's conservatism, he incorporates the sciences into a universalistic theory directed at protecting democracy and rationality from the threats lying within modernity.

Foucault identifies with Kant's Enlightenment *attitude* as exemplifying a certain form of reflection on the present. He sees this form of reflection within a stream of Enlightenment critique in which reason perpetually questions the present and hence itself. In linking his project to the Enlightenment, he adopts its ideals of critical reason and autonomy as the implicit grounds on which his critique of abusive forms of power and reason rest. However, he insists, we must not submit to the "blackmail" of the Enlightenment, the authoritarian demand to be either "for" or "against" it (PT, 109–10). Embracing the reflexivity of modern reason, Foucault rejects the notion that "one either recognises reason or casts it into irrationalism—as if it were not possible to write a rational criticism of rationality" (FL, 353). Since the Enlightenment is part of "the historical ontology of ourselves" that has determined what we are, any position for or against it remains inherently partial and conditioned by this determination. Enlightenment requires a philosophical *ethos* of permanent critique (PT, 109). The critical question today is "how can the growth of capabilities be disconnected from the intensification of power relations?" (PT, 116). This requires a process of becoming other than we are through the *agonic* use of reason, to call into question what is given as reason. By neither endorsing nor rejecting, but *suspending* the normative framework, Foucault adopts the stance of the radical observer who enquires into reason, not in terms of "its nature, its foundation, its powers and its rights", but in terms of its contingency, "its history and its geography . . . its immediate past and its present reality" (Foucault in Canguilhem 1991, 9). Foucault's histories show that what appears as the one and only possible reason has a history. He doesn't think we need a theory of what rationality *really* is, in contrast to what it takes itself to be. Social practices exist within certain "regimes of rationality", certain historical forms which cannot be viewed from a perspective outside all regimes.

Habermas is explicitly committed to the ideal of the universal decentred subject of modernity, towards which the learning process, the "logic of development", leads. In contrast, Foucault was critical of philosophies which rely on a universal subject, especially products of European modernisation that are "so universalising, so dominating with respect to others" (PT, 115–6). In his criticisms of humanism, Foucault not so much challenged its principles as its pretensions to universality and ahistoricity. "What I am afraid of about humanism is that it presents a certain form of ethics as a universal model for any kind of freedom. I think there are more secrets, more possible freedoms, and more inventions in our future than we can imagine in humanism"

(Foucault 1988c, 15). By rendering universals emerging in history less vulnerable to historicity, humanism stultifies the "undefined work of freedom" by which we transform ourselves. Today, philosophy's task is "to know how and to what extent it might be possible to think differently, instead of legitimating what is already known" (HS2, 9). Far more wary of reason's ahistorical universals than Habermas, he challenges particular forms of rationality in a way that his critics see as undermining reason per se.

As we have seen, each analytical strategy requires a particular framework to bring certain target objects into view, while occluding others. Both strategies can ultimately be characterised as moments of modern reason which we struggle to articulate from our positions within it. By employing this broader framework of modern reflexive reason, we can expose some limitations of both strategies and unravel some confusions.

NORMATIVE FOUNDATIONS AND CONFUSIONS

In chapter 6, I defended Foucault against the charge of being normatively confused because he fails to provide explicit normative foundations. This criticism is based on a misunderstanding that is worth revisiting, firstly because it bears on the role of science and secondly because it reveals the strategies of Habermas and Foucault as distinctly different intellectual genres. We have seen Habermas employ science to establish universal-necessary conditions to ground his normative developmental account. In contrast, Foucault's histories seem judgemental, yet offer no explicit basis for their judgements. According to Habermas, Foucault's judgements are "crypto-normative", resting on normative assumptions to which he is not entitled (PDM, 276). It appears that, because it does not articulate normative principles, Foucault's work can only be arbitrary.

Habermas endorses his ideal universal conditions as explicit principles or foundations, and he urges us to do likewise. Because these foundations are, though fallible, based on the best knowledge we have, we should accept them for the purposes of constructing social theories. If future science shows them to be wrong, we should drop them and adopt what science shows us to be the case. Habermas insulates his formal-transcendental account from any grounding in power relations by treating his rational reconstructions as "pure" forms of inquiry not shaped by any interest whatsoever. This opens a gap between the transcendental and empirical consistent with the structural split that runs throughout his work—between lifeworld and system, between the dynamics and the logic of development, and between theory and practice. While acknowledging that the critique of validity claims cannot be ultimately separated from the genetic question of the entanglement of power and validity, Habermas still insists that, at the level of theory, "the categorial distinction between power claims and truth claims is the ground upon which any theoretical approach has to be enacted" (PDM, 127). Habermas's strategy requires him to endorse the quasi-transcendental nature of what he insists must be necessarily and universally presupposed.

For Foucault, norms are not laid out in advance as foundations. Rather, normativity is generated in the context of, and for the sake of, transforming and reconstructing existing problems. The very idea of a normative basis laid out in advance is the object of his critique. The constitution of norms requires the constraints of an actually present field of struggle and counter-struggle. Rather than a stable, ethical self, guided by established principles, Foucault sees the subject immersed in inherently reciprocal intersubjective relations in which dialogue has no underlying logic and justificatory frameworks emerge from histories of struggle (STP, 3). Foucault would argue that we can still respond to norms without articulating foundations, although not like following tracks laid out in advance. What he is rejecting is the reliance on norms as prescriptions which determine in advance their application to future judgements and actions. Such prescription would stifle possible future freedoms and inventions beyond our current imaginations. Responding to specific contexts, there is no single way in which ethical agents arrive at a proper course of action, nor need there necessarily be stateable rules. We may employ something like *phronesis*, or a skill of knowing *how*, rather than of knowing *that*. Certainly we can give justifications, but when they reach an end, we can only say with Wittgenstein, "This is simply what I do" (Wittgenstein 2003, Remark 217).

To gain some clarity about these two very different approaches, we need to consider what we do with foundations. Foundations can be thought of as analogous to dictionaries, which have normative force for correct word usage but need not be given that authoritative role. Dictionaries are used to settle disagreements, not because their authority flows necessarily from their correctness, but because we have endorsed their authoritative use for these particular situations. The insistence that Foucault must articulate normative principles is based on a picture of principles as necessarily authoritative. However, such principles play the authoritative role that Habermas wants them to play only if we endorse that role. Foucault doesn't want to endorse any authoritative principle. If we do, he wants us to accept responsibility for this endorsement. His genealogies focus our attention on how universalist histories, anthropological foundations, and traditional emancipatory theories have been blind to their own dominating tendencies. Deploring "all the prophecies, promises, injunctions, and programs that intellectuals have managed to formulate over the last two centuries", Foucault thinks that "the role of the intellectual is not to tell others what to do" (PPC, 265).

Habermas thinks Foucault is caught in a performative contradiction, but this is not obviously the case, because Foucault's commitment to undermining foundations is itself a foundation. We cannot simply walk away from normativity. To attempt this would be to adopt a normative stance against norms. Foucault's lack of commitment to foundations is a *commitment* to this lack, a commitment which plays the role of a *foundation*.[8] This is not a matter of indifference for Foucault, but something he argues for doggedly, if at times obliquely. If he were to simply announce this foundation, he would immediately become embroiled in performative contradictions, much like commanding someone to be free. Rather than rejecting normativity per se, Foucault distances himself from regions of normativity by suspending normative force while describing the historical conditions in which certain norms have been socially constructed.

Foucault remains *in conformity with* his foundation, while *not relying on* it as, for example, a test in a decision procedure or a justification of the form "A, therefore B". He doesn't want to endorse his judgements as foundations compelling future communities because his judgements are responses to concrete experiences of oppression, either his own or those of others with whom we can identify, independently of philosophical justification. The difference between *conformity with*, and *reliance upon*, a foundation is a question of where we place ultimate authority. We can open the dictionary and declare, "Lets, for current purposes, stick to what the dictionary says" or "This is what it really means". Rather than leading to any determinate prescription, Foucault's normativity emphasises a *process*, a way of being which embodies a norm binding him to constantly question norms. Habermas, in contrast, insists that stable foundations are necessary for committed judgement and action. In his discourse ethics, Habermas formulates two principles, a universalising principle (U) and a discourse principle (D), that operate as unassailable foundations for moral argumentation because their negation cannot be asserted without performative contradiction (King 2009; MCCA,

---

[8] This point is seen in Foucault's account of three types of history of the sciences he discussed in AK: recurrent analysis, which is unreflexively normative and adopts a norm of scientificity coinciding with current science; epistemological history, which is reflexively normative and consciously employs current science as a norm for judging history without denying the historicity of that current science; and Foucault's archaeological history (AK, 208–11). Chimisso, however, points out that Foucault's account is itself a recurrent reconstruction of the history of historiography. "Just as the sciences, in Bachelard's and Foucault's view, reconstruct their history in a rational manner from the point of view of the present, and condemn to oblivion those theories and practices that cannot be integrated in this history, so Foucault's account of the historiography of science is simplified and progressive" (Chimisso, 300). Foucault presents his archaeology as an advance which gets at deeper and more general levels of epistemological history. My point is that Foucault *cannot avoid* a normative claim such as "this is the better way to do history of science". If he isn't making such a claim, his writing can only be understood as thoughts that have simply happened and to which he has no commitment, mere marks and noises.

65–6).[9] Habermas thinks he needs such foundations to safeguard the normative structure of the threatened lifeworld. He assumes Foucault requires such foundations because he sees Foucault as having the "serious intent of getting a science underway" that aspires to truly objective knowledge (PDM, 279).

To grasp the implications of Habermas's claim about Foucault's intention, we need to consider the function of science, and here I return to the functionalist account I drew upon in chapter 3. From this perspective, we see that although science involves an ongoing open-ended inquiry, its function as it enters the modern lifeworld is to provide a fixed point upon which we can all agree. It performs this role on the basis of its claims being accepted as not relying on the idiosyncratic predispositions of particular individuals or communities, but on how things are for everyone. In accepting current science, we grant it the role of anchoring discourse and engendering practical confidence in the lifeworld, thus providing a basis upon which to act, at least provisionally. Without such "foundations", we are compelled to argue interminably rather than solve problems with the sort of pragmatic agreements and outcomes that science generates. Habermas is suggesting that Foucault is engaging in a discourse which, like science entering the lifeworld, should serve in this foundational role. The principles of Habermas's discourse ethics provide examples of what he might accept as this sort of foundation, what could be called "science" in the broad sense of something authoritative where questioning must stop, in this case on pain of contradiction.

Habermas also thinks Foucault is "getting a science underway" because he assumes that he is what Koopman calls "normatively ambitious", attempting to discover determinate facts to definitively settle questions (Koopman 2013, 88–89). But as Koopman points out, Foucault's genealogy is normatively modest, not proving or disproving, supporting or subverting, but alerting us to dangers and encouraging us to be sceptical. Taken as a whole, his genealogy aims not to legislate the reconciliation of the real and ideal through the rule-bound use of reason, but to clarify and intensify awareness of present dangers. Modern power isn't necessarily bad, but it is problematic and demands serious attention. "This is a domain of very complex relations, which demand infinite reflection" (Foucault 1988a). Making a determinate judgement forecloses that reflection. Even Foucault's sympathetic interpreters, Rabinow and Dreyfus, have difficulty here, arguing that Foucault "owes us a criterion of

---

[9] See King (2009) for a more detailed account of the foundational status of these norms. (U) states that a norm is only valid if "*all* affected can accept the consequences and the side effects that the *general* observance can be anticipated to have for the satisfaction of everyone's interest (and these consequences are preferred to those of known alternative possibilities for regulation)". (D), which follows from (U), states "Only those norms can claim to be valid that meet (or could meet) with the approval of all affected in their capacity as participants in a practical discourse" (MCCA, 65–6, cited in King 2009).

what makes one kind of danger more dangerous than another" (Rabinow and Dreyfus 1983, 264). It is not clear to me that he does, especially because, since the demise of ancient philosophy's care of the self, an acceptable answer to such a question is always conceived in terms of a science of knowing the self, humanity, or society. The problem is that, while Foucault favoured "care" as a pragmatic, reflexive, and hermeneutic attitude of mutual adjustment and responsiveness to the self's own purposes, potentials, and meanings, only "knowledge" which entails a programmatic response to a determinate fixed object counts as an adequate answer for modern reason. Foucault wanted to separate our lives and ethics from knowledge and science.

Genealogy's radical perspectivism is clearly at odds with the conventional truth claims of science buttressed by foundations and reasons. Foucault wants to show that science's "value freedom" is itself a value that excludes other values. Genealogy takes its reflexivity seriously. There is no perspective outside power-laden discourses from which to speak truth. This doesn't mean genealogy isn't true, but rather that truth is not what we thought. The truth of genealogy is not anything like the truth that science sees itself pursuing. For Foucault, that truth is not enough. We need to ask about regimes of power, not to prove claims as true or untrue, but to uncover how truths have emerged. While Habermas builds his discourse ethics on foundations—points where questioning must necessarily end—these are precisely the necessities that Foucault would question. It may seem preposterous to challenge a performative contradiction, a point beyond which we can make no sense, but Foucault shows that what counts as making sense is never stable. His histories chart stranger things that our provincial imaginations can barely grasp—the Renaissance world where everything is connected by resemblances, the coming into being and the passing away of Man, the archaic truth ritual of the "ordeal". These are contingent universals or, as Foucault would say, historical a priori. Truth is not a context-transcendent realm but is of this world and only graspable within specific constellations of power/knowledge/subjectivity. He is acutely aware of how power creeps in under cover of hypostatised humanistic universals to bend them to its own ends. Science is complicit because it is never pure, but always embedded in specific constellations of power/knowledge.

Foucault does not, as Habermas claims he does, aspire to undertake science, by which a fixed subject seeks to discover and justify determinate conclusions. His critique is not "refusal and denial, but rather an investigative work that consists in suspending as far as possible the normative system which one refers to in order to test and evaluate it" (Foucault, cited in Lemke 2012, 61). The genealogy of forms of rationality follows "the course of a precarious and fragile history" (PPC, 37). This history "is sometimes more effective in unsettling our certitudes and dogmatism than is abstract criticism" (EW3, 323). Critique is linked to the space of concrete freedom by the possibility of transformation, not by argument backed by the authority of science, but by "unsettling our certitudes and dogmatism" to prompt a fresh way of seeing (EW3, 323; cf. Foucault 1981, 69–72).

We have seen that Foucault does have foundations—what we might characterise loosely as his "normative stance against norms"—that transcend the immediacy of his context. These "foundations" are suppressed though still operative. Foucault withdraws to become a mere observer, as though his descriptions allow history itself to unsettle our foundations. In contrast, Habermas as observer makes explicit what he has discovered as implicit in communication but then, as participant, endorses this discovery as the way we ought to be. However, behind such contrasts, we can see the entwinement of participant and observer, of power and validity, and of the strategies of discovery and transformation, reflecting the irreducible dualism of imbricated perspectives. For Foucault, any admission of foundations is ruled out by the observer stance that historicises the transcendental wherever he meets it. For Habermas, an admission of power relations is supressed by the participant stance that thematises context-transcendence.

WRAPPING UP THE DEBATE

Having characterised the two projects as being somewhat in tension though not in conflict, I want to return to the question of whether they can, or should, be reconciled within a combined standpoint. In taking the best from Foucault and Habermas, Koopman's strategy of delegation necessarily overlooks certain aspects in order to build its theoretical structure. I have sought to deconstruct these structures and make use of the overlooked residues, in particular Foucault's apparent lack of normativity and Habermas's apparent transcendentalism, to consider what Foucault and Habermas may have been tempted to say, but could not say because of the pressure that saying it would have put on what they did end up saying. Foucault's strategy of thematising the observer's perspective prevents him from admitting the fundamental normative commitments underlying his project. Habermas's strategy prevents him from thematising the constant vulnerability of the participant's normative stance to debunking by a hard-nosed observer.

We are tempted to think, on the basis of what could not be made fully explicit or what Foucault and Habermas did not own up to because they couldn't, that there is a disagreement which must be resolved one way or another because there is only one way the world is and thus only one true description of it. For scientific naturalism, this world is that which natural science describes. While the nuanced reinterpretations I have put forward are rightly wary of such ontological commitments, I suspect that more partisan participants in the Foucault/Habermas debate were not. Against such temptations, I would argue that we are not compelled to seek a synoptic view of a single world, mastered by a single vocabulary. We don't have to insist that either Habermas's or Foucault's strategies more closely address the world as it *really* is or really should be. As the Nietzschean Foucault would point out, such an insistence reflects a prior commitment to the existence of criteria for what we *count as* reality.

Both Habermas and Foucault would agree that there is a plurality of ways of sense making rather than a single true description of the world. Foucault's *epistemes*, or "regimes of truth", chart radically incommensurable positions that cannot be seen from a single viewpoint but require shifts in subjectivity. Habermas's discussion of freedom and determinism in BNR refers to "the *unnoticed* switch from the participant to observer perspective" that "can create the impression that the motivating of action by intelligible reasons forms a bridge to the determination of action by observable causes" (BNR 258). Habermas similarly refers to the unseen *switch* from "the participant who accuses or *justifies* to the observing analyst who ... *explains* the behaviour" (Habermas 2007b, 21). This switch amounts to a shift in subjectivity which opens one view of an object that we can't easily fit with another, much like Wittgenstein's duck/rabbit which we can only see either *as* a duck or *as* a rabbit (Wittgenstein II, xi). In the cases of minds or brains, language or marks and noises, normative force or behavioural regularity, it is not clear whether we are seeing *different* things or seeing the *same* thing differently, because what counts as same or different is up for grabs. In my view, we should not look to the world to settle such confusions, but rather acknowledge that these accounts reflect different ways to access the world which are not always easily fitted together.

The resistance of the two projects to being forced into alignment is heightened if we take seriously the self-reflexivity of modernity that both Habermas and Foucault recognise. It is not so much a question of how Foucault's strategy of self-transformation and Habermas's strategy of discovery enframe reality, but of how these strategies circle around to bear on themselves and each other in a never-ending "rational criticism of rationality". Given a prima facie resistance to reconciliation, the relationship between the strategies is an open question. In the methodological collaboration Koopman proposes, he tells us that "they cannot but fail to meet just insofar as the logic of question and response is not one of contradiction but rather one of reciprocal rhythm" (Koopman 2013, 226). However, Koopman's aim is to craft a viable critical theory, while mine is to understand Foucault and Habermas more in their own terms and to make use of the tensions arising to cast light on the question of our scientific modernity. Rather than "reciprocal rhythm", I am inclined to see a relationship of incommensurability, discontinuity, and even debunking between the two stances. Although perhaps disconcerting, this relationship is productive. Any object of discourse can be viewed through the lens of either perspective and either perspective can become an object of discourse for the other perspective. The imbrication of one latent perspective within another is the motor that generates the ongoing reflexive movement, modernity's craving for explanation, which Foucault describes as "a ceaseless task constantly to be undertaken afresh" to think the unthought (OT, 353).

The fact that Foucault and Habermas misunderstood each other shouldn't be surprising given their fundamentally different commitments and aims. Nor is it surprising that the substantive tension between rationality and power that seems endemic to social and political theorising shows up in the appearance of stubborn irreconcilability which appears to demand a decisive response. For an adequate account of our situation in the world we are tempted to say either that Foucault collapses truth into power or Habermas insulates truth from power, or perhaps Foucault collapses it *and* Habermas insulates it. However, we need not accept this framing of the debate. Rather than settling a dispute by seeing the two projects *saying* things that are either true or false, I regard the two strategies as *doing* different things more or less successfully or unsuccessfully.

Rather than seeing a problem with Habermas's quasi-transcendental tendencies, I see the problem as the ambitious reach of his comprehensive explanatory-prescriptive apparatus. As outlined in chapter 3, the broad strokes of his theoretical structures don't adequately capture modernity's complexity and opacity. Habermas seems to be aiming for a comprehensive explanatory-prescriptive account of modernity, an ambition that he falls short of by being trapped in *aporias*. However, concern about the over-reach of his theoretical structures can be tempered by looking to his discussions of concrete issues, such as liberal eugenics or the neuroscientific account of free will, in which he can draw upon these structures without needing to commit to an unyielding all-embracing quasi-scientific theory. And the concern about his quasi-transcendentalism can be tempered by understanding his abstract norms as requiring ongoing interpretation within actual contexts. This interpretation is reasonable because despite positing a strong categorical distinction between power and validity, Habermas actually does recognise that reason is never insulated from power (PDM, 323–4). By such interpretations, we are still able to draw judiciously from his framework to support incisive critiques that simply cannot be generated from Foucault's project.

Rather than Foucault collapsing knowledge into power or identifying power with knowledge, I think the problem is the misreading of Foucault (see PPC, 43). While knowledge and power are not the same, they are so entwined that we can only analyse them in relation to each other. Foucault's "power" need not undermine knowledge or freedom, because it is a condition of possibility which not only constrains but also enables knowledge and freedom. Rather than seeing a problem with Foucault's account of power/knowledge, we should see that reason and power aren't pure, but are always mixed. Rather than seeing a problem because Foucault didn't offer any prescriptive guidance, we should only see a problem if he set out to provide prescriptive guidance. He did not. Given these ways in which both projects can be interpreted, the nature of the debate does not demand an either/or response, but rather a response which criticises each project to the extent that it fails to achieve what it sets out to do. We need not buy into the debate on the terms in which it presents itself.

I now want to consider what each project yields, before drawing together the threads of arguments about Foucault, Habermas, and the nature of our modern scientific rationality. Enlisting the sciences to both discover and lend foundational authority to a progressive developmental account, Habermas's strategy of discovery yields what he considers the necessary and universal presuppositions of communicative practices. On this basis, he builds fallible foundations to guide us towards emancipation. While recognising the ambivalent legacy of modernisation in which "the communicative potential of reason has been simultaneously developed and distorted", Habermas can account for the dominance by instrumental reason in more concrete terms than his predecessors, specifically by the increasing influence of economic and administrative systems (PDM, 315). His theoretical framework brings modernity's pathologies into sharp relief by being brought to bear on contemporary social problems in determinate ways. Concepts like colonisation of the lifeworld, fragmentation of consciousness, and systematically distorted communication serve as analytical tools to address issues like the distortion of the public sphere by social power, the weaponisation of science or the dysfunctionality of modern democracies. His categories enable analysis and diagnosis from a standpoint outside the everyday lifeworld. In response to critics, Habermas can stress the fallibilist, formal, and procedural character of his normative foundations. His idea of a society based on agreements arrived at in free and equal exchanges is not the depiction of a concrete utopia, as Foucault once suggested (Foucault 1987, 18). Not only is Foucault wrong here, but so are the postmodernist critics who portray Habermas as authoritarian.[10] Habermas's formal and procedural idealisations and use of reconstructive sciences actually provide standards by which to recognise and criticise dogmatic authoritarianism. The pragmatic force of these context-transcendent ideals is that their regulative function opens reason to self-correction, enabling a foothold for criticism of what is found to fall short of these idealised presuppositions. Habermas's strategy of discovery, harnessing science as a vantage point outside the

---

[10] Lyotard (1988) and Cook (2001) criticise Habermas for authoritarianism.

everyday lifeworld, offers determinate tools not only to diagnose but also to begin thinking about remedies for the ills of modernity.

Foucault's strategy of self-transformation provides none of these tools and nothing like a remedy. It appears prima facie to yield only an amorphous *mélange* of power/knowledge/subjectivity. His genealogies, crammed with detail, yield no theoretical structures as diagnostic tools. Referring to his work as "a kind of toolbox others can rummage through to find a tool they can use however they wish in their own area", Foucault gives us no ideas about how to use those tools.[11] His analyses lead not to an account of a progressive logic of communication which has become distorted, but seemingly only to a crushing story of a modernity subjected to the inexorably normalising effects of discipline and biopower, backed by Enlightenment humanism. By lumping together all processes of socialisation and individuation as forces of normalisation and subjection and filtering out those aspects of institutional reform which represent unmistakable gains, Foucault makes it difficult to see how his critique could bear in any determinate way on contemporary social realities.

In his strongly non-normative reading, Mark Kelly defends Foucault against such criticism by arguing that the current threats to liberal democracies result from *too much* normativity and that Foucault's supposed non-normativity offers a remedy (Kelly 2018, 173). However, it seems to me that these threats call not for less norms, but better norms, better arguments, more communicative action governed by abstract regulative norms such as Habermas's ideal speech situation. The problem for Foucault is that by what he referred to as the "tactical polyvalence of discourse" his toolbox is open to serving ends that many of us would find problematic (HS1, 145). Some critics claim that the way in which Foucault takes up Kant's imperative to realise our maturity as a never-ending critique of critique has led to a solipsistic scepticism that undermines trust in institutions such as science and democracy by relativism and populism.[12] The best response is the one that Foucault himself prescribes—to adopt the Enlightenment attitude of critique, the norm of ceaseless questioning or norms, in order to "every day . . . determine which is the main danger", thus letting go the ambition of "ever acceding to a point of view that could give us access to any complete and definitive knowledge of what may constitute our historical limits" (EW1, 256; PT, 115). In other words, as we recognise dangers we should address them, yet we should never imagine that we have finally attained a view of things as they really are or

---

[11] Foucault (1994, 523), cited in *Foucault Studies* no. 10, Nov 2010, 1.
[12] Peter Harrison has linked Enlightenment scepticism with contemporary lack of trust in institutions such as science and democracy. See https://www.abc.net.au/religion/science-and-pervasive-scepticism-enlightenment-values-cannot-hel/10095952 (Retrieved 24/02/2021).

should be and that the task of attaining this view can be completed for all time.

If Foucault's seemingly totalising condemnatory comments are viewed in the context of the broader aims of his life's project, it becomes clear that his critique is not intended to bear directly on contemporary social problems in the same way as Habermas's work does. His aim is to open unforeseen possibilities by helping us see as problematic what was previously unproblematic. He highlights how frozen thought, captured by constellations of power/knowledge/subjectivity, is drawn into contexts where it unwittingly limits freedom. He provides a vantage point not only outside the everyday life-world, but also outside science, which he problematises, and truth, which he relativises. This is not to say that he doesn't believe or value science or truth. His project is to historicise the transcendental, the necessary, the universal, wherever he meets them, to show how much of our world is our "invention" and how even the most secure and obvious categories of scientific thought depend on us and would be different if we were different.

Foucault would want to show that what Habermas has discovered has also been "invented". He would emphasise the porosity of Habermas's categories and the negotiability of the moments of freedom and resistance. In a nominalist vein, his radical historicism would blur Habermas's categories, drawing out what is complex and hybrid and fails to fit within them. He could point out that Habermas's concept of communicative action, entailing interactions oriented towards intersubjective understanding, fails to grasp the operation of norms and communication as itself a form of power. He could criticise Habermas's theoretical framework not only for its abstractions, but also the universalism of the "universal intellectual" tradition that derives from the juridical tradition of the West (FR, 70). The problem he would point to is that "certain great themes such as 'humanism' can be used for any purpose whatsoever" (FR, 374). In contrast with Habermas's development of a comprehensive, explanatory analysis that aspires to grasp the outlines of the historical process as a whole, Foucault undertakes specific and fragmentary analyses of particular social institutions, informed by "local critique", which do not add up to a totality or "organic whole". "The whole of society is precisely that which should not be considered except as something to be destroyed" (cited in Best, 221). Foucault explicitly avoids the authoritative claims of science, but instead provides genealogical accounts which highlight science's constructed and contingent nature. Rather than seeking standards to be applied universally, he dramatises excesses of power in particular situations which arise from standards being applied universally. He acknowledges the limitations that his commitment to local partial inquiry imposes and thinks we should just accept them (PT, 115). Critique simply cannot tell us what to do. Nor can it tell us what is true or false like science does (GSO, 354). In fact, his genealogies are "anti-sciences" which "fight the power-effects of any discourse that is regarded as scientific" (SD, 9). While Habermas's theoretical structures aspire to an ambitious explanatory reach, Foucault withdraws from explanation into description. Rather than arguments that compel, his descriptions invite, prompt, and provoke readers to reflect on how claims of context-transcending validity can collude with repression by obscuring their own entanglements with power.

Habermas's concern is that Foucault denies us any "normative yardstick" as a basis for critique. Certainly, Foucault refuses to offer prescriptive guidance. However, his work is not norm-free. Foucault's entire project is governed by the sorts of unthematised situated norms I have discussed. He is, however, constrained from admitting this by his fear of drawing his entire project, governed by a hidden norm against norms, into a performative contradiction. Foucault's concern is that Habermas's theoretical structures, such as the ideal speech situation, can be drawn into constellations of power to become inexorable universal forces projecting across all contexts. Certainly, Habermas's theoretical structures risk hypostatisation if given metaphysical weight. Despite this, he seems tempted to discover some sort of quasi-scientific solution that responds to a determinate reality completely independent of us, in order to resist the performative contradiction he thinks Foucault finds himself in. Here he is drawn close to a performative contradiction of his own. When his abstract and formal formulations are not treated as sufficiently open to accommodate the interpretations of future communities, he is effectively telling others how they must be free. However, if we recall the abstract, formal, and procedural nature of the norms he urges us to accept, and understand their universal formulations as always interpreted and applied by particular communities in relation to specific problems and situations, he is not so far from Foucault.

It is by looking at the motivations of each project, or what each thinker wants to *do* as much as what each *says*, that we can see the two strategies within the broader frame of modern self-reflexive reason. It is by bringing them together in these ways that we can illuminate the *telos* of modern rationality as never-ending inquiry, including inquiry into itself. This reflexivity is heightened in the sciences. It is the basis of a neuroscience that inquires into rationality as though it weren't itself a particular form of rationality, of an economic science that inquires into how society values as though it were itself value-free, and of a science of sociology that itself inquires into science. Just as in the struggles between the strategies of discovery and self-transformation, no-one has the last word. In modernity's restless movement, there is always more to be said. We will conclude by considering how the two strategies are reflected in Foucault's and Habermas' accounts of the development of science and its relations to other forms of reason.

CONCLUDING REFLECTIONS

Consistent with the Frankfurt School approach, Habermas harnesses science for his philosophising. Empirically plausible universalistic theories are brought to bear on contemporary crises. Thus, we see Habermas's social evolutionary account of a progression of developmental milestones yield a normative standpoint towards modernity. This account includes a progressive developmental account of science itself. While the sciences have their risks and setbacks, there is a logic to their development. This "logic" is firstly (in KHI) a logic of interests. Here the empirical sciences are guided by a technical interest in dealing with the objective world, while the human sciences are guided by a practical interest in the social world. In his mature theory, this logic is vested in the development of communicative reason, which makes increasingly finer distinctions between objective, intersubjective, and subjective "worlds", separating out what is in the objective world from what is in us and what is between us. Habermas counters the reductive imperatives of scientific naturalism initially by admitting the fundamental reality of three separate and irreducible interests, and then later by distinguishing three distinct and irreducible worlds. While the development of communicative action constitutes progress, the main threat, both in his earlier Kantian theory in KHI and later in TCA, is the threat to balance. In both cases, the developmental communicative logic is skewed by contingencies of power that usher in a limited form of reason which is to be taken as reason per se. This view of a developmental logic thwarted by the contingencies of power which infect reason is consistent throughout Habermas's work, from his early critiques of positivism to his lifeworld colonisation thesis and his more recent defences of lifeworld structures against reductive scientific naturalism.

Foucault's genealogies chart the course of concepts and discourses on their way to becoming sciences, as they interact with each other and with the non-conceptual and non-discursive under different truth regimes. His histories reveal the play of power/knowledge/subjectivity by abstracting from truth, to deal only with the contingencies of what is taken-as-true, so revealing the fine grain of events beyond everyday awareness. In chapter 7, we saw Foucault account for the emergence and dominance of a particular form of truth, the demonstrative scientific truth of the "Cartesian moment" (HM, 550–74). In chapter 6, we saw him sketch the genealogy of the natural sciences as a particular form of rationality emerging from judicial techniques of inquiry, the power-effects of which they have largely escaped. The human sciences grow out of a later technology of power—the examination, linked to observation and normalising judgement—which becomes pervasive in modernity. These sciences form regions modelled on the empirical sciences and, after their emergence in asylums, schools, prisons, and hospitals, diffuse throughout society, still inextricably linked to power. Medicine establishes a normative basis for the entire human sciences and, from the nineteenth century, scientific knowledge becomes linked to new forms of governmentality and is increasingly directed towards life at the level of populations. Human and natural sciences are linked to each other, and to ideologies and discourses and practices like law and morality, to constitute networks of power/knowledge.

In his 1979 lectures, Foucault relativises scientific truth in relation to a broader frame provided by his concept of "regimes of truth". Rather than accept "truth" as an obvious good to be sought at all costs, Foucault's historicism problematises truth and its technologies of power. He employs past *epistemes* to throw our current situation into sharp relief by revealing that what is taken as natural, universal, and necessary is local and contingently constructed. Foucault doesn't convey a story of either progress or regress. Nor does he find anything which points the way to progress. His genealogies constitute an experience, an *ascesis* that can change us and through us can change social practices and relations, though not in any way that can be specified in advance. Foucault wants us to feel the contingency of our present as being just as strange as that of the earlier *epistemes* from which it accidentally developed. His densely factual histories of discrete areas of the human sciences—sex, mental illness, criminality—refuse systematic theories. In a Nietzschean spirit, Foucault would insist that, while social evolution has occurred, it need not be viewed as progress. Certainly we have become more able to distinguish validity claims, but is this better? Like Habermas's use of science, Foucault's genealogy establishes a distance from which to question what was previously unquestionable. However, Habermas wants conclusions and so draws a line at what must be necessarily presupposed. Foucault is more interested in questioning what seems necessarily presupposed as being presupposed only by creatures such as Man. Habermas's worry is that Foucault's critique stops nowhere, is totalising, and pulls the rug out from under its own feet. Foucault's worry is that Habermas's theory nails the rug to the floor and exempts itself from modernity's "rational criticism of rationality" (Foucault 1989a, 353).

Neither Habermas nor Foucault dismiss religion like the positivists did, assuming that it would fade away under the clear light of an expanding scientific worldview. Certainly Habermas affirms the priority of science and secular discourse. Secular discourse informed by science not only is, but should be, the *lingua franca* of modernity. Natural science represents a rational advance by its exclusive focus on the differentiated validity sphere of truth to grapple with the external objective world. Natural scientific discourses are unproblematic when they don't overstep their domain to be drawn into constellations of power where their authority is co-opted to legislate in other validity spheres. Habermas thinks that scientifically informed secular language, aspiring to transcend provincial perspectives by granting communicative action its proper scope, must have the primary place in politics. However, he is acutely aware that "religious language is the bearer of semantic content that is inspiring and even indispensable" and that its absence leaves a loss which it seems we can only lament. He thinks we should attempt to translate the residue of meaning locked in religious language into secular language. He acknowledges that at times this eliminates the intended substance and leaves irritations (FHN, 110). Habermas wants non-believers to shoulder equal responsibility for undertaking the cooperative translation of religious language into secular language (BNR, 131). Although not a straightforward task, such translation may enable divergent traditions to find a common language separate from coercive forms of economic, political, or social power.

For Foucault, science and religion are different facets of power/knowledge/subjectivity which coalesce and concatenate to relay its effects. In PP, he describes how the Christian confession migrated into the psychoanalytic confession and the fear of monsters mutated into the search for instincts. Foucault is not interested in religion as a system of metaphysics or form of knowledge, but rather in its relations with other discourses and practices and the power effects that flow from these relations. He is concerned with "spirituality", a term with a specific sense not involving commitments to religious concepts, but rather to "the search, practice and experience through which the subject carries out the necessary transformations on himself in order to have access to the truth" (HS, 15). Unlike science, where the truth is open to all, the spiritual subject's access to truth is grounded on *ascesis*, on transforming her mode of being a subject, of seeing and inhabiting the world, by certain exercises amounting to a sort of conversion (HS, 15). This notion of a critical-experimental practice of self-transformation is essentially political. "There is no first or final point of resistance to political power other than in the relationship one has to oneself" (HS, 252).

In accounting for these two responses, we can detect the biographies of both thinkers. Habermas wants a degree of certainty, to identify and call out the deceptions and power imbalances which fed the criminal totalitarianism he witnessed in his teenage years. He believes that science, by its "external" stance, gives us at least some purchase on our situation. Its counter-intuitive perspective infuses the traditional lifeworld with fresh perspectives that aspire to universality, a role that modernity can't leave to religion. However, he is all too aware of the imperialistic tendencies of science which, albeit compelling to the modern mind, occlude other dimensions of validity and meaning, some of which reside uniquely in religion. Foucault wants to alert us to the dangers of authoritarianism which he sees in science, especially the human sciences. He is particularly concerned, for example, about the psychiatric/psychological power that tells us what we are by nature and hence must be. To that end, his more radical observer's view recruits spirituality as self-transformation in a series of overcomings of whatever is presented as necessary.

It is philosophy that must traverse these disparate domains to make science at home in the modern lifeworld while acknowledging and drawing on religion's semantic potentials. While drawing on science as a resource providing an "external" view, many fundamental questions still require philosophy's capacity to step back from the empirical data and draw it into a normative framework. What Foucault and Habermas both value about philosophy is its anarchic nature, unconstrained by scientific method. Like Habermas, Foucault thinks that philosophy occupies a somewhat contested place within the particular social and historical formation of modernity, from where it seeks to understand, diagnose, and change the present by a critique of modernity's paradigmatic form of reason, science.

Habermas draws on the authority of science to buttress a progressive account of reason, yielding an explanatorily powerful diagnosis of modernity, as well as conceptual tools of analysis and prescription. His account of the development of science stresses reason's autonomy. Despite constant entanglement with power and frequent setbacks, reason tentatively traces a path of progress towards Enlightenment. His fallibilism admits that the world could always be some way that our descriptions fail to capture, thus leaving room for foundational principles to be revised. Foucault suspends science's authority and adopts agnosticism regarding its progress or truth, thus bringing it into view as one form of knowledge amongst others. Reason is not autonomous but entwined with power which flows through subjects, bodies, and objects like the panopticon. He suspends not only truth, but the "truth regime", the preconditions for what can be taken as true or false. Though disconcerting, this stance brings to awareness the unseen endorsement granted to science as the most authoritative knowledge, an endorsement not necessarily flowing from its purported correctness. This exposure of our constitutive activity reminds us of our epistemic responsibilities and of the dangers of scientific reason being drawn into circuits of power. It follows that what we take for truth shifts not only in response to scientific discovery but also in response to shifts in subjectivity.

We have seen the restless movement of modernity's reflexive reason between the imbricated perspectives of participant and observer found within strategies of discovery and self-transformation. Clearly, we need to posit context-transcending ideals, yet we also need to unmask their status as illusions rooted in power-laden contexts. We need a rational framework to protect freedoms from anonymous systems of surveillance and discipline, yet freedom is already lost if this framework is beyond question. To this end, Habermas attempts to discover some solid ground to build the modern project out of itself, while Foucault urges an ongoing process of self-transformation of ourselves and our possibilities. Both projects are relevant to today's problems. Whether we are talking about climate change denialism, the dissolution of the public sphere, the erosion of trust in science and democracy, new forms of subjectivity, or further radicalisations of neoliberal governmentality, we find science and technology driving changes far beyond our awareness. Science also provides the means to understand these changes, not simply by understanding science's claims or theories, their truth or falsity, but by viewing science in its relations to power and subjectivity as constituting, maintaining, and legitimating systems that operate behind our backs, reshaping subjects and social relations.

By addressing science in these terms, Habermas and Foucault can help us recognise opportunities and points of resistance. This common *telos* reflects that strand of the Enlightenment tradition which culminates in critical theory, aiming at both understanding and changing society by liberating human beings from entrapment in systems of dependence or domination, both internal and external. Both Habermas and Foucault also offer responses to contemporary ideological distortions induced by metaphysical scientism. Although reductive scientific naturalism cannot be refuted by any proof, it can be problematised by broader contextualisation. In this regard, Habermas and Foucault examine scientific reason, not in isolation, but in relation to its history and to other forms of reason. Adopting a normative stance, implicit in Foucault and explicit in Habermas, they both display the power-effects that flow from the unconsidered endorsement of scientific authority. For both thinkers, this is a political project to change consciousness by revealing it as conditioned by a vast range of historical and social factors. Both recognise the need for a more reflexive perspective and see philosophy as able to articulate social problems not visible within the specialised perspectives and limited scope of science.

Bibliography

BOOKS BY FOUCAULT

Books are shown by date of original French publication in square brackets.
2006 [1961]. *History of Madness.* Oxford: Routledge.
1994 [1963]. *The Birth of the Clinic.* New York: Vintage Books.
2002 [1966]. *The Order of Things.* Oxford: Routledge Classics.
1989 [1969]. *The Archaeology of Knowledge.* Oxford: Routledge Classics.
1995 [1975]. *Discipline and Punish.* New York: Vintage Books.
1979 [1976]. *History of Sexuality Vol. 1: The Will to Know.* London: Penguin.
1990 [1984]. *History of Sexuality Vol. 2: The Use of Pleasure.* London: Penguin.
1992 [1984]. *History of Sexuality Vol. 3: The Care of the Self.* London: Penguin.

FOUCAULT INTERVIEWS, LECTURES, ESSAYS

1980. *Power/Knowledge Selected Interviews and Other Writings.* C. Gordon (ed). New York: Vintage Books.
1981. "The Order of Discourse." In *Untying the Text: A Post-Structuralist Reader.* R. Young (ed). Boston, MA: Routledge.
1983. "*Fearless Speech* 6 Lectures at Berkeley Online Audio Resource." http://www.lib.berkeley.edu/MRC/foucault/parrhesia.html. Retrieved 25/02/2021.
1984. *The Foucault Reader.* P. Rabinow (ed). Middlesex, UK: Penguin.
1987. "The Ethics of Care for the Self and a Practice of Freedom." In *The Final Foucault.* J. Bernauer, D. Rasmussen (eds). Cambridge, MA: MIT Press, originally appearing in *Philosophy and Social Criticism* 1987 12 (2–3).

1988(a). "Power, Moral Values, and the Intellectual: An Interview with Michel Foucault." In (San Francisco 3/11/1980) *History of the Present*. M. Bess (ed). 4 (Spring 1988), 1–2, 11–13. http://www.critical-theory.com/read-me-foucault-in terview-in-a-sense-i-am-a-moralist/. Retrieved 25/02/2021.

1988(b). *Michel Foucault: Politics, Philosophy Culture Interviews and Other Writings 1977–1988*. L. Kritzman (ed). London: Routledge.

1988(c). "Truth Power and Self: An Interview with Michel Foucault." In *Technologies of the Self*. L. Martin, H. Gutman, and P. Hutton (eds). Amherst, MA: University of Massachusetts Press.

1989. *Foucault Live, Collected Interviews, 1961–84*. S. Lotringer (ed). New York: Semiotext(e).

1993. *Foucault, M., and Blasius, M*. "About the Beginnings of the Hermeneutics of the Self: Two Lectures at Dartmouth." *Political Theory* 21(2): 198–227.

1997(a). *Essential Works Volume 1 Ethics, Subjectivity and Truth*. P. Rabinow (ed). New York: The New Press.

1997(b). *The Politics of Truth*. S. Lotringer (ed). New York: Semiotext(e).

2000. *Essential Works Volume 3 Aesthetics, Method and Epistemology*. J. Faubion (ed). London: Penguin.

2001 [1983]. *Fearless Speech*. J. Pearson (ed). Los Angeles: Semiotext(e).

2002. *Essential Works Volume 2 Power*. J. Faubion (ed). London: Penguin.

FOUCAULT'S LECTURE COURSES AT COLLEGE DE FRANCE

2003(a). *Society Must Be Defended: Lectures at the College de France 1975–76*. M. Bertani and A. Fontana (eds). New York: Picador.

2003(b). *Abnormal: Lectures at the College de France 1974–5*. V. Marchetti and A Salomoni (eds). New York: Picador.

2005. *The Hermeneutics of the Subject: Lectures at the College de France 1981–82*. F. Gros (ed). New York: Picador.

2007. *Security, Territory and Population: Lectures at the College de France 1977–78*. M. Senellart (ed). New York: Picador.

2008(a). *Psychiatric Power: Lectures at College de France 1973–74*. J. Lagrange (ed). New York: Palgrave Macmillan.

2008(b). *The Birth of Biopolitics: Lectures at the College de France 1978–79*. M. Senellart (ed). New York: Picador.

2011. *The Government of Self and Others: Lectures at the College de France 1982–83*. F. Gros (ed). New York: Picador.

2012. *The Courage of Truth: Lectures at the College de France 1983–84*. F. Gros (ed). New York: Picador.

2013. *Lectures on the Will to Know College de France 1970–71*. D. Defert (ed). Hampshire, New York: Palgrave Macmillan.

2014. *On the Government of the Living: Lectures at College de France 1982–83*. M. Senellart (ed). New York: Palgrave Macmillan.

BOOKS BY HABERMAS

The date of the original publication is shown in square brackets.

1971(a) [1963, Introduction 1971]. *Theory and Practice.* New York: Beacon Press.
1971(b) [1968]. *Towards a Rational Society.* London: Heinemann Educational Books Ltd.
1972 [1968]. *Knowledge and Human Interests.* New York: Beacon Paperback.
1976 [1969]. *The Positivist Dispute in German Sociology* (essays by Habermas, J., Adorno, T., Albert, H., Dahendorf, R., Pilot, H., Popper, K.) G. Adey and D. Frisby (trans). London: Heinemann.
1979 [1976]. *Communication and the Evolution of Society.* T. McCarthy (trans). Boston: Beacon Press.
1983 [1981]. *The Theory of Communicative Action Vol. 1: Reason and the Rationalistion of Society.* T. McCarthy (trans). Boston, Ma: Beacon Press.
1988(a) [1967]. *On the Logic of the Social Sciences.* Cambridge, MA: MIT Press.
1988(b) [1973]. *Legitimation Crisis.* Cambridge: Polity Press.
1989(a) [1962]. *The Structural Transformation of the Public Sphere.* Cambridge, UK: Polity Press.
1989(b) [1981]. *The Theory of Communicative Action Vol. 2: Lifeworld and System: A Critique of Functionalist Reason.* T. McCarthy (trans). Boston, MA: Beacon Press.
1990(a) [1983]. *Moral Consciousness and Communicative Action.* Cambridge, MA: MIT Press.
1990(b) [1985]. *The Philosophical Discourse of Modernity.* Cambridge, UK: Polity Press.
1992 [1988]. *Postmetaphysical Thinking.* Cambridge, MA: MIT Press.
2001 [1984]. *On the Pragmatics of Social Interaction.* Cambridge, MA: MIT Press.
2002(a) [1981]. *Religion and Rationality.* Cambridge, MA: Polity Press.
2002(b) [1999]. *On the Pragmatics of Communication.* Cambridge, MA: Polity Press.
2003(a) [1998]. *Truth and Justification.* B. Fultner (trans). Cambridge, MA: MIT Press.
2003(b) [2001]. *The Future of Human Nature.* Cambridge, UK: Polity Press.
2009 [2005]. *Between Naturalism and Religion.* C. Cronin (trans). Cambridge, UK: Polity Press.
2010 [2008]. *An Awareness of What is Missing; Faith and Reason in a Post-Secular Age.* Cambridge, Ma: Polity Press.

HABERMAS' INTERVIEWS, LECTURES, ESSAYS

1953. "Mit Heidegger gegen Heidegger denken. Zur Veröffentlichung von Vorlesungen aus dem Jahre 1935." *Frankfurter Allgemeine Zeitung.* July 25, 1953.
Habermas, J., and Luhmann, N. 1971. *Theorie der Gesellschaft oder Socialtechnologie.* Frankfurt.

1980 [1970]. "Hermeneutics Claim to Universality." In *Contemporary Hermeneutics*. J. Bleicher (ed). London: Routledge and Kegan.
1975. "A Postscript to Knowledge and Human Interests." *Philosophy of the Social Sciences* 3: 157–189.
1996 [1981]. "Modernity: An Unfinished Project." In *Habermas and the Unfinished Project of Modernity*. M. Passerin d'Entreves and S. Benhabib (eds). Cambridge: Polity Press.
1982. "A Reply to My Critics." 219–283. In *Habermas: Critical Debates*. J. Thompson and D. Held (eds). London: The MacMillan Press.
1983. "The German Idealism of the Jewish Philosophers." In *Philosophical-Political Profiles*. Cambridge, MA.
1986 [1981]. "The Dialectics of Rationalisation." In *Autonomy and Solidarity: Interviews*. P. Dews (ed). London: Verso.
1990. "Jürgen Habermas: Morality, Society and Ethics: An Interview with Torben Hviid Nielsen." *Acta Sociologica* 33(2): 93–114. SAGE Publications.
1992. "Taking Aim at the Heart of the Present." In Kelly (ed): 149–154.
1994. *The Past as Future*. Interviews Max Pensky (ed). Lincoln, London: Nebraska University Press.
2007(a). "Interview." https://www.youtube.com/watch?v=jBl6ALNh18Q. Retrieved 25/12/2021.
2007(b). "The Language Game of Responsible Agency and the Problem of Free Will." *Philosophical Explorations* 10(1): 14–45.

LITERATURE CITED

Adorno, T., and Horkheimer, M. 1997 [1944]. *Dialectic of Enlightenment*. London: Verso Books.
Allen, A. 2008. *The Politics of Our Selves: Power, Autonomy and Gender in Contemporary Critical Theory*. New York: Columbia University Press.
Anderson, J. 2005. Review: "The Future of Human Nature." *Ethics* 115(4): 816–821.
Argyrous, G. 2011. "Cumulative Causation." In *Readings in Political Economy: Economics as a Social Science*. G. Argyrous and F. Stilwell (eds). Prahran, VIC: Tilde University Press.
Ashenden, S., and Owen, D. 1999. *Foucault Contra Habermas*. London: SAGE Press.
Austin, J. 1962 [1955]. *How to Do Things With Words: The William James Lectures Delivered at Harvard University*. J. Urmson and Marina Sbisà (eds). Oxford: Clarendon Press.
Ayre, A. J. 1952. *Language, Truth and Logic*. New York: Dover Publications.
Baghramiam, M. 2008. "Relativism about Science." In *The Routledge Companion to the Philosophy of Science*. S. Psillos and M. Curd (eds). New York: Routledge.
Bateson, P. "The Corpse of a Wearisome Debate." *Science* 297(5590): 2212–2213. DOI: 10.1126/science.1075989. Retrieved 24/02/2021.
Benhabib, S. 1986. *Critique, Norm and Utopia: A Study of the Foundations of Critical Theory*. New York: Colombia University Press.

Berger, P., and Luckmann, T. 1966. *The Social Construction of Reality: A Treatise in the Sociology of Knowledge*. Garden City, NY: Anchor Books.

Bernstein, R. 2010. *The Pragmatic Turn*. Cambridge, UK: Polity Press.

Best, S. 1995. *The Politics of Historical Vision: Marx, Foucault, Habermas*. New York: The Guilford Press.

Bloor, D. Barnes, and Henry, J. B. 1996. *Scientific Knowledge: A Sociological Approach*. Chicago: University of Chicago Press.

Bostrom, N. 2005. "In Defence of Posthuman Dignity." *Bioethics* 19(3): 202–214. http://www.nickbostrom.com/ethics/dignity.html.

Boudrey, M., and Pigliucci, M. (eds). 2017. *Science Unlimited? The Challenge of Scientism*. Chicago: University of Chicago Press.

Brandom, R. 2004. "Knowledge and the Social Articulation of the Space of Reason." In *Epistemology: An Anthology*. E. Sosa and K. Jaegwon. Oxford: Blackwell. www.scheuer.estranky.cz/file/19/brandomknowledge---space-od-reason.pdf.

Butler, J. 2001. "What is Critique? An Essay on Foucault's Virtue." https://transversal.at/transversal/0806/butler/en. Retrieved 25/02/2021.

Canguilhem, G. 1988. *Ideology and Rationality in the History of the Life Sciences*. Cambridge, MA: MIT Press.

Canguilhem, G. 1991. *The Normal and the Pathological*. New York: Zone Books.

Canguilhem, G. 1994. *A Vital Rationalist: Selected Writing from George Canguilhem*. New York: Zone Books.

Canguilhem, G. 2012. *Writings on Medicine*. New York: Fordham University Press.

Carson, R. 2002 [1962]. *Silent Spring*. New York: First Maritime Books.

Chimisso, C. 2003. "The Tribunal of Philosophy and Its Norms: History and Philosophy in Georges Canguilhem's Historical Epistemology." *Studies in History and Philosophy of Biological and Biomedical Sciences* 34(2): 297–327.

Churchland, P. 1984. *Matter and Consciousness*. Cambridge, MA: MIT Press.

Cook, D. 2001. "Two Faces of Liberal Democracy." *Philosophy Today*.

Cooke, M. 2006. *Representing the Good Society: Philosophical Issues in Critical Theory*. Cambridge, MA: MIT Press.

Daston, L. 2000. *Biographies of Scientific Objects*. Chicago: Chicago University Press.

Dawkins, R. 1976. *The Selfish Gene*. London: Granada.

Dawkins, R. 1997. 'Is Science a Religion?' First Published in *The Humanist*, January/February 1997. https://www.skeptical-science.com/essays/science-religion-richard-dawkins. Retrieved 24/02/2021.

De Caro, M., and MacArthur, D. (eds). 2004. *Naturalism in Question*. Cambridge, MA: Harvard University Press.

Descartes, R. 1968. *Discourse on Method and the Meditations*. London: Penguin Classics.

Dews, P. 1999. *Habermas: A Critical Reader*. Oxford: Blackwell Publishers.

Dupré, J. 2003. "Making Hay with Straw Men." In *Review of Steven Pinker, The Blank Slate: The Modern Denial of Human Nature*. Research Gate. Retrieved 24/02/2021.

Durkheim, E. 1974. *Sociology and Philosophy*. New York: Free Press.

Edgar, A. 2005. *The Philosophy of Habermas.* Montreal: McGill-Queens University Press.
Elden, S. 2001. *Mapping the Present: Heidegger, Foucault and the Project of a Spatial History.* London: Bloomsbury Continuum.
Elgin, C. 1997. *Between the Absolute and the Arbitrary.* Ithaca: Cornell University Press.
Eribon, D. 1991. *Michel Foucault.* Cambridge, Ma: Harvard University Press.
Eriksen, T. H. 2007. "Tunnel Vision." *Social Anthropology* 15(2): 237–243. DOI: 10.1111/j.0964-0282.2007.00015.x. Retrieved 24/2/2021.
Feenberg, A. 1999. *Questioning Technology.* New York: Routledge.
Feyerabend, P. 2010. *Against Method: Outline of an Anarchist Theory of Knowledge.* New York: Verso Books.
Fisher, D., and Freudenburg, W. 2001. "Ecological Modernization and Its Critics: Assessing The Past And Looking Toward The Future." *Society and Natural Resources* 14(8): 701–709.
Fleck, L. 1979. *Genesis and Development of a Scientific Fact.* Chicago: University of Chicago Press.
Frankfurt, H. 1988. *The Importance of What We Care About.* Cambridge, UK: Cambridge University Press.
Fraser, N. 1981. "Foucault on Modern Power: Empirical Insights and Normative Confusions." *Praxis International* 3: 272–287.
Fraser, N. 1989. *Unruly Practices: Power Discourse and Gender in Contemporary Social Theory.* Cambridge, UK: Polity Press.
Freud, S. 1985. *Civilisation, Society and Religion: Group Psychology, Civilisation and Its Discontents and Other Works.* J. Strachey (trans). Harmondsworth, UK: Penguin.
Gadamer, H.-G. 2004 [1960]. *Truth and Method.* London: Continuum.
Gadamer, H.-G. 1990. "Reply to My Critiques." In *The Hermeneutic Tradition: From Ast to Ricoeur.* G. Ormiston and A. Schrift (eds). New York: SUNY Press.
Galison, P., and Stump, D. 1996. *The Disunity of Science.* California: Stanford University Press.
Ghosh, A. 2016. *The Great Derangement.* London: Penguin.
Giladi, P. 2020. "Foucauldian Critique of Scientific Naturalism: 'Docile Minds.'" *Critical Horizons* 21(3): 264–286.
Giladi, P. 2021. "Habermas and Liberal Naturalism in (Forthcoming)." M. De Caro and D. Macarthur (eds). *The Routledge Handbook of Liberal Naturalism.* New York: Routledge.
Gilligan, C. 1982. *In a Different Voice.* Cambridge, MA: Harvard University Press.
Gould, K., Pellow, D., and Schnaiberg, A. 2016. *Treadmill of Production: Injustice and Unsustainability in the Global Economy.* Oxford: Routledge.
Greene, B. 2000. *The Elegant Universe.* New York: Vintage Books.
Grimes, D. Bauch, C and Ionnidis, J. 2018 "Modelling Science Trustworthiness under Publish or Peril Pressure." Royal Society Publishing. https://doi.org/10.1098/rsoc.171511 retreived 10/5/2021
Grumley, J. 1989. *History and Totality: Radical Historicism from Hegel to Foucault.* London: Routledge.
Grumley, J. 1992. "Two Views of the Paradigm of Production." *Praxis International* 12(2): 192–210.

Gutting, G. 1989. *Michel Foucault's Archaeology of Knowledge.* Cambridge, UK: Cambridge University Press.
Gutting, G. (ed). 2005. *The Cambridge Companion to Foucault.* Cambridge, UK: Cambridge University Press.
Haack, S. 2003. *Defending Science Within Reason: Between Scientism and Cynicism.* Amherst, New York: Prometheus Books.
Haack, S. 2008. *Putting Philosophy to Work: Inquiry and Its Place in Culture.* Amherst, New York: Prometheus Books.
Haack, S. 2011. "Six Signs of Scientism." *Logos and Episteme* 3(1): 75–95.
Hacking, I. 1999. *The Social Construction of What?* Cambridge, MA: Harvard University Press.
Hacking, I. 2002. *Historical Ontology.* Cambridge, MA: Harvard University Press.
Hadot, P. 1995. *Philosophy as a Way of Life.* Oxford, UK: Blackwell Publishing.
Hadot, P. 2006. *The Veil of Isis: An Essay on the History of the Idea of Nature.* M. Chase (trans). Cambridge, MA: The Belknap Press of Harvard University Press.
Han, B. 2002. *Foucault's Critical Project: Between the Transcendental and the Historical.* California: Stanford University Press.
Han, B. 2005. "Is Early Foucault a Historian? History, History and the Analytic of Finitude." *Philosophy and Social Criticism* 31(5–6): 586–608.
Hanson, N. 1958. *Patterns of Discovery: An Inquiry into the Conceptual Foundations of Science.* Cambridge, UK: Cambridge University Press.
Harrington, A. 2001. "Dilthey, Empathy and Verstehen: A Crticial Reappraisal." *European Journal of Social Theory* 4(3): 311–329.
Harrington, A. 2019. *The Mind Fixers.* London: W. W. Norton & Co.
Hawking, S. and Mlodinow, L. 2010. *The Grand Design.* London: Bantam Press.
Hayek, F. 2008. *The Road to Serfdom.* London: Routledge.
Heidegger, M. 1977. *The Question Concerning Technology and Other Essays.* San Francisco: Harper Row.
Hollis, M., and Lukes, S. (eds). 1982. *Rationality and Relativism.* Oxford, UK: Basil Blackwell.
Honneth, A., and Joas, H. (eds). 1991. "A Reply." 214–265 In *Communicative Action Essays on Jurgen Habermas' Theory of Communicative Action.* Cambridge, Ma: MIT Press.
Horkheimer, M. 2002. "Traditional and Critical Theory." In *Critical Theory.* New York: Continuum Press.
Hoy, D., and McCarthy, T. 1994. *Critical Theory.* Cambridge, MA: Blackwell Publishers.
Iliopoulis, J. 2017. *The History of Reason in the Age of Madness: Foucault's Enlightenment and a Radical Critique of Psychiatry.* London: Bloomsbury.
Ingram, D. 2010. *Habermas: Introduction and Analysis.* Ithaca: Cornell University Press.
Jeffries, S. 2017. "Habermas by Stephan Muller-Doohm." *The Guardian.* February 15, 2017. https://www.theguardian.com/books/2017/feb/15/habermas-biography-stefan-muller-doohm-review. Retrieved 24/02/2021.

Jiang, F., Jiang, Y., Zhi, H., et al. 2017. "Stroke and Vascular Neurology." https://svn.bmj.com/content/svnbmj/2/4/230.full.pdf. Retrieved 24/2/2021.

Johnson, P. 2009. *Habermas: Rescuing the Public Sphere*. London: Routledge.

Kalecki, M. 1971. "Costs and Prices." In *Selected Essays on the Dynamics of the Capitalist Economy*. Cambridge, UK: Cambridge University Press.

Kant, I. 1963. *Introduction to Logic*. T. K. Abbott (trans). Westport, Connecticut: Greenwood Press.

Kant, I. 1987. *Critique of Judgement*. W. S. Pluhar (trans). Indianapolis: Hackett Press.

Kant, I. 1998. *Critique of Pure Reason*. P. Guyer and A. Wood (trans). Cambridge: Cambridge University Press.

Keen, Steve. 2020. "The Appallingly Bad Neoclassical Economics of Climate Change?" *Globalizations*: 1–29. Retrieved 25/2/2021.

Kelly, M. (ed). 1992. *Critique and Power Recasting the Foucault/Habermas Debate*. Cambridge, MA: MIT Press.

Kelly, M. G. 2009. *The Political Philosophy of Michel Foucault*. Oxon, UK: Routledge.

Kelly, M. G. 2013. "Foucault, Subjectivity, and Technologies of the Self." In *A Companion to Foucault*. C. Falzon, T. O'Leary, and J. Sawicki (eds). London: Blackwell Publishing.

Kelly, M. G. 2018. *For Foucault: Against Normative Political Theory*. New York: SUNY Press.

King, J. E. 2011. "Arguments for Pluralism in Economics." In *Readings in Political Economy: Economics as a Social Science*. Melbourne, VIC: Tilde University Press.

King, M. 2009. "Clarifying the Foucault-Habermas Debate: Morality, Ethics and 'Normative Foundations." *Philosophy & Social Criticism* 35(3): 287–314.

Kohlberg, L. 1971. "From is to Ought." In *Cognitive Development and Epistemology*. T. Mischel (ed). New York: Academic Press.

Kolakowski, L. 1972. *Positivist Philosophy: From Hume to the Vienna Circle*. New York: Pelican Books.

Kompridis, N. 2009. "Technology's Challenge to Democracy: What of the Human?" *Parrhesia* 8: 20–33. www.parrhesiajournal.org.

Koopman, C. 2010. "Revising Foucault: The History and Critique of Modernity." *Philosophy and Social Criticism* 36(5): 545–565.

Koopman, C. 2013. *Genealogy as Critique: Foucault and the Problems of Modernity*. Indiana, USA: Indiana University Press.

Krauss, M. (ed). 1989. *Relativism Interpretation and Confrontation*. Indiana, USA: University of Notre Dame Press.

Kripke, S. 1980. *Naming and Necessity*. Cambridge, MA: Harvard University Press.

Kuhn, T. 1996 [1962]. *The Structure of Scientific Revolutions*. Chicago: University of Chicago Press.

Lakoff, G., and Johnson, M. 2003. *Metaphors We Live By*. Chicago: University of Chicago Press.

Latour, B. 1992. "Where Are the Missing Masses? The Sociology of a New Mundane Artifacts." In *Shaping Technology/Building Society: Studies in Sociocultural Change*. W. Bilker and J. Laws (eds). Cambridge, MA: MIT Press.

Latour, B. 1993. *We Have Never Been Modern.* C. Porter (trans). Cambridge, MA: Harvard University Press.
Latour, B., and Woolgar, S. 1979. *Laboratory Life: The Construction of Scientific Facts.* California: SAGE Press.
Lecourt, D. 1969. *Marxism and Epistemology: Bachelard, Canguilhem, Foucault.* Ben Brewster (trans). London: Verso.
Lecourt, D. 2008. *George Canguilhem it la Vie Humaine.* Paris: Presses Universitaires de France.
Lemke T. 2012. *Foucault, Governmentality and Critique.* Boulder, Colorado: Paradigm Publishers.
Lewontin, R. 1991. *Biology as Ideology: The Doctrine of DNA.* Canada: Anansi Press.
Lovelock, J. 2000. *Gaia a New Look at Life on Earth.* Oxford: Oxford University Press. Lyotard, J.-F. 1988. *The Differend: Phrases in Dispute.* Minneapolis: University of Minneapolis Press.
Macey, D. 1995. *The Lives of Michel Foucault.* New York: Vintage Books.
Marcuse, H. 1964. *One Dimensional Man.* UK: Sphere Books.
McCarthy, T. 1978. *The Critical Theory of Jurgen Habermas.* Cambridge, MA: MIT Press.
McCarthy, T. 1991. *Ideals and Illusions*: *On Reconstruction and Deconstruction in Contemporary Critical Theory.* Cambridge, MA: MIT Press.
McCarthy, T. 1992. "The Critique of Impure Reason." M. Kelly (ed): 243–282.
McGushin, E. 2007. *Foucault's Askesis.* Illinois: Northwestern University Press.
Mead, G. H. 1934. *Mind Self and Society.* Chicago: Chicago University Press.
Mendieta, E. 2003. "Communicative Freedom and Genetic Engineering." *Logos* 2(1): 124–140.
Merton, R. K. 1979. *The Sociology of Science.* Chicago: University of Chicago Press.
Midgley, M. 1992. *Science as Salvation: A Modern Myth and Its Meaning.* London: Routledge.
Miller, J. 1993. *The Passion of Michel Foucault.* Cambridge, MA: Harvard University Press.
Mirowski, P. 1989. *More Heat Than Light: Economics as Social Physics, Physics as Nature's Economics.* New York: Cambridge University Press.
Muller-Doohm, S. 2016. *Habermas: A Biography.* Cambridge: Polity Press.
Myrdal, G. 1944. *An American Dilemma: The Negro Problem and Modern Democracy.* New York: Harper.
Nordhaus, William D. "Revisiting the Social Cost of Carbon." *Proceedings of the National Academy of Sciences* 114(7): 1518–1523. Retrieved 25/2/2021.
Nowotny, H., Scott, P., and Gibbons, M. 2001. *Re-Thinking Science: Knowledge and the Public in an Age of Uncertainty.* London: Polity Cambridge.
Oels, A. 2005. "Rendering Climate Change Governable: From Biopower to Advanced Liberal Government?" *Journal of Environmental Policy and Planning* 7(3): 185–207.
Olson, E. 2021. "Personal Identity." In *The Stanford Encyclopedia of Philosophy.* E. Zalta (ed). forthcoming. https://plato.stanford.edu/archives/spr2021/entries/identity-personal/.

Orr, H. "Darwinian Storytelling." *New York Review of Books* 50(3). Retrieved 24/02/2021.
Outhwaite, W. 2009. *Habermas: A Critical Introduction*. California: Stanford University Press.
Patton, P. 1985. "Michel Foucault: The Ethics of an Intellectual." *Thesis 11* 10–11(1): 71–80.
Patton, P. 2013. "From Resistance to Government: Foucault's Lectures 1976–1979." In *A Companion to Foucault*. C. Falzon, T. O'Leary, and J. Sawicki (eds). London: Blackwell Publishing.
Piaget, J. 1965. *The Moral Development of the Child*. New York: Free Press.
Pinker, S. 2002. *The Blank Slate*. New York: Viking Press.
Popper, K. 1963. *Conjectures and Refutations: The Growth of Scientific Knowledge*. London: Routledge.
Price, H. 2004. "Naturalism Without Representationalism." In De Caro and MacArthur (eds). *Naturalism in Question*. Cambridge, Mass: Harvard University Press.
Putnam, H. 2002. *The Collapse of the Fact/Value Dichotomy and Other Essays*. Cambridge, MA: Harvard University Press.
Rabinow, P., and Dreyfus, H. 1983. *Michel Foucault Beyond Structuralism and Hermeneutics*. Chicago: University of Chicago Press.
Rajchman, J. 1985. *Michel Foucault: The Freedom of Philosophy*. New York: Columbia University Press.
Rajchman, J. 1988. "Foucault's Art of Seeing." *October* 44: 88–117. MIT Press.
Ramachandran, V. S. 2004. *A Brief Tour of Human Consciousness*. London: Profile Books.
Rasmussen, D. 1990. *Reading Habermas*. London: Basil Blackwell.
Rehg, W. 2009. *Cogent Science in Context: The Science Wars, Argumentation Theory, and Habermas*. Cambridge, Mass: MIT Press.
Robinson, J. 1962. *Economic Philosophy*. UK: Penguin Books.
Rorty, R. 1981. *Objectivity, Relativism and Truth*. Cambridge, UK: Cambridge University Press.
Rorty, R. 1998. *Truth and Progress: Philosophical Papers Vol. 3*. Cambridge, UK: Cambridge University Press.
Rorty, R. 2000. "Response to Brandom." In *Rorty and His Critics*. Brandom (ed). Oxford: Blackwell.
Rose, N. 2007. *The Politics of Life Itself: Biomedicine, Power and Subjectivity in the Twenty-First Century*. Princeton: Princeton University Press.
Rosenberg, A. 2011. *The Atheist's Guide to Reality*. New York: W. W. Norton & Co.
Rouse, J. 1987. *Towards a Political Philosophy of Science*. New York: Cornell University Press.
Rouse, J. 1993. "Foucault and the Natural Sciences." Wesleyan University Faculty Publications Paper 33. http://wesscholar.wesleyan.edu/div1facpubs/33. Retrieved 25/2/2021.
Searle, J. 1969. *Speech Acts: An Essay in the Philosophy of Language*. Cambridge: Cambridge University Press.

Searle, J. 2005. "What is an Institution?" *Journal of Institutional Economics* 1(1): 1–22.
Sellars, W. 1963. "Philosophy and the Scientific Image of Man." In *Science, Perception, and Reality*. New York: Humanities Press.
Shapin, S., and Schaffer, S. 1985. *Leviathan and the Air-Pump: Hobbes, Boyle, and the Experimental Life*. Princeton: Princeton University Press.
Smith, A. 1776. "The Wealth of Nations." http://www.bibliomania.com/2/1/65/112/frameset.html. Retrieved 25/2/2021.
Smith, R. E. 2019. *Rage Inside the Machine*. London: Bloomsbury Press.
Specter, M. 2010. *Habermas: An Intellectual History*. New York: Cambridge University Press.
Stockwell, D. 2013. "A Sea Change for Climate Science?" *Quadrant Online*, October. https://quadrant.org.au/opinion/doomed-planet/2013/10/sea-change-climate-science/. Retrieved 24/2/2021.
Stretton, H. 1999. *Economics: A New Introduction*. Sydney: UNSW Press.
Szakolczai, A. 1998. *Max Weber and Michel Foucault: Parallel Life-Works*. Oxford: Routledge.
Szasz, T. 1974. *The Myth of Mental Illness: Foundations of a Theory of Personal Conduct*. New York: Harper.
Talcott, S. 2014. "Errant Life, Molecular Biology, and the Conceptualization of Biopower: Georges Canguilhem, François Jacob, and Michel Foucault." *History and Philosophy of the Life Sciences* 36(2): 254–279.
Talcott, S. 2019. *Georges Canguilhem and the Problem of Error*. New York: Springer International Publishing.
Thompson, J., and Held, D. 1982. *Habermas: Critical Debates*. London: The MacMillan Press.
Thornton, T. 2017. *From Economics to Political Economy: The Promise, Problems and Solutions of Pluralist Economics*. London: Routledge.
Tiles, M. 2005. "Technology, Science and Inexact Knowledge: Bachelard's Non-Cartesian Epistemology." In *Continental Philosophy of Science*. G. Gutting (ed). London: Blackwell Publishing.
Tulley, J. 1999. "To Think and Act Differently: Foucault Four Reciprocal Objections to Habermas Theory." In *Foucault Contra Habermas*. S. Ashenden and D. Owen (eds). London: SAGE Press.
Watters, E. 2010. *Crazy Like Us: The Globalisation of the American Psyche*. New York: Free Press.
Webb, D. 2013. *Foucault's Archaeology: Science and Transformation*. Edinburgh: Edinburgh University Press.
Weber, M. 1978. *Economy and Society: An Outline of Interpretive Sociology*. G. Roth and C. Wittich (ed). Berkeley: University of California Press.
Weber, M. 2002. *The Protestant Ethic and the Spirit of Capitalism*. P. Baehr and G. Wells (trans and ed). London: Penguin.
Weber, Max. 2004. "Science as Vocation." In *The Vocation Lectures*. R. Livingstone (trans); D. Owen and T. Strong (eds). Illinois: Hackett Books.

Whyte, J. 2019. *The Morals of the Market: Human Rights and the Rise of Neoliberalism.* New York: Verso.
Williams, B. 1985. *Ethics and the Limits of Philosophy.* London: Fontana.
Wilson, E. O. 1998. *Consilience: The Unity of Knowledge.* New York: Vantage.
Wittgenstein, L. 2003. *Philosophical Investigations.* Oxford: Blackwell.
Wynne, B. 2001. "Creating Public Alienation: Expert Cultures of Risk and Ethics on GMOs." *Science as Culture* 10(4): 445–481. DOI: 10.1080/09505430120093586. Retrieved 25/4/21.
Young, A. 1997. *The Harmony of Illusions: Inventing Post-Traumatic Stress Disorder.* Princeton: Princeton University Press.

Index

Page numbers followed by "n" refer to footnotes

actor network theory (ANT), 14, 92n13
Adorno, T., 13, 23, 47, 65, 72, 83;
   Dialectic of Enlightenment, 22
aesthetics of the self, 217–18
Alcibiades, 222
Allen, A., 95, 96
the analytic of finitude, 157–60
ANT. *See* actor network theory (ANT)
anti-psychiatry, 140–41
archaeology, 144–50; and genealogy,
   168; and history of science, 144–50
The Archaeology of Knowledge
   (Foucault), 138, 144–49, 166
artificial intelligence, 3n5, 65, 65n14
ascesis, 32, 216, 216n9, 218, 220, 225,
   231n19, 235, 262, 263
Austin, J. L., 24, 67, 69
avant-garde literature, 142

Bachelard, G., 27, 28, 30, 139, 144,
   145
Baudelaire's aesthetic modernity, 218
Becker's theory of human capital,
   197n12
Behrent, M., 205
belief, Pierce's view of, 52–54
Benhabib, S., 76, 96

Between Facts and Norms (Habermas),
   25, 101–9
biological determinism, 183, 184, 255
biological essentialism, 107, 183
biological normality, 183
biological norms, 130
biology, emergence of, 153–54
biopower, 178, 191–94, 208, 226, 227;
   resistance to, 215
biotechnology, 114–15, 117
The Birth of the Clinic (Foucault), 138,
   148–49
the "blackmail" of the enlightenment,
   248
Bostrom, N., 118
Boyle, R., 89
Bush, G. W., 6, 20
Butler, J., 203

Canguilhem, G., 26–30, 144, 183,
   189, 230; biological norms, 130;
   epistemology, 27; error, 29–30;
   history of the concepts, 28, 145,
   146, 148, 156; *The Normal and the
   Pathological*, 27, 29, 30; norms, 29;
   scientific ideology, 29
capitalism, 22, 65, 82, 83, 92

the carceral archipelago, 186, 189, 190, 201
care of the self, 215–17, 219, 221, 222, 224, 226, 227
Carson, R. *Silent Spring*, 47
the "Cartesian moment", 207, 220, 224–27, 261
Charcot, J.-M., 170–72; hysterics, 180, 181
Chomsky, N., 112
Christian asceticism, 223
circular and cumulative causation, 181n5, 189, 199
Classical *episteme*, 151, 153
climate change, 7, 8n12, 48–49, 105, 106, 198; contrarianism, 14; denial, 49, 106
Cogito-unthought double, 159
cognitive interests: theory of, 49–62
Communication and the Evolution of Society (Habermas), 77
communicative action, 68, 70, 72, 73, 76, 79, 98–99, 115, 246, 247, 258, 259, 261, 263
communicative competence, 68, 77, 78, 90, 240, 245
communicative power, 102
competition: neoliberalism and, 197
the concepts of the normal and the abnormal: emergence of, 176
confession: disciplinary technology of, 182
connaissance, 147n16
the constitution of the subject, 33, 179–81, 211, 214
context transcendence, 95, 96, 98
Cooke, M., 96
Cooper, D., 141
counter-sciences, 138, 138n5, 156, 163, 164
crises: lived and social, 76, 77
critical sciences: Habermas's idea of, 51n7, 57–60
critical theory, 20, 22, 23, 25, 34n38, 234, 241

criticisms of BFN, 103
critique, 133–38; and limits, 137
crypto-normativism, 200
cultural impoverishment, 80–81
culture: evolution of, 128
the Cynics, 221–22, 236

danger as a condition of liberalism, 195
Dawkins, R., 3, 14
death of Man thesis, 160–63
debate between Habermas and Foucault, 35–36
deliberative democracy, 25, 45, 102–4; science and, 104–9
the delinquent, 178, 180, 181, 202
democratic public sphere, 108, 130
Derrida, J., 15
Descartes' meditations, 213, 224, 225
developmental psychology, 50, 68, 78, 241
developmental stages, 78
development of systems, 79
Diagnostic and Statistical Manual of Mental Disorders (DSM), 172
Dialectic of Enlightenment (Horkheimer and Adorno), 22
the diffusion of power/knowledge, 176
Dilthey, W., 23; hermeneutics, 55
disciplinary power, 173–76, 179, 185, 187, 188, 190, 191
discourse and communicative action, 71
discursive formations, 146–49, 146n15; and sciences, 147n16
dispositif, definition, 168
distantiation: Piaget's concept of, 80
documentation and the human sciences, 178
Dreyfus, H., 26, 32, 164, 252–53
DSM. *See Diagnostic and Statistical Manual of Mental Disorders* (DSM)
dualism of epistemic perspectives, 127, 128
Durkheim, E., 68, 112
the dynamics of development, 78, 83

economics, 75, 105, 153, 193–200
economism, 130
efficiency and technology, 63
Elgin, C., 180, 181
empirical-analytic science, 51n7, 52–54
the empirical sciences, 138n5, 153–54
"empirico-transcendental doublet", 157
the Enlightenment, 135, 136, 179, 247–48; in France, 229–30; in Germany, 229–30; Kant, I., 58, 134, 135, 227–30, 247–48
the *episteme*, 144, 146n15, 149, 156, 163, 165
epistemological ruptures, 144, 144n12
ethics, 209, 211, 215–18, 215n7, 221, 234–36
ethnocentrism, 95, 97
etho-poesis, 211n5, 218
events: role in Foucault's genealogy, 169
the examination: disciplinary technology of, 176, 178, 185
experiences and arguments: relation to self-transformation, 236

the "fabrication" of subjects, 180
fact/value distinction, 15, 17, 44
faith, 3; and knowledge, 120–22
Feenberg, A., 63, 91
formal and informal discourses within the public sphere, 102
formal pragmatics, 68–73
formal world concepts, 70, 74
Foucault, M.: account of his work as "fictions", 162; account of hysteria, 170–72; account of psychiatry, 170–71; account of the archaic notion of truth, 246; account of the philosophical-scientific standpoint, 246; account of the subject's genealogy, 220–28; accounts of the "carceral archipelago", 190, 201; aesthetics, 217–18; analyses of liberalism, 193–95, 198; analyses of neoliberalism, 8, 195–98; analysis of power relations, 168, 172–79, 184, 204, 205; analytic of finitude, 157–60; analytic of power, 173; anti-essentialism, 183; archaeology, 138, 144–50, 164; (and genealogy, 168–72); *The Archaeology of Knowledge*, 138, 144–50, 166; biography, 26, 241; biopower, 178, 189, 191–94; *The Birth of the Clinic*, 138, 148–49; and Canguilhem, 26–30; care for oneself, 211; the "Cartesian moment", 207, 220, 224–27, 261; collapsing truth into power, 256; concept of governmentality., 208–9; constructivism, 186; criticism of humanism, 136, 248–49; critique, 134–36, 229–31; "crypto-normative" judgements, 249; death of Man thesis, 160–63; disciplinary power, 173–75, 179, 187, 190, 191; and empirical sciences, 153–54; and Enlightenment, 11, 30; Enlightenment *ethos* of ongoing critique, 136; ethics, 209, 215–18, 236; "experience" of order, 235; foundations, 251, 254; genealogical problematisations, 234–36; governmentality, 134, 134n1, 194, 198, 199; and Heidegger, 31; historical critique, 135; *The History of Madness*, 138–44, 148, 161; history of philosophy, 212, 222; *The History of Sexuality*, 32; and humanism, 136; and human sciences, 154–57; Kant's Enlightenment and, 227–30, 247, 248; local critique, 259; modernity, 34–36, 238, 239; modern power/knowledge, 172–79; the natural sciences, 185–88; and Nietzsche, 15, 31, 238, 255, 262; nominalism, 168, 183; normative "confusions", 200–206; normative modesty, 252; normative principles, 250; normativity, 135–36, 203, 251, 254; *The Order of Things*, 138, 150–63, 165; *parrhesia*, 221–24, 226,

227, 236; periodisation of works, 31; power/knowledge, 209, 212, 228, 256–59, 261, 263; (and the natural sciences, 186–87); radicalisation of critique, 133–38; reconcilability with Habermas, 234–37; regimes of truth, 212–14, 255, 262; science and religion, 263; and science as social institution, 17; science's "value freedom", 253; scientific knowledge, 168, 174; spirituality, 218–19; strategy of self-transformation, 242–49, 255, 258, 265; and the transcendental, 135, 161; truth, 218; and truth of a science, 141; truth-telling, 221; turn from power to the subject, 208–12; understanding of the modern sciences, 152; view of human nature, 162; *What Is Enlightenment?*, 137–38; works, 27, 31–33

fragmentation of consciousness, 81
Frankfurt, H., 15
the Frankfurt School, 228–30
Fraser, N., 86–87, 200
freewill and determinism, 122–29
Freud, S., 23, 28, 112, 140; diagnosis of "communal neurosis", 59; psychoanalysis, 58–59
Friedman, M., 196
The Future of Human Nature (Habermas), 114–22

Gadamer, H.-G., 55, 57; hermeneutics, 56, 69, 73
Gauss, C., 68
gender identity and Habermas' proceduralism, 107
gender orientation, 184
genealogy, 234–35; archaeology and, 168–72; and political engagement, 205
general sense of critique, 134–35
genetic intervention, 114–16, 118–20
Ghosh, A., 106
Giladi, P., 92, 104, 143

governmentality, 134, 134n1, 194, 198, 199
Government by truth, 208
the Great Confinement, 139, 142
Gutting, G., 28, 142, 161

Habermas, J.: analogy of premodern cultures, 94; analysis of technocratic consciousness, 46, 48–49, 63; analysis of the new politics, 85; "authoritarianism", 237, 257; authority of science, 264; colonization of the lifeworld, 83–85, 87, 99, 130; *Communication and the Evolution of Society*, 77; communicative action, 244, 246, 247, 258, 259; communicative rationality, 98; concept of lived "crisis", 76; contingency and complexity, 245; critical theory, 20, 22, 23, 25, 96; criticism of technocratic thought, 45; debate between Gadamer and, 55–57; deliberative democracy, 103–9; and democracy, 20, 22, 24, 25, 34, 83; developmental stages, 78; diagnosis of modernity, 80–85; dialectic of potential and will, 43–44, 49; Dilthey's hermeneutics and, 55; epistemic perspectives, 127–28; *Between Facts and Norms*, 25, 101–9; fallibilism, 246, 257, 264; formal pragmatics, 68–73; and freewill and determinism, 122–29; *The Future of Human Nature*, 114–22; and Heidegger, 19; ideal speech situation, 72, 72n5, 260; "interaction" and "labour", 46, 49–51, 53, 65; Kantian Enlightenment tradition, 247; *Knowledge and Human Interests*, 10, 23, 49–62, 90, 94; *Legitimation Crisis*, 34n38, 68; life biography, 18–21, 23–25; lifeworld and system, 9, 73–77; and limits of natural sciences, 137; logic of development, 245, 248; modernity,

34–36, 80–85, 238, 239, 248, 256, 257; natural science and technology, 64; normativity, 236–37; Peircean account of belief, 53; Popper and, 41; power and ideology concepts, 59; quasi-transcendentalism, 96, 97; rationalisation of society, 77–79; and reasons and causes, 125; reconcilability with Foucault, 234–37; reinterpretation, 94–99; on religious language, 120–22; response to positivism, 40–49; response to technocracy, 43–46; social theory, 77, 86; soft naturalism, 129; strategy of discovery, 240–49, 255, 257–58; *The Structural Transformation of the Public Sphere*, 20–21, 23, 25, 102, 103, 108; theoretical partitions, 62–66; *The Theory of Communicative Action*, 10, 24, 34n38, 59, 67–69, 73, 86, 88, 95, 102, 104, 108, 109, 120, 261; treatment of science and technology, 47–48; use of evolutionary theories, 77–79; use of "foundations", 241, 249, 251–53; validity claims, 70, 72, 74, 76, 77, 79, 81, 82, 90; "weakly transcendental" approach, 97, 130; *What Is Universal Pragmatics?*, 68; works, 20–25
Hacking, I., 180, 181, 199
Hadot, P., 217, 218, 223
Han, B., 152
Harrington, A., 171–72
Harris, S., 14
Hawking, S., 7, 14
Hayek's "spontaneous order", 196
Hegel, G. W. L., 23, 25, 228
Heidegger, M., 5, 19, 31, 39, 240; historicism, 69
hermeneutics, 55–58, 69, 73, 97
hierarchical observation: disciplinary technology of, 175
the historical *a priori*, 144, 144n11, 151, 152, 157, 165

historical-hermeneutic sciences, 51n7, 54–57
historical ontology, 32
The History of Madness (Foucault), 138–44, 148, 161
The History of Sexuality vol. 1 (Foucault), 32, 182–83, 191–93
Hobbes, T., 89
the homosexual, 178, 181, 202
Horkheimer, M., 13, 19, 65, 229; critical theory, 20; critique of instrumental reason, 83; *Dialectic of Enlightenment*, 22
human monster: Foucault's genealogy of, 176
human nature, 9–10, 16, 17n25, 107, 160, 162, 163, 183, 230
the human sciences, 138n5, 154–57, 175, 178, 185, 186
Husserl's phenomenology, 73
the hysteric, 170–72, 180, 202

ideal speech situation, 72, 72n5
ideology, 46, 60, 72, 86, 98n14, 147–48, 168, 214
Iliopoulis, J., 141
impact science, 105. See also production science
informal collective discourses, 21, 108
informal public opinion formation, 102
the inquiry and Foucault's genealogy of natural science, 185–86
Insel, T., 172
instability of the human sciences, 156
institutions, 1, 20, 22, 33; science as a social institution, 9, 10, 17
institutionalisation of technological knowledge, 188–90

juridification, 84

Kant, I., 25, 50, 98, 157, 158, 160, 258, 261; aesthetics, 217; anthropology, 135, 160; conditions for representation, 152; critical theory,

23; dualism, 161; Enlightenment, 58, 134, 135, 227–30, 247–48; epistemological critique, 228–29; epistemology, 159; philosophy and science after, 228–31; "regulative ideas", 72; transcendental critique, 133–37, 152; transcendentalism, 51, 135, 179, 227–29
Kelly, Mark, 200, 258
Kelly, Michael, 200
knowledge: as an "invention", 169; as struggle and compromise, 170
Knowledge and Human Interests (Habermas), 10, 23, 49–62, 90, 94, 261; critical sciences, 57–60; empirical-analytic science, 52–54; historical-hermeneutic sciences, 54–57
Kohlberg, 68, 77
Koopman, C., 96–97, 143, 235, 255; modernity, 202; normatively ambitious, 252; "strategy of delegation", 234, 236, 254
Kuhn, T., 13, 36, 53; discontinuity, 145–46

Laches, 222
Laing, R. D., 141, 142
Latour, B., 13, 64, 92, 184
law and morality, 103
Legitimation Crisis (Habermas), 68
liberal eugenics, 114, 116, 119
liberalism, 103, 193–95, 198; and republicanism, 103
liberal naturalism, 17
liberal paradigm, 106–7
lifeworld, 68, 73–74, 87–88, 95, 240; colonisation of the, 45, 80, 82–83, 86, 87, 90, 92; cultural impoverishment, 80–81; definition, 9n14; development of, 78; power in the, 86–88; rationality, 79; structural components, 74, 78, 82; and system, 73, 75, 88, 99; (distinction between, 87, 98); "unity of reason", 82
limits to self-objectification, 127

logical positivism, 13, 40; Haber's critique of, 40–43, 57, 65
the logic of development, 83, 95
the "looping" effects of interactive kinds, 180, 184, 199
Lorenzer, A., 59
Luhmann's social systems theory, 49, 69, 73
Lyotard, J.-F., 15

Madness: Foucault's history of, 139–44
Mannheim, K., 203
Marcuse, H., 45, 63; Habermas dispute with, 46–47, 49, 91
market economy, 75, 79, 196, 199
Marx, K., 23, 28, 153; historical materialism, 59; lifeworld and system, 80; scientific-technological rationality, 51
Marxism, 112, 219, 220, 226
mass media, 108–9
McCarthy, T., 18, 93–95, 97, 98, 209
McGushin, E., 220, 235
Mead, G. H., 68, 112
mental causation and free will, 124
mental illness, 171–72; as an object of scientific inquiry, 139–40
Merton, R., 81
metaphorical structure of understanding, 106n5
Mind Fixers: Psychiatry's Troubled Search for the Biology of Mental Illness (Harrington), 171–72
modern *episteme*, 151–53, 155–57
modernity, 34–36, 111, 143, 216, 219, 228, 247; diagnosis of, 80–85; reflexivity, 34, 35, 113, 238, 248, 264
modern myths of liberation, 136–37
modern power/knowledge: emergence and dissemination, 172–79
modern psychiatry and biology, 171
modern scientific worldview: emergence of, 207, 213, 225
modes of veridiction/truth telling, 221

myth of scientific objectivity in 19th century psychiatry, 140
myths informing science, 106n5

natality: Arendt's concept of, 117
the natural sciences, 40, 52–54, 62–65, 185–88
nature society distinction, 6
neoclassical economics, 76, 106n5
neoliberalism, 195–98, 205
neurological description, 123, 127
neurology, 170
neuroscience, 126, 128, 129
neuroscientific debunking of free will, 126
neutrality of the sciences, 44
new technologies, 109, 119, 120
Nietzsche, F., 15, 31, 134, 163, 166, 168, 169, 228, 238, 244, 255, 262
Nordhaus, W., 105
The Normal and the Pathological (Canguilhem), 27, 29, 30
normalisation of society, 188
normalizing judgement: as a technology of disciplinary power, 175
normative commitments: Foucault and Habermas, 135
normative "confusions", 200–206
normative foundations, 202, 241, 252n9, 257; and confusions, 249–54
normativity, 236; definition, 237; expanded sense, 237
norms: entwinement of social and technological, 64, 189; two uses of the term, 177

observer/participant perspectives, 124–25, 127–29, 237–38; inextricable interlocking, 127, 238
Oels, A., 197
order, conceptual structures and, 150
The Order of Things (Foucault), 138, 150–53, 165; analytic of finitude, 157–60; death of Man thesis, 160–63; empirical sciences, 153–54; human sciences, 154–57

parrhesia, 221–24, 226, 227, 236
Patton, P., 205
Peirce, S., 23; belief, 52–53
philosophical life, 223
philosophy, 7, 34, 36; and critique, 135, 227–28, 230, 231; Habermas and Foucault in relation to, 264; and its contexts, 12–18; modernity, science, and, 33–37; nature of, 16; and politics, 219; and science, 109–13, 228–31; as a theoretical discipline, 223; as a way of life, 223, 228
Piaget, J., 68, 77
political economy/economics, 193–94, 197, 198, 199, 199n15
the politics of truth, 205, 228
Popper, K., 12–13, 23, 28, 40, 41, 54
positivism, 10, 23, 23n29, 25, 30, 65, 214; dominance, 50; Habermas's response to, 40–49
post-traumatic stress disorder (PTSD), 184n7
power: domination distinguished from "games between liberties", 204
power in the lifeworld, 86–88
power/knowledge: constitution of the subject, 179–85; emergence and dissemination, 172–79
PP. *See* psychiatric power (PP)
pragmatism, 17, 37
prescientific commitments, 105–7
problematisations, 134n2, 163, 234
problems with "reflection" in KHI, 61
proceduralist paradigm, 107
production science, 105
progress and reason, 93–94
psychiatric power (PP), 170, 172
psychiatry, 171–72, 176–78
psychoanalysis, 58–59, 112, 156, 164, 171, 219, 220, 226
psychology, 142, 154–55
public opinion: democratic formation of, 102–4
purification of concepts and practices within modernity, 143

Putnam, H., 17

Quine, W. V., 14

Rabinow, P., 26, 32, 164, 252–53
Rajchman, J., 142, 162
the rationalisation of society, 77–79
rational reconstruction, reconstructive sciences, 61, 61n12, 68
reasons and causes, 125
reconciliation between Foucault's and Habermas's projects, 234–37
reflexivity: its modern radicalization, 34–35, 237–39
regimes of truth, 212–14, 255, 262
religion, 120–22, 129; Habermas and Foucault in relation to, 262
religious language, 120–22
Renaissance *episteme*, 151
repressive hypothesis, 143, 182, 220
republicanism, 103
resistance to biopower, 204, 215
resistance to subjection, 209
the "retreat of the state", 196
Ricardo's view of economic history, 153
Rorty, R., 16, 110–11
Rose, N., 119
Rouse, J., 187, 188

savoir, 147n16
Schaffer, S., 13, 17, 186
Schnaiberg, A., 105
science: analysing as social institution, 9, 10, 17, 33; authority of, 7, 264; and belief, 53; crisis in, 3, 34n39; definition, 2; and deliberative democracy, 104–9; dissemination of, 6, 17; faith in, 3; hierarchy and organization of, 204–5, 188; history of, 144–50; ideology and, 29, 148; and lifeworld and system, 84; philosophy and, 109–13; social function of, 90; scepticism towards, 3; and social power, 4–5; and

society, 1–7; and technology, 46–47, 63, 64, 90–93; (distinctions between, 62, 62n13); truth and, 213–14; unification of, 2
scientific ideologies, 28–29
scientific naturalism, 14–18, 65, 110, 121, 130, 133, 143, 254, 261, 265
scientific technologies, 45–48, 51, 65, 91, 91n10
scientism, 2, 3n4, 4n8, 14, 14n20, 92, 97
Searle, J., 24, 67, 69, 126
self-transformation, 212, 215, 226, 228, 235, 236
Sellars, W., 14, 17, 124
sex: concept of, as a pseudo-object, 183
sexuality, 177–78; and sex, 182–85
Shapin, S., 13, 17, 186
Silent Spring (Carson), 47
Slocum, R., 197
Smith, A., 194–96
social and biological norms: their inextricable links, 183
social evolutionary theory, 77–79
the social function of scientific authority, 88–90
social media algorithms, 7, 8
social norms within technology, 91–92
social power, 107, 108; science and, 4–5
social science and positivism, 40–41
social theory, 73, 75–77
sociology: emergence of, 154–55
sociology of scientific knowledge (SSK), 13
sovereign power, 173, 176. *See also* disciplinary power
the specific intellectual, 205
speech acts: illocutionary component of, constative, regulatory and expressive, 70–71; theory, 24
spirituality, 218–19, 218n12, 223, 227, 263
SSK. *See* sociology of scientific knowledge (SSK)

"strong program" of SSK, 13
The Structural Transformation of the Public Sphere (Habermas), 20–21, 23, 25, 102, 103, 108
struggle between technical knowledges, 170
"subjectivation", 216, 216n8
subjugated knowledges, 198, 204, 204n19
surveillance and disciplinary power, 173, 187
systematically distorted communication, 72
system(s), 75–79, 95; definition, 9n14
systems theory, 67, 68, 68n1, 76, 77
Szasz, T., 141; *The Myth of Mental Illness*, 4n7

tactical polyvalence of discourses, 181, 193, 258
Talcott, S., 27
technical efficiency, 63
technical knowledge, 44
technocracy, 23, 43–45, 47
technocratic consciousness, 44–46, 48–49, 56, 63, 65
technological "normalisation" of society, 189
technologies of the self, 210
technology, 64, 91–93; and science, 4, 5
the *telos* of modern rationality, 260, 265
Thatcher, M., 6, 14, 195
theory-ladenness of observation, 148n17
theory of cognitive interests, 49–62

The Theory of Communicative Action (Habermas), 10, 24, 59, 67–69, 73, 86, 88, 95, 102, 104, 108, 109, 120, 261
transhumanism, 118n8
trauma: pseudo-concept of, 170, 183
trust in institutions, 91, 258
truth, 53, 213–14, 218, 219, 253; and knowledge, 169; power and, 14
Tulley, J., 200
two-track model of deliberative democracy, 102, 104–9

the universal intellectual, 205, 205n20, 236, 259

validity claims, 70, 72, 74, 76, 77, 79, 81, 82, 90

Weber, M., 12–13, 68, 80, 112, 228; account of religion, 120; theory of modernity, 143; thesis of the separation of value spheres, 81–82
welfare paradigm, 106, 107
welfare state, 83–84
What Is Enlightenment? (Foucault), 134, 137–38, 227
What Is Universal Pragmatics? (Habermas), 68
Williams, B, 16
the will to truth, 167
Wilson's "evolutionary ethics", 16n24
Wittgenstein, L., 69, 250, 255; "perspicuous presentation", 235
women's rights, 107

About the Author

John McIntyre is a research affiliate at the University of Sydney. He has lectured at the University of Sydney and the Melbourne School of Continental Philosophy. His research is focussed on science, technology, and their relationship to society and draws on philosophical thought from across both the analytic and continental philosophical traditions. Prior to commencing formal studies in philosophy, John McIntyre worked as horticultural development advisor in Papua New Guinea, followed by a career as an environmental planner within state and local agencies.

www.ingramcontent.com/pod-product-compliance
Lightning Source LLC
Chambersburg PA
CBHW021847300426
44115CB00005B/46